The Kolsky-Hopkinson Bar Machine

Ramzi Othman
Editor

The Kolsky-Hopkinson Bar Machine

Selected Topics

Springer

Editor
Ramzi Othman
Faculty of Engineering, Department
of Mechanical Engineering
King Abdulaziz University
Jeddah
Saudi Arabia

ISBN 978-3-030-10130-5 ISBN 978-3-319-71919-1 (eBook)
https://doi.org/10.1007/978-3-319-71919-1

Softcover re-print of the Hardcover 1st edition 2018

Printed on acid-free paper

This Springer imprint is published by the registered company Springer International Publishing AG part of Springer Nature
The registered company address is: Gewerbestrasse 11, 6330 Cham, Switzerland

Contents

About the Editor

Ramzi Othman received M.S. in Engineering from Ecole Polytechnique de Tunisie in 1998. In 1999, he also completed an M.S. in Structural Engineering from Ecole Centrale Paris. In 2002, he received his Ph.D. from Ecole Polytechnique (France). In 2005 he became an Associate Professor at Ecole Centrale Nantes (France). Since 2012, he has been on leave from Ecole Centrale de Nantes and working as Associate Professor in King Abdulaziz University, Jeddah, Saudi Arabia, where he is promoted in February 2018 to the rank of Full Professor. His major research interests are characterization and modeling of the impact behavior of materials and structures. He is author and co-author of about 100 publications in refereed journals and conference proceedings.

Introduction

The Kolsky-Hopkinson bar machine is a well-established experimental technique widely used to characterize materials and structures under dynamic, impact, and explosion loads. Since the pioneer work of Bertram Hopkinson (1914), continuous improvements have been undertaken to extend the range of use of this machine and to improve the quality of measurements. For more than one hundred years, this technique has been used in several engineering fields (automobile, aeronautical, manufacturing, biomedical, military, etc.).

Several publications (book, review articles, etc.), written by prominent scientists, were dedicated to this technique yet. The purpose of this book is not to compete with these past publications. However, this book aims at giving the opportunity to several worldwide prominent scientists to share herein their viewpoints and expertise on special topics related to the Kolsky-Hopkinson bar technique. Indeed, the Kolsky-Hopkinson bar machine is not a simple experimental device. It is rather a philosophical approach to solve the problem of measuring impact events. This field is attracting an increasing interest each year, and new improvements are continuously proposed. Thus, this book aims at supporting engineers, R&D professional, graduate students, scientists, and scholars by new viewpoints on some topics related to the Kolsky-Hopkinson bar machine.

Chapter 1 went back to "The Origins of the Hopkinson Bar Technique". He takes the reader in a time travel through the nineteenth and twentieth centuries, where a review of the first interests in measuring the stress–strain curves of materials is presented. It also explains the motivation of Bertram Hopkinson in developing a testing device measuring dynamic pressures induced by explosion and bullet impacts. This chapter depicts the impact of Bertram Hopkinson's works on detonators and fuses design in World War I. It reviews the main improvements that lead to the nowadays-used double Hopkinson bar or more commonly named split Hopkinson pressure bar. It gives also a noticeable interest in the life of Bertram Hopkinson.

The most used configuration of this technique is the split Hopkinson pressure bar (SHPB) system. It is also called Kolsky bars system due to the works of Herbert Kolsky in 1949. This consists in two long elastic bars supported so to allow a guided motion in the axial direction of the bars. A very short specimen is sandwiched between them. A striker is accelerated, mostly by a gas gun, impacts a first bar or input bar. A compressive incident wave is then generated within the input bar. The incident wave propagates till the bar–specimen interface. Because of the impedance mismatch between the bar and the specimen, a part of the incident wave is reflected back into the incident bar as a tensile reflected wave. A second part of the wave is transmitted through the specimen to the second bar or output bar. One strain gauge station is cemented on each bar. The input bar's strain gauge measures together the incident and reflected waves, whereas the output bar's strain gauge measures the transmitted wave. The wave duration, which is equal to the time needed by the wave to make one round-trip in the striker bar, is assumed short enough so that the incident wave ends up before that the reflected wave arrives at the input bar strain gauge. Thus, it is possible to measure separately the incident, reflected, and transmitted waves. Once this is done, it is possible to deduce the force and velocities at the specimen–bar interfaces. Subsequently, it is possible to calculate the strain rate, the strain, and the stress properties of the specimen.

The SHPB is appropriate to measure stress–strain curves of metallic materials under uniaxial compression and over high strain rate, i.e., strain rates ranging between 500 and 5000 s^{-1}. This is because of some limitations of this device. Indeed, a striker impact is used to induce a pressure wave in the input bar, and this induces a compression deformation of the specimen. Several improvements have been proposed to test materials at high strain rates under tensile, shear, etc., loadings using a Kolsky-Hopkinson bar type machine.

Chapter 2 is interested in "Tensile Testing Using the Kolsky-Hopkinson Bar Machine". It starts with a recall of the advantages, basics, assumptions, processing and limitations of static tensile tests. The remaining parts of this chapter are dedicated to the dynamic tensile testing. First, the one-dimensional elastic wave propagation theory was reviewed. This recalls the framework of the processing and analyzing of Kolsky-Hopkinson bar machine tests in general and particularly the tensile Kolsky-Hopkinson bar machine tests. Second, the different techniques, proposed in the literature, to test materials under tensile loads using Kolsky-Hopkinson bar machines, are discussed. Two types of solutions were mainly used. There are techniques that are based on the classical SHPB set-up. A careful design of the specimen is then required. The second type of solutions is designed so to induce a tensile wave in the incident bar. To do so, one of the solutions is to use a pre-tensile load in the incident bar. This alternative attracts the attention of the author in the fourth section where he details its design and gives its analysis procedure. The chapter is closed by two sections dedicated to the direct measurement of deformation and measuring material properties at high temperature. These two issues are very important not only to dynamic tensile testing but more generally in dynamic testing of materials under any types of load.

Tensile dynamic tests are widely used and of a valuable interest. However, they have some drawbacks. The shear testing can overcome some of these shortcomings. Chapter 3 is interested in "Shear Testing Using the Kolsky-Hopkinson Bar Machine". First, this chapter depicts the motivation to dynamic shear testing.

Mainly, it explains that shear testing largely avoids localized necking encountered in tensile testing. Thus, the fracture observed by a shear test is a material fracture, which opposes geometrical/structural fracture observed in a tensile test after necking. Second, this chapter draws the two main shear deformation modes: simple shear and pure shear modes. Third, it reviews shear testing of sheet specimens using Kolsky-Hopkinson bar technique. This includes a discussion on the design of the shear specimen in order to avoid premature failure at the edges, to achieve homogeneous stress and strain fields within the gauge parts, and to reduce specimen and grip inertias. Next, the chapter reviews shear testing of bulk materials using Kolsky-Hopkinson bar technique. Mainly, torsion Hopkinson bars are presented and discussed. A careful design of the different parts of the machine and torsion samples is depicted. Moreover, some other shear testing solutions, based on the Kolsky-Hopkinson bar machine, are also described including using hat-shaped specimens. At last, this chapter closes with a section dealing with the formulation of shear strains mainly when large deformations are involved.

Tensile and shear dynamic testing introduced in the second and third chapters is more suitable for ductile materials and more appropriately ductile metallic and polymeric materials. For brittle geomaterials, such as rocks and concrete, different techniques are proposed. Direct uniaxial tensile tests, using classical dumbbell or dogbone-shaped specimens, are extremely hard with rocks or concrete. The Brazilian test is typically used to determine the tensile strength of brittle materials. Chapter 4 is then dealing with the "Dynamic Brazilian Test Using the Kolsky-Hopkinson Bar Machine". First, the chapter starts with an Introduction where the authors depict the assets and success of the static Brazilian in measuring the tensile strength and, in some cases, the fracture toughness, the tensile elastic modulus, and the bi-modulus of rocks, concrete, or more generally brittle materials. They also introduce the dynamic Brazilian test using the Kolsky-Hopkinson bar machine. Mainly, they emphasize the use of non-contact optical strain measurements. In the second section, the authors present the principle of the indirect-tensile/Brazilian test. They give the equations for the stress and displacement fields as derived for planar static elastic deformation. A first main advantage is that the static and dynamic stress distributions are alike. A second important advantage is that the stress distribution is independent of the specimen's material parameters. The third section of this chapter deals with the non-contact full-field strain measurement. Namely, the authors gave a brief overview of the principle of the digital image correlation method. Subsequently, they give some useful information of the high-speed photography technique. The fourth section is dedicated to some practical aspects of the experimentation. They illustrate the assets of the dynamic Brazilian test on Al_2O_3 ceramic, granite, and PMMA.

The dynamic Brazilian test is mainly used to determine the tensile strength of brittle materials. It can also be used (the flatted Brazilian disk), among other techniques, to measure the fracture toughness of materials, which is a key material property in fracture mechanics. Thus, Chap. 5 is interested in a "Comparative Study of the Dynamic Fracture Toughness Determination of Brittle Materials Using the Kolsky-Hopkinson Bar Machine". This chapter's introduction focuses on the difficulty of measuring such parameter for brittle materials compared to ductile materials. It emphasizes on the fact that this material parameter is loading rate-dependent. It lists four possible techniques to measure it. In the second section, the authors compare three techniques used with the split Hopkinson bar to measure the dynamic fracture toughness: the flatted Brazilian disk test, the three-point bending test, and the notched semicircular bending test. For each technique, the authors give the basic equations. They explain how these techniques work together with a Kolsky-Hopkinson bar machine. The third section focuses on the strain and displacement fields measured by digital image correlation technique and high-speed photography for the types of specimens: the flatted Brazilian disk sample, the three-point bending specimen, and the notched semicircular bending specimen. It is also interested in comparing crack-opening displacement curves obtained with the three techniques. The fourth section deals with the determination of the dynamic fracture toughness from stress intensity factor curves. A brief micromorphologic analysis is presented in the fifth section. This chapter is closed with a discussion on the three reviewed techniques, where the authors motivate their preference to the flatted Brazilian disk test.

In Chaps. 2–5, the main goal is how to extend the use of split Hopkinson bar to test materials under dynamic tensile (ductile materials), shear tensile, dynamic tensile (brittle materials), and fracture toughness, as the conventional Kolsky-Hopkinson bar machine is mainly dedicated to the test materials under uniaxial compression. These chapters depict several solutions to the problem of how to induce a specific stress state within the specimen that is different from uniaxial compression. The presented works tackled the problem from two different sides. First, it is possible to modify the loading device in order to generate different stress pulses within the bars such as: tensile or torsion waves. Second, it is possible to modify the specimen geometry in order to transform the stress wave to a desired stress state within the specimen. Once this is done, the last difficulty is to establish the equations that give the strain and stress in terms of the force and displacements at the specimen-bar interfaces.

Independently of the final use of the Kolsky-Hopkinson bar machine: compression, tensile, shear, etc., they are equipped with long (visco)-elastic bars which are basically dynamic force and dynamic velocity transducers. These bars/transducers work as waveguides. There are three main waves: the incident, reflected, and transmitted waves. They are recorded through two strain gauge stations that are cemented one on each bar. Knowing these three waves and assuming one-dimensional non-dispersive wave propagation within the bars, it is possible to deduce the force and velocity at the specimen-bar interfaces. However, this type of processing is only straightforward and simple in the case of thin metallic (elastic) bars and within a certain range of strain rate, that is the high strain rate interval. As soon as, the used bars are not elastic or their diameter is important compared to the involved wavelengths, dispersion occurs. Moreover, working in the intermediate strain rate range lengthens the wave pulses. The three elementary waves, the incident, reflected, and transmitted waves, cannot be separately determined by only one gauge station on each bar. In the two above cases, it is laborious to determine the force and velocity at the specimen-bar interfaces.

Chapter 6 is dealing with the "Wave Dispersion in Kolsky-Hopkinson Bar Machine". After a brief Introduction, this chapter depicts, in the second section, the analytical approaches to assess the wave dispersion in three-dimensional viscoelastic rods. Indeed, wave dispersion can be caused by either geometrical or material effects. The author writes the viscoelastodynamic equations for an infinite-length rod with a given diameter. He derives the Pochhammer-Chree equation and gives some techniques to solve this nonlinear equation. The author discusses the multi-mode solution and emphasizes the fact that the second and higher modes propagate only at very high frequencies. The third section of this chapter focuses on the experimental techniques to measure the wave dispersion in viscoelastic rods. The author divides these techniques into two families. The first family of techniques uses the first round-trip of waves in the bars. They are named the techniques using short-duration signals. The second family allows multiple reverberations of waves in the bars, and they are named techniques using long-duration signals.

The measurement of forces and velocities at the specimen-bar interfaces assumes that the incident and reflected waves can be measured separately with one strain gauge station on the input bar. It also assumes that the strain gauge station on the output bar only measures the transmitted wave and that no reflection comes from the output bar end. These assumptions are only valid at high strain rates, where the loading wave pulse is short. Mainly, this is possible when the striker is roughly shorter than the half the length of the incident bar and roughly shorter than the output bar. At intermediate strain rates, longer loading pulses are needed. Consequently, wave separation techniques are needed and they are reviewed in Chap. 7: "Wave Separation Techniques". After a brief introduction, this chapter deals first with the wave separation techniques based on the one-dimensional wave equation written in the time domain. The solutions presented in this section are derived using the characteristics diagram of waves. They mainly neglect wave dispersion. As soon as wave dispersion is important and has to be considered, the wave separation techniques should use the wave equations written in the Fourier/frequency domain. This is the objective of the third section of this chapter. The author insisted on the importance of using redundant measurements in order to reduce noise amplification and list some application using Kolsky-Hopkinson bar-like machines.

After applying the right loading condition (Chaps. 2–5), and after correctly measuring the force and velocity at the specimen-bar interfaces (Chaps. 6 and 7), the experimenter faces another difficulty, which is how to deduce the stress and strain properties of the specimen using the force and velocity boundary measurements. To this aim, a certain stress state is assumed for the specimen. Moreover, the stress and strain fields are considered homogeneous based on an equilibrium hypothesis. However, important accelerations are required in high strain rate experiments, which can lead to significant inertia effects. Because of these inertia effects and also friction effects in compression tests, the measured stress is only an apparent/structural stress which is sometimes much higher than the material real stress. Chapter 8 deals then with "Inertial and Frictional Effects in Dynamic Compression Testing". This chapter is organized in ten sections. It starts with an introduction. Five sections follow dealing with the inertia effects. It closes with four sections interesting in the friction effects. In the second section, the authors detail the two main approaches on how to calculate inertia effects, namely the energy-based approach and the stress-based approach. This section is closed by a comparison in terms of the application of the equations resulting from the two approaches. The third section is dedicated to giving some order of magnitudes about the structural stress due to inertia effects. It is amazing to know that the inertia stress can be as high as 1GPa at very high strain tests (10^6 s^{-1}). The following fourth section is emphasizing the impact of inertia effects on the design of the specimen geometry. The authors discussed this separately for three classes of materials: metals and polymers, rubbers, and concrete. Later, the authors reviewed the numerical works investigating inertia effects which are mainly undertaken using finite element analysis. To close the discussion about inertia effects, the sixth section gives an overview on the inertia effects in tensile experiments. The authors emphasize the fact that the inertia can influence metal ductility. They also give lower and upper bounds of the inertia stress in tensile specimens. In the seventh section, the authors move to the second topic of this chapter: friction effects. They are mainly classified into three major effects: shear at specimen/anvil interface, barreling, and cones of non-deforming material. The effects of friction on the bulge factor are also presented. Some solutions to overcome/cancel the friction effects are also reviewed in this section, mainly compression in small stages and compression between two cones. Though friction effects exist also in static tests, their effect is more important in dynamic tests. The authors close the seventh section by presenting techniques to study the flow fields in barreling cylindrical specimens. In the eighth section, the authors go back to the origins of the friction. They mainly recall the friction law. They also give an insight into the new techniques to undertake fundamental studies of friction. The ninth section deals with the measurements techniques of friction, mainly the focus on ring tests and the aspect ratio tests. The closing section is dedicated to the lubrication under non-conventional conditions and mainly at non-ambient temperatures.

The inertia effects increase as the strain rate increases. This makes Kolsky-Hopkinson bar tests are extremely difficult at very high strain rates (higher than 5000 s^{-1}). Chapter 9 deals with using Kolsky-Hopkinson bar machine in the "Very high strain rate range". This chapter starts with a brief introduction giving mainly a definition of the strain rate. It is then stated that to increase strain rate and mainly to achieve very high strain rate it is possible to either decrease the dimensions of the specimen or increase the impact velocity. Consequently, two classes of solutions are proposed to overcome the very high strain rate limitations: the miniaturized Kolsky-Hopkinson bar and the direct-impact Kolsky-Hopkinson bar. These two techniques are reviewed in the second section. Subsequently, the authors focus on the design of the direct-impact Kolsky-Hopkinson bar machine. They mainly discussed three design ratios which are the bar length-to-diameter ratio, the specimen length-to-diameter ratio, and the specimen diameter-to-bar diameter ratio. Though these ratios are discussed for the direct-impact Kolsky-Hopkinson bar machine, most of the constraints can also be considered for a conventional Kolsky-Hopkinson bar machine or a split Hopkinson pressure bar. In the fourth section, the authors discuss some criteria that involve first the specimen and the projectile together, second the specimen and the bar, and at last the projectile and the bar. Moreover, the authors discuss the problem of wave dispersion as shorter wavelengths are involved in the very high strain rate testing range. The fifth section summarizes the constraints discussed in the third and fourth sections to build a design procedure based on an optimization problem. The sixth chapter is interested in the processing of direct-impact Kolsky-Hopkinson tests. Mainly, the authors present techniques to deduce stress, strain, and strain rate from the direct measurements. The use of digital image correlation technique is also introduced. The chapter is closed by a conclusion.

Chapter 10 of this book gives an exotic application of the Kolsky-Hopkinson bar machine. It deals with "Testing of Adhesively Bonded Joints by Split Hopkinson Bar Technique". The chapter starts with an introduction where the author relates the impact rate, the testing machine, and the industrial application. In the second section, the split Hopkinson pressure bar and the clamped Hopkinson bar are presented. In the third section, the author reviews researches that measured the strength of adhesive joints using the Kolsky-Hopkinson bar machine. The chapter is closed by a conclusion.

Ramzi Othman

Chapter 1
The Origins of the Hopkinson Bar Technique

Stephen M. Walley

In the last 20 years or so, several reviews have been published of the split Hopkinson bar technique for measuring the dynamic stress-strain curves of materials [1–8]. The purpose of this chapter, therefore, is not to repeat what was written in those publications but to take a more historical approach, and in particular to give an account of the man behind the technique as far as is possible to do some 97 years since Professor Bertram Hopkinson tragically died in a plane crash. I performed much of the research for this chapter when preparing the introduction to the theme issue 'Shock and Blast: Part 1" published in 2014 in the *Philosophical Transactions of the Royal Society Series A* [9].

That themed issue was one of a pair [10, 11] compiled to commemorate the centenary of the publication in the same journal in 1914 of one of Bertram Hopkinson's most famous papers [12]. In it he introduced a momentum trapping technique (Fig. 1.1) whereby the shape of the pressure-time pulse produced by the impact of a bullet or by an explosion could be roughly determined, the first time (as far as he knew) this had ever been done [13]. In his own words quoted from the beginning of his 1914 paper [12]: "The determination of the actual pressure produced by a blow such as that of a rifle bullet or by the detonation of high explosives is a problem of much scientific and practical interest but of considerable difficulty. It is easy to measure the transfer of momentum associated with the blow, which is equal to the average pressure developed, multiplied by the time during which it acts, but the separation of these two factors has not hitherto been effected. The direct determination of a force acting for a few hundred-thousandths of a second present difficulties which may perhaps by called insuperable, but the measurement of the other factor, the duration of the blow, is more feasible. In the case of impacts such as those of spheres or rods moving at moderate velocities the time of contact can be

S. M. Walley (✉)
SMF Group, the Cavendish Laboratory, The University of Cambridge,
J.J. Thomson Avenue, Cambridge CB3 0HE, UK
e-mail: smw14@cam.ac.uk

R. Othman (ed.), *The Kolsky-Hopkinson Bar Machine*,
https://doi.org/10.1007/978-3-319-71919-1_1

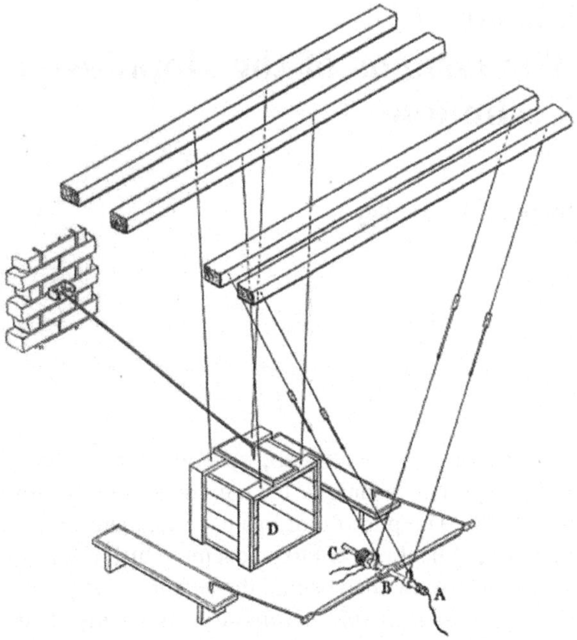

Fig. 1.1 Bertram Hopkinson's original apparatus for measuring the shapes of stress pulses produced by bullet impact or by explosions [12]

determined electrically with considerable accuracy. The present paper contains an account of a method of analysing experimentally more violent blows and of measuring their duration and the pressures developed."

So who was Bertram Hopkinson and why did he develop this technique? For answers to the first part of this question, I am indebted to a number of published accounts (listed in Table 1.1), to two of Hopkinson's descendants (Alexander Hopkinson-Woolley and Adam Woolley) and to Professor John Young who up until 2015 was the holder of the Hopkinson & ICI Chair of Applied Thermodynamics at the Engineering Department of the University of Cambridge. Alexander Hopkinson-Woolley provided the transcript (see Appendix) made by his mother of a talk about Hopkinson given on the BBC by Sir Henry Tizard on March 5, 1937. The BBC told me that sadly no audio recording exists of this broadcast as at that time radio programmes were broadcast live and not often recorded as the technology (the Blattnerphone) for doing so at that time was both expensive and dangerous to use.

When he died, Bertram Hopkinson was a Fellow of the Royal Society of London, a Fellow of King's College, Cambridge, Professor of Mechanism and Applied Mechanics in the University of Cambridge, a Colonel in the Royal Engineers, and a Companion of the Most Distinguished Order of Saint Michael and Saint George (CMG). A photograph and a portrait of him are given in Fig. 1.2.

Bertram Hopkinson was born into a distinguished Anglo-German family of engineers. Both his mother and his wife were German. His father was John Hopkinson, who was Professor of Electrical Engineering at King's College,

Table 1.1 List of published articles and books about Bertram Hopkinson and his family.

Obituaries of Bertram Hopkinson
1919, Ewing, J. A. "Bertram Hopkinson, 1874–1918" *Proc. R. Soc. Lond. A* **95** xxvi–xxxvi
1920, Hill, A. V. "An appreciation [of Bertram Hopkinson]" *The Alpine Journal* **32** 353–357
1921, "Bertram Hopkinson, 1874–1918" *Min. Proc. Inst. Civil Engrs* **212** 425–426
1927, Ewing J. A. "Bertram Hopkinson, 1874–1918", in "The Dictionary of National Biography 1912- 1921", ed. H. W. C. Davis & J. R. H. Weaver, pp. 268–269, publ. London, Oxford University Press
1929 "Bertram Hopkinson", in "Who Was Who. Vol. 2: 1916–1928", p. 397, publ. A&C Black, London
Published information about Bertram Hopkinson and his family
1928, Hopkinson, E. "The Story of a Mid-Victorian Girl", printed privately by the Cambridge University Press
1931, Inglis C. E. "Cambridge as a place of education", *Proc. Inst. Mech. Engrs* **121** 15–22
1945, Ricardo, H. R. "Applied research" *Proc. Inst. Mech. Engrs* **152** 143–148 1965, Clark, R. W. "Tizard", pp. 37–48, publ. London, Methuen
1967, Hilken, T. J. N. "Bertram Hopkinson: Professor of Mechanism and Applied Mechanics, 1903–1918", in "Engineering at Cambridge University 1783–1965", pp. 129–144, publ. Cambridge, Cambridge University Press
1974, Charlton, T. M. "Professor Bertram Hopkinson, C. M. G., M.A., B.Sc., F.R.S. (1874–1918)", *Notes Records R. Soc. Lond.* **29** 101–109

Fig. 1.2 **a** Photograph of Bertram Hopkinson first published in [14]; **b** Portrait of Bertram Hopkinson (courtesy of the Cambridge University Engineering Department)

London. Although born into a life of privilege, he must have been hit hard by a terrible tragedy that occurred in 1898 when his father and three of his five siblings were killed in a mountaineering accident on the Petite Dent de Veisivi, Switzerland

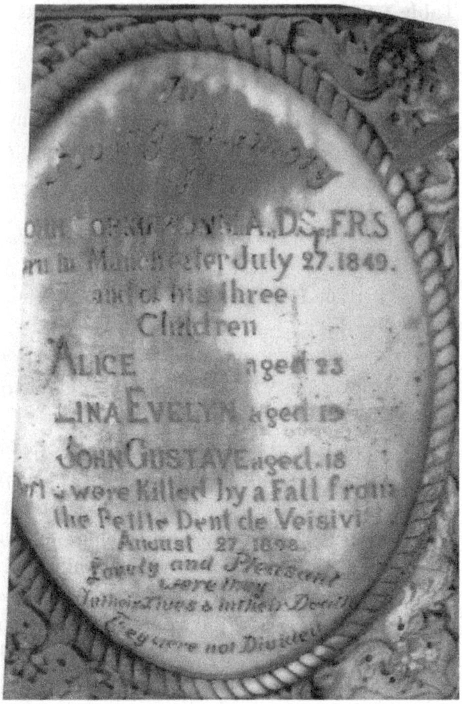

Fig. 1.3 Gravestone of John Hopkinson, two of his daughters, and one of his sons in a graveyard in Territet, Vaud, Switzerland

(Figs. 1.3 and 1.4). Although his mother Evelyn lived on until 1933, her account of her life (printed privately in 1928; [15]) ends with a heart-rending cry of grief[1]. She also had to nurse one of her other sons whose body was shattered in the Great War until he died in Cambridge in 1917. Only one of her six children outlived her. The Great War must also have torn this Anglo-German family apart in many other ways, but none of the published documents about the family mention this. Rather they emphasise Hopkinson's scientific and engineering contribution to the British war effort [14, 16–22]. Despite these tragedies and his early death in 1918, Bertram Hopkinson had many descendents (in the mid-1970s, the family knew of about 75 who were still alive).

What was the context within which John and Bertram Hopkinson worked as Professors of Engineering with research interests in the dynamic properties of metals? The major applications of iron and steel during the 19th century were bridges [23–28], buildings [29], boilers for steam engines (which were prone to explode [30–36]), armour for naval ships [37–46], and guns [47–59]. There was also considerable concern about gas explosions in coalmines [60–66].

[1]Then came the End: my life cut short as by a sword, my task seemed done, finished with this brief half of the story. He who reads may perchance gather what made the continuity of the after years possible.

Fig. 1.4 Plaque on Free School Lane, Cambridge, commemorating John Hopkinson and his son John Gustave Hopkinson who both died in 1898 in an Alpine climbing accident

Early pioneers of materials testing include de Réaumur in France [67, 68], Thomas Young (of Young's modulus fame) in England [69], Stodart, Faraday, Tredgold, and Hodgkinson (again in England) [70–73] and Wade in the United States [74]. But it was the Scotsman, Kirkaldy, who in London started the first known materials and structures testing business in the middle of the 19th century [75–77]. His Universal Testing Machine still exists in the building in which it was installed in 1874 (99 Southwark Street, London, SE1 0JF) and which is now a museum (www.testingmuseum.org.uk).

This early materials research was magisterially summarised for students of engineering and practising engineers by Rankine in two massive books [78, 79]. Nineteenth century French research on the plasticity of metals was summarised in two papers published in English by Tresca [80, 81]. The application of optical microscopy to the study of the microstructure of steels was pioneered in the late 19th century by two Frenchmen, Osmond [82–84] and Sauveur [85–88], who emigrated to the United States. The reason why it took around 250 years from the invention of the microscope [89] to its application in microstructural investigations of metals was the need first to develop good polishing and etching techniques [90]. The state of knowledge at the beginning of the 20th century concerning the plasticity and brittleness of materials (and methods of obtaining this information) was summarised by Guest in 1900 [91].

In his 1914 Royal Society publication [12], Hopkinson reported a mechanical waveguide method for the measurement of the shapes of dynamically-produced elastic pulses. The method consisted of using a ballistic pendulum (Fig. 1.1) to determine the momentum trapped in a short steel rod of known length (later termed the time piece [92]) attached either magnetically [12] or with grease [92] to the far end of a long steel rod whose other end is subjected either to impact by a lead bullet or to the blast produced by the detonation of gun cotton. By repeating the experiment using short rods of various lengths, it proved possible to build up a picture of the pulse shape (Fig. 1.5). He had previously presented many of these ideas at the Royal Institution in 1912 [13] and to the North East Coast Institution of Engineers and Shipbuilders [93]. Bertram Hopkinson had for some years before this had an interest in studying explosions in coalmines [94–96] as well as the dynamic properties of metals [97], a topic that his father (John Hopkinson) had also been interested in [98, 99]. His experiments to determine the shapes of elastic pulses produced in steel by ballistic impact and by explosions enabled him to explain the phenomenon of spall in ship and tank armour [13, 93] (see also Fig. 1.6). This kind of spall was later called 'Hopkinson fracture' by Kolsky [100].

According to two papers published by Robertson in 1921, the Hopkinson bar technique proved invaluable to the British during the First World War [102, 103], enabling, for example, improvements to the design of detonators and fuses through giving information on the impulse delivered and 'the necessity of avoiding gaps in the train of detonation on account of the very rapid falling off in violence of the blow when even a small air-gap is introduced' [103]. Its first major use, therefore,

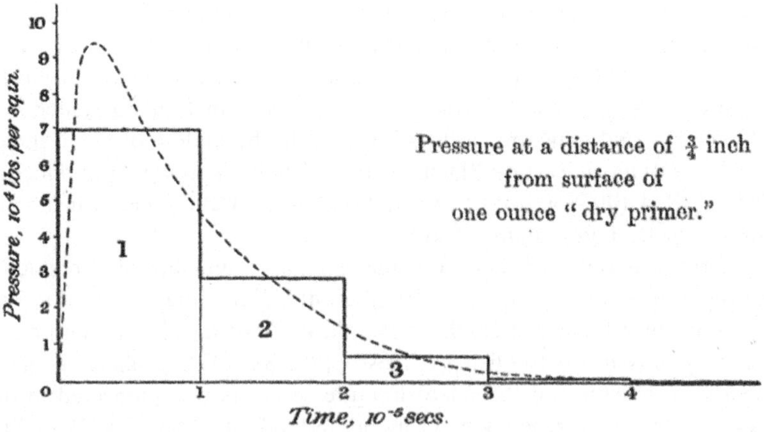

Fig. 1.5 Graph showing the shape of a pulse produced in a steel rod by the detonation of an explosive primer at one end [12]

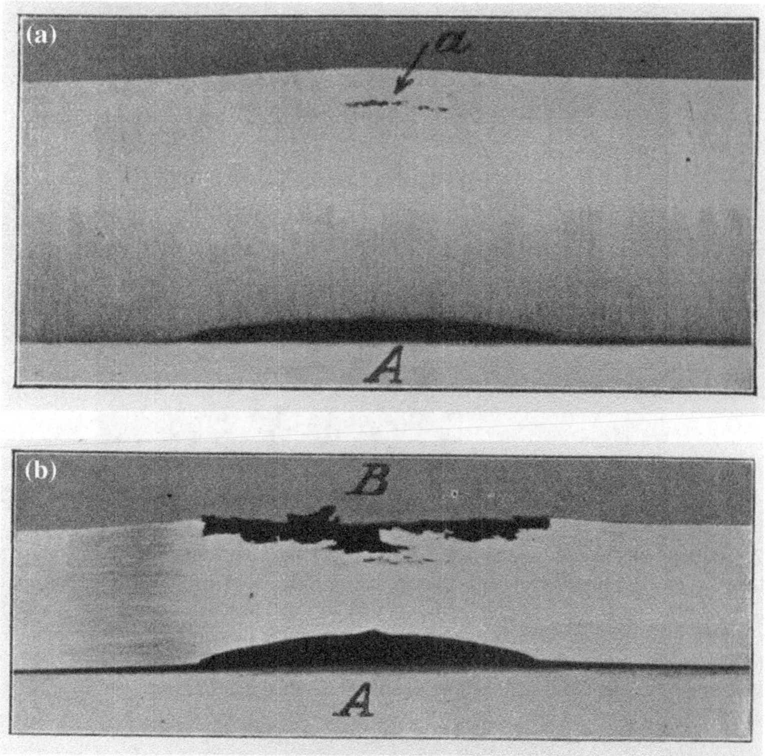

Fig. 1.6 Sections through mild steel plates of **a** 0.75 inch (19 mm) and **b** 1.25 inch (31.8 mm) thicknesses, both of which had been exposed to the detonation of guncotton in contact with side A. From [101]

was in the study of explosives and explosive devices. Its use was revived for this purpose during the Second World War [104, 105] and there have been occasional papers on this use of it right up to the present time [106, 107]. The apparatus was first named in print the 'Hopkinson pressure bar' in a paper by Landon and Quinney published in 1923 [92].

The paper on the technique that bears his name was the last one he himself published (although several of his previously unpublished reports were published posthumously in his collected works [14]). The reason for this was that he died on August 26, 1918 while piloting his own plane. He learnt to fly in 1917 because he had a senior military research role during the First World War (see Fig. 1.7 for a photograph of him in uniform with his colleagues at Orfordness) and flying was the quickest and easiest way of travelling between research stations. He also wished to gain the respect of the pilots he was performing research for (see Appendix). Notable obituaries appeared in the *Proceedings of the Royal Society Series A* [16] and in *The Alpine Journal* [17] as, like his father and siblings, he was a keen mountaineer. An assessment of his life and work was published many years later

Fig. 1.7 Top Row: R. Bourdillon, Holder, W. Barrett, Howarth., Middle Row: Air Vice-Marshal Oxland, R. Collett, Gribble, Vernon Brown., Seated: Prof. B. Hopkinson, J. P. Cooper., Bourdillon was a chemist at Oxford. Holder, with Saundby, was awarded the MC for destroying a Zeppelin at Leiston in 1917. Collett was the first man to make a parachute descent from an aeroplane in 1917. Prof. Hopkinson, Professor of Engineering, Cambridge was killed flying in 1918. Cooper was the first Inspector of Accidents in 1920. Photograph and original caption from [108]

(1974) in *Notes and Records of the Royal Society of London* [22]. The obituary in *The Alpine Journal* contains the intriguing statement that when he was an undergraduate at Trinity College, Cambridge his fellow students believed Bertram Hopkinson would be the first man to reach the South Pole.

His funeral took place in King's College Chapel on August 30, 1918 where his name is inscribed on the Fellows' War Memorial. His name is also listed on the War Memorial of Trinity College. His final resting place is St Giles Cemetery, Cambridge (now called the Ascension Parish Burial Ground) where his mother and one of his brothers are also buried close by.

The use of Hopkinson pressure bars as a method of measuring the dynamic properties of materials in compression was first mentioned in the open literature by Taylor in 1946 [109] (although in secret wartime research performed in 1942 Taylor and Davies had used a single Hopkinson bar to measure the dynamic properties of cordite [110]; Figs. 1.8 and 1.9). The apparatus consisted of two Hopkinson bars supported by strings (see Fig. 1.10). It was developed to study the high rate (*ca.* 600 s^{-1}) properties of soft materials such as plastics and rubbers (Fig. 1.11). The published measurements showed that the dynamic flow stress of polyethylene was about twice the quasistatic value. The design was Enrico Volterra's who published his own paper on the technique in Italian in 1948 [111]. Volterra stated in his obituary of Davies published in *Nature* in 1958 [112] that

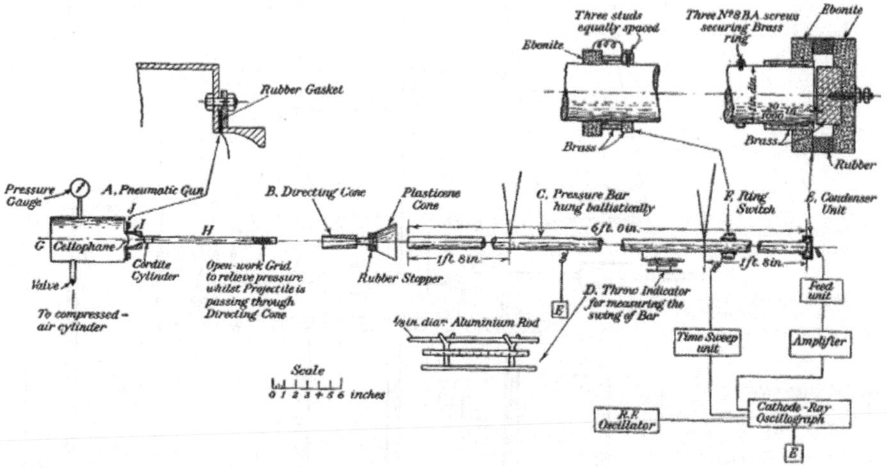

Fig. 1.8 Schematic diagram of the apparatus used by Taylor and Davies in 1942 to measure the dynamic compressive strength of cordite. From [110]

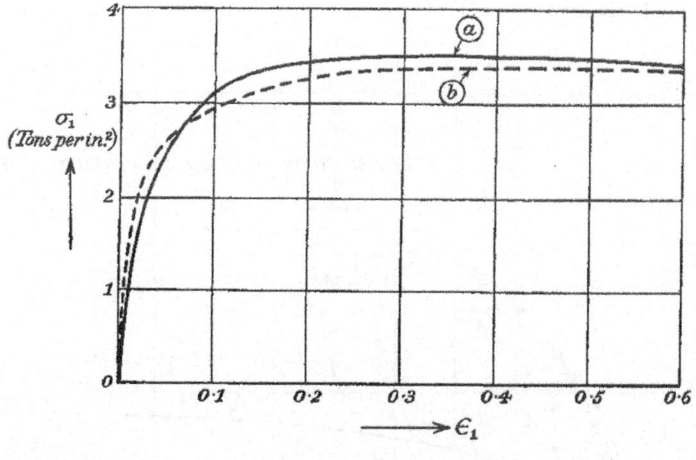

Fig. 1.9 Dynamic true stress-true strain curves for cordite obtained using the single Hopkinson bar apparatus shown in Fig. 1.7 and calculated in two different ways. From [110]

Hopkinson's pressure bar had been largely neglected until Davies resurrected it during World War 2).

This pendulum arrangement could not produce stresses great enough to deform metals. So Herbert Kolsky designed a split-bar system that was dynamically loaded at one end using a detonator [113] (Fig. 1.12). This method of explosive loading was only used once more in a compression split Hopkinson pressure bar system, namely the one he helped set up at Fort Halstead in the UK in the late 1950s [114–116]. A gun-loaded system had previously been developed and installed at the

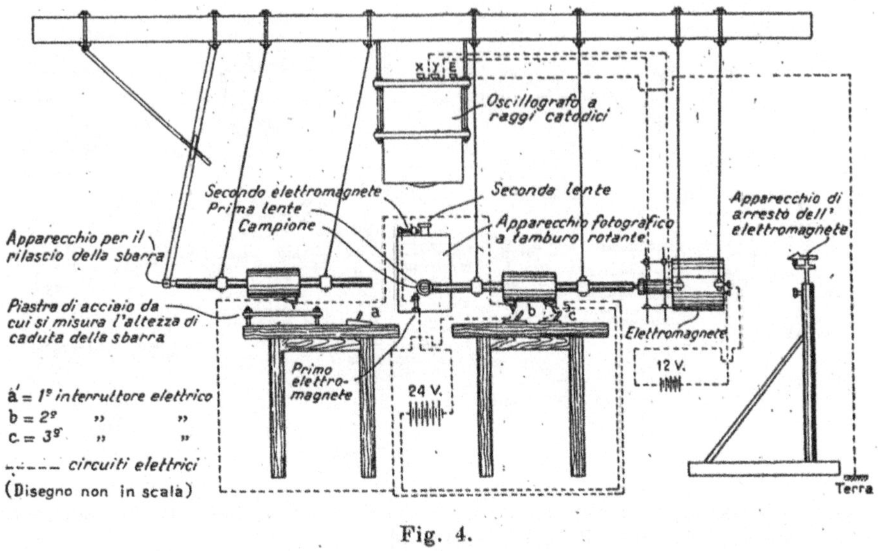

Fig. 1.10 Enrico Volterra's double Hopkinson bar apparatus for measuring the dynamic properties of polymers and rubbers [111]

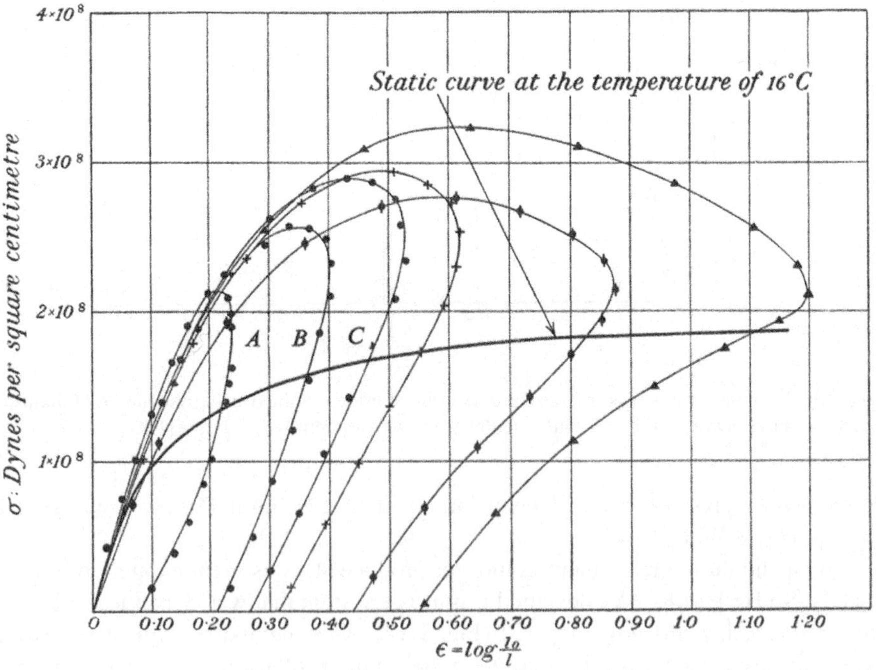

Fig. 1.11 First known compression high strain rate curves (for polyethylene) obtained using a double Hopkinson pressure bar apparatus [109, 111]

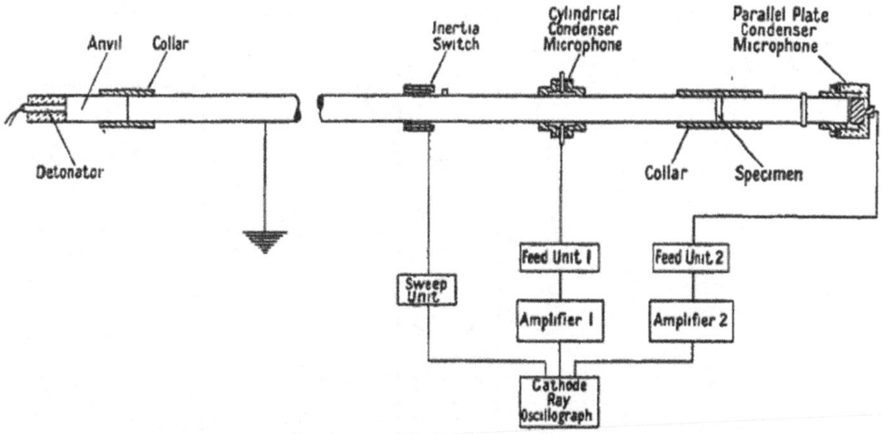

Fig. 1.12 Double Hopkinson bar apparatus reported by Kolsky in 1949 for obtaining dynamic stress-strain curves. From [113]

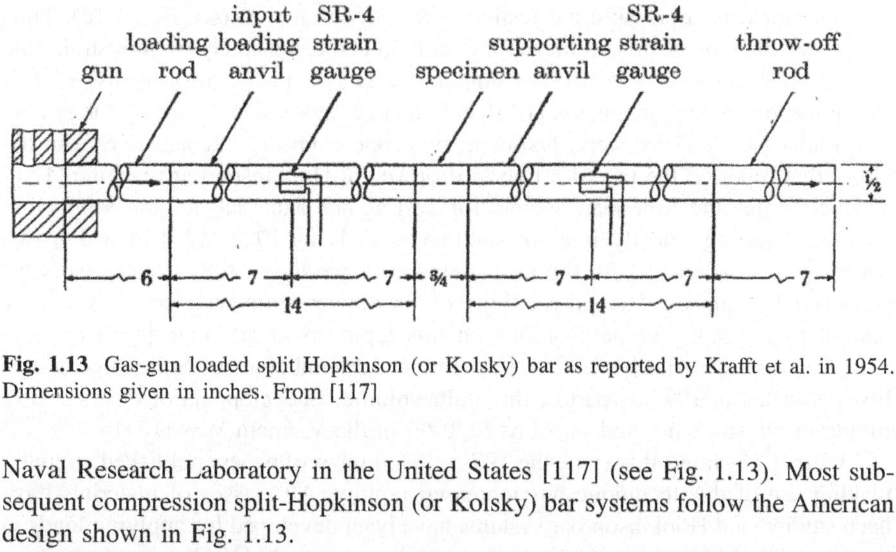

Fig. 1.13 Gas-gun loaded split Hopkinson (or Kolsky) bar as reported by Krafft et al. in 1954. Dimensions given in inches. From [117]

Naval Research Laboratory in the United States [117] (see Fig. 1.13). Most subsequent compression split-Hopkinson (or Kolsky) bar systems follow the American design shown in Fig. 1.13.

Kolsky reported high strain rates results for polyethylene, rubber, Perspex, copper and lead, but he only compared the dynamic and quasistatic stress-strain curves for three of them, namely polyethylene, perspex and copper. All three of these materials supported substantially greater stresses under dynamic loading (e.g. Fig. 1.14). Note that Kolsky did not quote a value for the strain rates. As far as I have been able to ascertain, the only previous researcher to have compared the static and dynamic compressive strengths of a metal (copper) was Dunn in 1897 (Fig. 1.15) [118, 119] although Thurston in his comments on Dunn's 1897 paper [120] stated that he had previously found that the work needed to deform copper in a dropweight machine increased with the impact velocity.

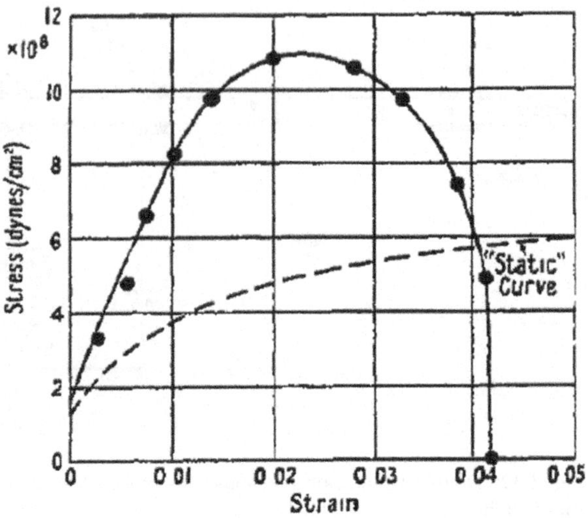

Fig. 1.14 Comparison of dynamic and static stress-strain curves for copper as reported by Kolsky in 1949. From [113]

Split Hopkinson pressure bar testing was slow to take off (see Fig. 1.16). Thus the proceedings of the first conference ever held dedicated solely to high strain rate testing (in London in 1957) [121] contains only two papers making use of this technique, one loaded in tension [122] and one in compression loaded by the impact of a bullet [123]. There were, however, two (one from the UK and one from the USA) that reported the use of a hybrid dropweight-Hopkinson bar machine [124–126]. Note that the American Society for Testing and Materials held a symposium on impact testing (mostly fracture studies) as early as 1922 [127] in which two papers were concerned with the measurement of pressures [128] and forces [129] produced by impact. The proceedings of this symposium also contains a very helpful bibliography of publications on this topic up to that date [130] as does J. F. Bell's magisterial book 'The Experimental Foundations of Solid Mechanics' first published in 1973 as part of the multi volume 'Encyclopedia of Physics' as a tribute to his son who died on May 27 1969 in the Vietnam War [131].

Figure 1.16 shows that since the 1970s, the number of papers published annually making use of this technique has increased rapidly. All classes of materials have been studied and Hopkinson bar systems have been developed for applying loads in tension [122, 132], torsion [133–135], as well as biaxially [136].

To summarise, although Bertram Hopkinson had a research interest in the dynamic strengths of materials, the bar technique he published in 1914 was developed to roughly determine the shapes of stress pulses produced by bullet impact and explosions. His pressure bar proved useful to the British explosives and armour industry during the Great War allowing them to improve their munitions. It also allowed the phenomenon of spall of metal plates produced by ballistic impact or by explosions to be understood, but I do not know whether this led to improved designs of tank armour or battleships during that conflict. Another greater war had to occur before Hopkinson's pressure bar was applied by Taylor, Davies and

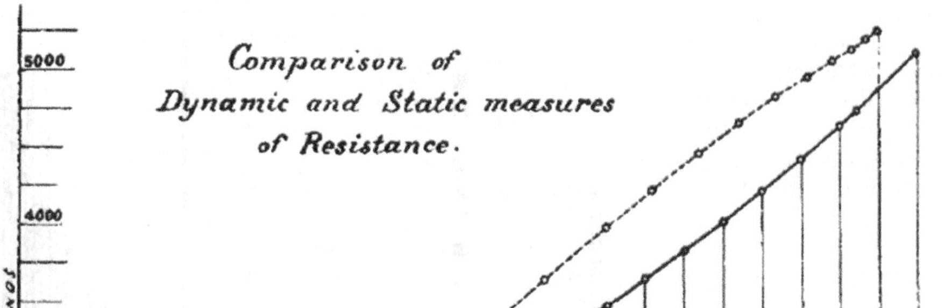

Fig. 1.15 Comparison of dynamic and static compressive strengths of copper as reported by Dunn in 1897. From [118]

Volterra in the UK to the measurement of the dynamic compressive properties of explosives and polymers. Unlike after the First World War, why this research was thought important to the British armed forces during the Second World War was not explained in the declassified reports published once the war was over.

Soon after the end of the Second World War, Kolsky made a number of improvements to the double Hopkinson bar system including using detonators to substantially increase the stresses that could be applied to specimens. This made possible for the first time the accurate determination of the dynamic compressive stress-strain curves of metals. Detonators are, however, impractical for all but defence labs to obtain and use. So an alternative method of loading was needed, and this was developed by Krafft and co-workers at the Naval Research Laboratory in Washington DC in the early 1950s. Their design used a gun to launch a projectile rod against one end of the double bar system and this American method eventually became standard in all laboratories that use the split Hopkinson pressure bar technique. However, it was not until the end of the 1970s that split Hopkinson bars

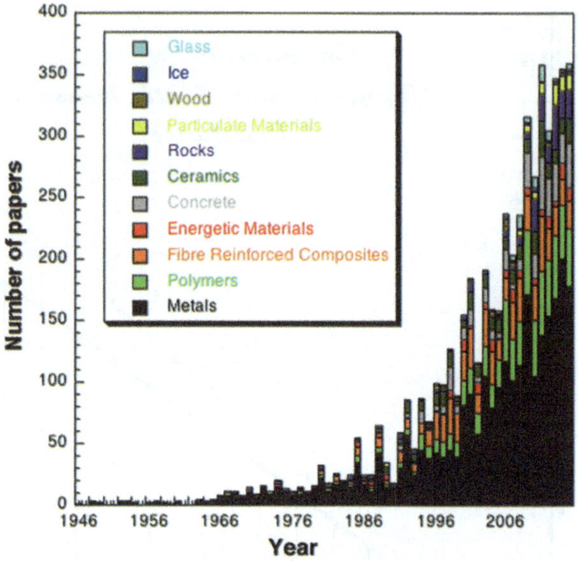

Fig. 1.16 Histogram showing the annual number of papers published between 1946 and 2015 that report results for various classes of materials obtained using the Hopkinson pressure bar

began to be built in laboratories in an ever increasing number of countries around the world. For example, the first international DYMAT conference held in Paris in 1985 [137] contained 21 Hopkinson bar papers of which 10 originated from France, 3 from the USA, 3 from the UK, 2 from Italy, 1 from China, 1 from Czechoslovakia, and 1 from Sweden. By contrast, in the most recent DYMAT conference held in 2015 at Lugano, Switzerland [138] 73 papers involving the use of the Hopkinson bar were presented by researchers from 22 countries.

Acknowledgements I would like to thank Bradley Dodd and Lewis J. Lea for their comments on this chapter.

Appendix

Transcript of a radio talk for sixth-formers about Professor Bertram Hopkinson broadcast on the BBC on March 5, 1937 by Sir Henry Thomas Tizard. Original transcript typed by Mariana Hopkinson-Woolley.

When I agreed to be one of the speakers to talk to you about someone who had influenced me in my life, I began to wonder whom I should choose among so many possible choices. When you come to my mature age, and look back on your life, you will realise how many people have influenced you in one way or another. Everyone is influenced by books; that is why it is a good thing, especially when you're young, to read widely whatever takes your fancy; it helps to bring out what good and what powers you have in you. But on the whole I think that most of us are far more influenced by people we have known than by people we have never seen, however celebrated their words my be. Even schoolmasters have an influence on us, odd though that may appear to you who are still at school. I know that I remember some of my schoolmasters with much greater respect than I ever showed them when they had the depressing job of teaching me. One of them, who is dead, I remember with special affection. He used to teach English to what he called the young barbarians of the Science Sixth, and stung by this doubtless well-meaning contempt, we used to retaliate by… no, perhaps I'd better not say what we did. Anyway he managed, I don't know how, to inspire us all with a love of English literature: perhaps he was helped by not having to work us up for an examination.

Even the greatest of scientists has but a passing influence on men. His work, it's true, may profoundly affect the growth of civilisation, but it gets lost in the growth, as much of the seed gets lost in the tree. The essays of Thomas Henry Huxley still live, and by reading them we can form a clear idea of his character, but his scientific work is rapidly disappearing in an ever-growing mass of unread scientific papers. The number of people who study the works of Isaac Newton must be negligible compared to the number who read and discuss the views of Socrates. How many of you, I wonder, know anything of Isaac Newton as a man? The fact is that the growth of scientific knowledge is so rapid that by comparison art, literature and morals seem to stand still, and the personal influence of a great man of science disappears almost effectively in a century or so as that of the artists who first made wall paintings on the caves of paleolithic man. The scientist exerts his greatest influence on his contemporaries, and greatest of all on those who work with him and know him well.

For all these reasons, reinforced by the somewhat astonishing news that the three other speakers in this series of talks were each going to talk to you about eminent foreigners, long since dead, I thought that I would tell you about a great Englishman, whose name probably conveys nothing to most of you, and whom I knew intimately in the Great War. His name is Bertram Hopkinson. He was educated at St Paul's School and Trinity College, Cambridge, and he was Professor of Engineering at Cambridge University from 1903 until his death in 1918.

But first, let me tell you a little about his father, John Hopkinson. He was one of the great engineers of the latter part of the last century. Only three years after he became Senior Wrangler at Cambridge he was busily engaged in revolutionising the optical system of lighthouses and introducing the flashing light system which is so familiar today. Then he became a consulting engineer and devoted himself mainly to the problem of the young electrical industry. The great developments that are so familiar to us today were then only at their very beginnings, and electric lighting was just a scientific curiosity. Dynamos and motors were crude affairs, designed by hit-and-miss methods. John Hopkinson changed all that; and by studying the principle of design, developing methods of testing, and in many other ways, put the young industry on its feet. He must have been a remarkable man; among other thing he broke the record for the mile at Cambridge three weeks before his final examination. He was passionately fond of climbing and was killed in the Alps along with two daughters and a son in 1898.

When this appalling tragedy happened, Bertram Hopkinson, his eldest son, was on the way to Australia. Like his father, he had taken a mathematical degree, but had then become a barrister. He came back from Australia immediately and decided to become an engineer and go on with his father's work. That must have required some courage, for he had never had any training as an engineer, apart from helping his father in experiments. However, he soon succeeded, and five years later was made Professor of Engineering at Cambridge, when he was only 29 years old. The electors to that professorship deserve to have their names written in gold in the Engineering Laboratory.

Let me try to describe him to you, so that you may get some kind of picture of him, even if it is very imperfect. He was a big man—perhaps not so very big now I come to think about it—but he seemed big. There was nothing small about him. He had a big voice, a big laugh, a stride, and a big mind and a big soul—and a big moustache. He was gruff in manner, and perhaps a little forbidding to younger men when they first met him; but that soon wore off. I am told by those who knew him before the war that he was rather unapproachable, as if he had a poor opinion of humanity; but if so, that must have been due to a form of shyness, and the war must have done him a lot of good, as it did to so many of us, by bringing us into close contact with all sorts and conditions and ages of men, all working with a common object. I am sure that you would not have thought him unapproachable if you had seen him on guest-night in an officers' mess, when strained nerves relaxed and everyone was thoroughly childish. He must have been very like his father in many ways; he had the same rare combination of physical and mental powers. I don't think I can better describe him than in the words of a friend of his and of mine, who said: "He was the kind of man you would have chosen to lead you on a Polar expedition. He had the courage, the strength, the stamina, and the intelligence."

I am not going to tell you anything about his scientific work before the war. I had better confine myself to what he did when I knew him. I first met him early in 1916 or late in 1915, I forget which, when I was one of a small group of men stationed at Upavon, and doing experimental work for the Royal Flying Corps, as it was then. Hopkinson had been engaged mainly on work for the Admiralty up to then, but someone, I don't know whom, had the sense to see that he was given charge of the experimental work for the Flying Corps. He was exactly fitted for this. You must remember that it was then only some ten years since Orville Wright made his first flight, and the use of aeroplanes in war was quite new. The Royal Aircraft Factory existed at Farnborough, and a good deal of experimental work on the design of aeroplanes and engines had been done there and elsewhere, but practically nothing on the development of the aeroplane as a weapon. At the beginning of the war there was no armament in aeroplanes, and English and German pilots used to wave greetings to each other as they passed upon their errands. But soon a Navy of the air came into being, and problems arose and there was little experience to help in their solution. The oddest ideas prevailed. Clouds were thought to be magnetic, because pilots lost all sense of direction in them, and the compass consequently lost its head. No-one knew how to aim a bomb. I remember pilots who argued that if eggs were dropped from a sufficient height they would be going so slowly when they reached the ground that they would not break. One of them actually made the experiment, but as the eggs were never found, the problem remained unsolved.

Looking back, it seems to me that some of us young scientists who were engaged on the more serious problems, under great difficulties, were really only playing with them until Hopkinson came along and put fresh life into us. One of his first acts was to found a station at Orfordness on the east coast of England, for experimental work on air aviation, gunnery and bombing and many other distasteful tasks that had to be done then, and still have to be done. Then he founded at Martlesham Heath, near Woodbridge, another large experimental station for the testing of all new aeroplanes, and for providing designers with information to help them. Orfordness was practically given up after the war, but Martlesham remains, a monument to his work.

You can imagine, perhaps, the effect that all these developments had on young men like me, whose experiments had been confined to the laboratory, and had little or nothing to do with practical affairs. It was not so much the scale of the work that was impressive as the way it was done. It is just as easy to make mistakes on a large scale as on a small—and much more expensive. What was impressive was that there were so few mistakes; so few that I cannot think of any worth mentioning. Hopkinson seemed to combine all the best qualities of a great scientist, the great soldier and the great organiser. If he had merely been an organiser, the scientific work of the stations would have been left without a leader, and would have been largely misdirected; if he been only a great scientist he would never have gained the complete confidence of practical men. As it was, I doubt if anywhere else could have been found a group of men, so different in their outlook and their interests, and yet so happily welded together by the influence of one man. Everyone had confidence in him, and he never failed them.

But the greatest influence he had on me came from his power to go to the very heart of a problem. In all engineering, in all scientific research, indeed one might say in everything in life, there are always so many things that might be done; the difficulty is to know what best to do. In war, to waste time over non-essentials may mean the unnecessary loss of many lives; may mean indeed the difference between victory and defeat. In peace, in all national affairs, it may mean the difference between prosperity and unrest. Hopkinson swept aside non-essentials with unerring instinct, and taught one to concentrate on things that mattered. I believe that this is one of the real hallmarks of genius; you will find it in the great statesman, the great doctor, the great teacher, as well as in the great scientist. It is not a gift that can be easily passed on or we should all be geniuses. But to work in close association with someone who had it to such a supreme degree, and who carried it into everyday life, was an education not merely in engineering and organisation, but in living.

In the midst of all this work he decided to learn to fly. He had then passed the venerable age of forty. Somehow that age does not now seem to me to be so advanced as it did then; it has more of a retreated air. But even now it is unusual for men of over forty years of age to learn to fly, and it was very different then when the art was still quite new, when engine failures were far more frequent, and when there were few reliable instruments to assist the pilot in bad weather. Some of us tried to dissuade Hopkinson from learning, but only half-heartedly. It was not as if he did it for the pure love of adventure—although he had more of that than most people; I do not think he was the kind of man to risk his life unnecessarily when there was important work to be done. The real object and advantage of his flying was that it improved his sense of perspective; it helped him to realise the things that matter in the air. For things that seemed easy for other people to do or use in the air when one was sitting safely on the ground assumed quite a different aspect when one was in the cramped and cold cockpit of an aeroplane, cumbered with heavy clothing—quite apart from the fact that other misguided people might be trying to shoot one down. Quite a lot a time and money was wasted in the early stages of experimental work through an insufficient knowledge of practical conditions. Perhaps it still is.

Hopkinson set us all a new standard, and by showing that he was prepared to share the risks that others took, inspired everyone who worked under him. There can be no doubt that it was right of him to fly; but in the end what we all feared happened, and he was killed in an accident when flying alone back to London in bad weather, twenty years, to within a day, after his father had been killed in the Alps.

Shortly before his death, foreseeing the early end of the war, he had been engaged on a plan for aeronautical research in peacetime. Much has been done since then, but his full scheme never came off, because there was no-one good enough to take his place. It is seldom that one can say that truthfully of anyone, but it was true of him. Had he lived he would have done for the aircraft industry all, and perhaps more than all, that his father did for the electrical industry; as it was, he did more for his country in a few short years than most people do in a lifetime. I do not like to feel that his name is forgotten; that is why I have told you something about him.

It is barely nineteen years since Hopkinson was killed: there must be few boys now at school who were born before he died. But I have a very vivid memory of him. I can now see him quite plainly, sitting in a chair in a characteristic attitude, brooding over some technical problem; or dominating by sheer force of intellect and character a committee of senior officers rudely called 'brass hats'; or even playing a flippant game of bridge, at which he was a singularly bad performer, and hesitating which card to play until I said, "Better play the ace", when out it would come with a laugh and a surprised, "How did you know I'd got it?"

But best of all I can remember him in the open air. I can see myself struggling to keep up with him when walking across the park—though I was over ten years younger—and hoping he did not observe the effort. I can see him standing by an aeroplane whose engine was being run up, telling a story in a stentorian voice which carried above the roar of a propeller, and coming to the point just at the moment when the engine was throttled down, with disturbing results on the bystanders. Most vividly of all I seem to see him sitting in front of me as my passenger in an aeroplane when we were flying back one Sunday from Martlesham Heath to London. We had been having an argument at lunch about the effect of thunderstorms on aeroplanes. Somebody had said that when flying near a thundercloud the wires of an aeroplane would glow with electrical discharges, and we had said that that was very unlikely. We started off from Martlesham in beautiful clear weather, with a few clouds about in the distance. Before we had gone halfway the clouds in the east had increased in number and size, and looked dark and threatening. I was looking backward at one which was fairly close and saw a flash of lightning pass between it and another cloud. I am sorry to say that I experienced no violent desire to investigate the matter further, and kept steadily upon my course. But soon afterwards Hopkinson began looking round and, as luck would have it, saw another lightning flash behind us. He pointed eagerly in its direction, and I registered suitable astonishment. Then he passed me back a note with "Let us go and see if the wires glow". So we turned, and flew towards the thundercloud, and round its edges. Then we went inside for a short time. It was quite dark inside, and the cloud was full of hailstones, as I discovered by incautiously putting my head out from behind the shelter of the windscreen. We got out all right without observing any glow from the wires; but when we eventually landed we found that the hailstones had chipped the propeller, and frayed the fabric on the leading edges of the wings so much that the wings had to be re-covered. However, we did not know that while we were in the air, so we were quite happy. I don't suppose the risk was really very great, as aeroplanes have often since been struck by lightning without, so far as I know, any serious results, although it must be alarming at the time. It was an interesting experience, which I should certainly have missed if I had not been with Hopkinson. He seemed to lose all sense of danger when there was something interesting to be done.

All my memories of him are happy ones; there is no-one who served under him in the war who is ever likely to forget him. I hope that you, too, will have the great good fortune to know early in your life someone like him, so vigorous in mind and body, and yet so unself-seeking; so full of the joy of living, yet with so little fear of dying; who will teach you almost unconsciously to do whatever job comes to hand as well and as thoroughly as you can, and who will help you by his own example not to worry over trifles, but to spend your lives on things that matter.

References

1. Field JE, Walley SM, Bourne NK, Huntley JM (1994) Experimental methods at high rates of strain. J Phys IV France 4(C8):3–22
2. Field JE, Walley SM, Bourne NK, Huntley JM (1998) Review of experimental techniques for high rate deformation studies. In: Proceedings of the Acoustics and Vibration Asia '98, Acoustics and Vibration Asia 98 Conference, Singapore, pp 9–38
3. Gray GT III (2000) Classic split-Hopkinson pressure bar testing. In: Kuhn H, Medlin D (eds) ASM Handbook, vol 8. Mechanical testing and evaluation. ASM International, Materials Park, pp 462–476
4. Field JE, Proud WG, Walley SM, Goldrein HT (2001) Review of experimental techniques for high rate deformation and shock studies. In: Nowacki WK, Klepaczko JR (eds) New experimental methods in material dynamics and impact. Institute of Fundamental Technological Research, Warsaw, Poland, pp 109–177

5. Field JE, Walley SM, Proud WG, Goldrein HT, Siviour CR (2004) Review of experimental techniques for high rate deformation and shock studies. Int J Impact Eng 30:725–775
6. Gama BA, Lopatnikov SL, Gillespie JW Jr (2004) Hopkinson bar experimental technique: A critical review. Appl Mech Rev 57:223–250
7. Chen W, Song B (2011) Split Hopkinson (Kolsky) Bar: design, testing and applications. Springer, New York
8. Alves M (2014) Material behavior at a wide range of strain rates. In: Hashmi S (ed) Comprehensive materials processing, vol 2. Materials modeling and characterization. Elsevier, Amsterdam, pp 95–110
9. Walley S, Eakins D (2014) Introduction to the theme issue 'Shock and blast: celebrating the centenary of Bertram Hopkinson's seminal paper of 1914'. Philos Trans R Soc A 372:20130220
10. Walley S, MacGillivray H, Field J, Eakins D, Pierron F, Siviour C (eds) (2014) Shock and Blast: celebrating the centenary of Bertram Hopkinson's seminal paper of 1914 (Part 1) (Philos Trans R Soc A 372, no. 2015). The Royal Society, London
11. Walley S, MacGillivray H, Field J, Eakins D, Pierron F, Siviour C (eds) (2014) Shock and Blast: celebrating the centenary of Bertram Hopkinson's seminal paper of 1914 (Part 2) (Philos Trans R Soc A 372, no. 2023). The Royal Society, London
12. Hopkinson B (1914) A method of measuring the pressure produced in the detonation of high explosives or by the impact of bullets. Philos Trans R Soc Lond A 213:437–456
13. Hopkinson B (1921) The pressure of a blow. In: Ewing JA, Larmor J (eds) The scientific papers of Bertram Hopkinson. Cambridge University Press, Cambridge, pp 423–437
14. Ewing JA, Larmor J (eds) (1921) The scientific papers of Bertram Hopkinson. Cambridge University Press, Cambridge
15. Hopkinson E (1928) The story of a Mid-Victorian Girl. Cambridge University Press, Cambridge
16. Ewing JA (1919) Bertram Hopkinson, 1874–1918. Proc R Soc Lond A 95:xxvi-xxxvi
17. Hill AV (1920) Obituary of Bertram Hopkinson. Alpine J 32:353–357
18. Anon (1921) Obituary of Bertram Hopkinson, 1874–1918. Min Proc Inst Civil Eng 212:425–426
19. Clark RW (1965) Tizard. Methuen, London, pp 37–48
20. Ricardo HR (1945) Applied research. Proc Inst Mech Eng 152:143–148
21. Hilken TJN (1967) Bertram Hopkinson: professor of mechanism and applied mechanics, 1903–1918. Engineering at Cambridge University 1783–1965. Cambridge University Press, Cambridge, pp 129–144
22. Charlton TM (1974) Professor Bertram Hopkinson, C.M.G., M.A., B.Sc., F.R.S. (1874–1918). Notes Records R Soc Lond 29:101–109
23. Anon. (1817) On chain bridges. Philos Mag (Ser 1) 49:371–375
24. Hodgkinson E (1831) On the chain bridge at Broughton. Mem Liter Philos Soc Manchester (Ser 2) 5:384–397, 545–553
25. Hodgkinson E (1831) A few remarks on the Menai Bridge. Mem Liter Philos Soc Manchester (Ser 2) 5:398–406
26. Barlow P (1837) A treatise on the strength of timber, cast iron, malleable iron, and other materials, with rules for application in architecture, construction of suspension bridges, railways etc. John Weale, London
27. Paxton RA (1979) Menai Bridge (1818–26) and its influence of suspension bridge development. Trans Newcomen Soc 49:87–110
28. James JG (1990) Some steps in the evolution of early iron arched bridge designs. Trans Newcomen Soc 59:153–185
29. Tredgold T (1822) Practical essay on the strength of cast iron, and other metals; intended for the assistance of engineers, iron masters, architects, millwrights, founders, and others engaged in the construction of machines, buildings, &c. Containing practical rules, tables, and examples; founded on a series of new experiments, with an extensive table to the properties of materials. J Taylor, London

30. Anon (1815) Dreadful accident occasioned by the explosion of a boiler at Mr. Constant's sugar-house in Well-street on the 15th of November. Philos Mag (Ser 1) 46:372–375
31. Taylor J (1827) On the accidents incident to steam boilers. Philos Mag (Ser 2) 1:126–133
32. Earle T (1831) On the causes of some explosions of steam boilers. J Franklin Inst 11:154–156
33. Airy GB (1863) On the numerical expression of the destructive energy in the explosions of steam-boilers, and on its comparison with the destructive energy of gunpowder. Philos Mag (Ser 4) 26:329–336
34. Marten EB (1870) On the conclusions derived from the experience of recent steam boiler explosions. Proc Inst Mech Eng 21:179–218
35. Mollison J (1927) Historical references to the progress in the use of high-pressure steam. Trans Inst Eng Shipbuilders Scotland 70:783–803
36. Winship IR (1991) The decline in locomotive boiler explosions in Britain 1850–1900. Trans Newcomen Soc 60:73–94
37. Ford J (1862) On the manufacture of armour plates. Trans Inst Naval Architects 3:144–146
38. Russell JS (1863) On the present state of the question of issue between modern guns and iron-coated ships. Trans Inst Naval Architects 4:20–30
39. Heath LG (1867) On the construction of the sides of armour-plated ships. Trans Inst Naval Architects 8:117–128
40. Barnaby N (1879) Armour for ships. Trans Inst Naval Architects 20:27–45
41. Robinson RS (1879) On armour-plating ships of war. Trans Inst Naval Architects 20:1–26, plate I
42. Russell JS (1880) On the true nature of the resistance of armour to shot. Trans Inst Naval Architects 21:69–95, plate IV
43. FitzGerald CP (1889) On the possible effect of high explosives in future designs for warships. Trans Inst Naval Architects 30:130–136
44. Barba J (1891) Recent improvements in armour plates for ships. Trans Inst Naval Architects 32:138–163
45. Ellis CE (1911) Armour for ships (1860 to 1910). Trans Inst Naval Architects 53(2):338–350
46. Watts P (1911) Warship building (1860–1910). Trans Inst Naval Architects 53(2):291–337, plates XXXIV-XLI
47. Thompson B (1798) Experiments to determine the force of fired gunpowder. Philos Trans R Soc Lond 87:222–292
48. Bagnold ME (1826) On the Indian methods of working iron and steel, for the damascus gun-barrels and sword-blades. J Franklin Inst 2:1–3
49. Greener WW (1835) The gun: or a treatise on the various descriptions of small fire-arms. Longman, London
50. Wilkinson H (1841) Engines of war: historical and experimental observations on ancient and modern war-like machines and implements, including the manufacture of guns, gunpowder, and swords, with remarks on bronze, iron, steel & c. Longman, London
51. Rodman TJ (1861) Reports of experiments on the properties of metals for cannon, and the qualities of cannon powder, with an account of the fabrication and trial of a 15 inch gun etc. H.C. Baird, Philadelphia, PA
52. Whitworth J (1866) On the proof of guns by measurement with description of the instrument employed. Proc Inst Mech Eng 17:105–114
53. Merrifield CW (1879) On naval guns. Trans Inst Naval Architects 20:46–58
54. Samuda JD (1882) Armoured ships and modern guns. Trans Inst Naval Architects 23:1–12
55. Cubillo L (1888) On the manufacture and treatment of steel for field guns. J Iron Steel Inst 33:123–150
56. Cubillo L (1888) On steel castings for the manufacture of guns. J Iron Steel Inst 33:151–170
57. Roberts-Austen WC (1898) The action of the projectile and of the explosives on the tubes of steel guns. J Iron Steel Inst 54:233–237, plates XXIII–XXV

58. Reynolds O (1900) On the bursting of the gun on board 'The Thunderer'. Papers on mechanical and physical subjects, vol 1. Cambridge University Press, Cambridge, pp 399–402
59. Braid DHP (1986) The armament of naval ships in the nineteenth century. Trans Newcomen Soc 56:111–132
60. Clanny WR (1813) A method of procuring a steady light in coal mines without the danger of explosion. Phil Trans R Soc Lond 103:200–205
61. Davy H (1816) On the fire-damp of coal mines, and on methods of lighting the mines so as to prevent its explosion. Phil Trans R Soc Lond 106:1–22
62. Little G (1850) Preventing explosions in coal mines. J Franklin Inst 49:8–10
63. Galloway W (1875) On the influence of coal dust in colliery explosions. Proc R Soc Lond 24:354–372
64. Freire-Marreco A, Morison DP (1878) Results of some experiments on the effect of coal dust in colliery explosions. Trans Chesterfield Derbyshire Inst Min Civil Mech Eng 5:267–277
65. Atkinson WN, Atkinson JB (1886) Explosions in coal mines. Longmans, Green & Co., London
66. Stuart DMD (1894) Coal dust as an explosive agent. E. & F.N, Spon, London
67. de Réaumur RA (1722) L'Art de Convertir le Fer Forgé en Acier et l'Art d'Adoucir le Fer Fondu. Michel Brunet, Paris
68. de Réaumur RA (1956) Réaumur's memoirs on steel and iron (translated by AG Sisco). University of Chicago Press, Chicago, IL
69. Young T (1807) A course of lectures on natural philosophy and the mechanical arts: vol 1, pp 144–148. Joseph Johnson, London
70. Stodart J, Faraday M (1820) Experiments on the alloys of steel made with a view to its improvement. Philos Mag (Ser 1) 56:26–35
71. Hodgkinson E (1824) On the transverse strain and strength of materials. Mem Liter Philos Soc Manchester (Ser 2) 4:225–289
72. Tredgold T, Hodgkinson E (1842) Practical essay on the strength of cast iron, and other metals; Containing Practical rules, tables, and examples; Founded on a series of new experiments, with an extensive table of the properties of materials. The Fourth Edition to Which are Added Experimental Researches on the Strength and Other Properties of Cast Iron with the Development of New Principles; Calculations Deduced from Them; and Inquiries Applicable to Rigid and Tenacious Bodies Generally. John Weale, London
73. Hadfield RA (ed) (1931) Faraday and his metallurgical researches. Chapman & Hall, London
74. Wade W (1856) Hardness of metals. In: Craig HK (ed) Reports on experiments on the strength and other properties of metals for cannon with a description of the machines for testing metals, and of the classification of cannon in service. Henry Carey Baird, Philadelphia, PA, pp. 259–275, 313–314, plate XVI
75. Kirkaldy D (1862) Results of an experimental enquiry into the comparative tensile strength and other properties of various kinds of wrought iron and steel etc. Bell & Bain, Glasgow, UK
76. Kirkaldy D (1866) Experiments on wrought-iron and steel, 2nd edn. D. Kirkaldy Publishing, London
77. Smith D (1982) David Kirkaldy (1820–1897) and engineering materials testing. Trans Newcomen Soc 52:49–65
78. Rankine WJM (1858) A manual of applied mechanics. Richard Griffin and Company, London
79. Rankine WJM (1862) A Manual of civil engineering. Griffin, Bohn and Company, London
80. Tresca H (1867) On the flow of solids with the practical application in forgings etc. Proc Inst Mech Engrs 18:114–150
81. Tresca H (1878) On further applications of the flow of solids. Proc Inst Mech Engrs 30:301–345
82. Osmond F (1893) Microscopic metallography. Trans AIME 22:243–265

83. Osmond F, Stead JE (1904) Microscopic analysis of metals. Charles Griffin, London
84. Rosenhain W (1912) Floris Osmond. Nature 89:454–455
85. Sauveur A (1893) Microstructure of steel. Trans AIME 22:546–590
86. Sauveur A (ed) (1912) The metallography and heat treatment of iron and steel. McGraw-Hill, New York
87. Sauveur A (1920) Improvements in metallurgical microscopes. Trans Farad Soc 16:181–182
88. Sauveur A (1937) Metallurgical reminiscences. American Institute of Mining and Metallurgical Engineers, Englewood CO
89. Hooke R (1665) Micrographia: or some physiological descriptions of minute bodies made by magnifying glasses with observations and inquiries thereupon. The Royal Society, London
90. Osmond F, Cartaud A (1913) The science of polishing. In: Stead JE, Sidney LP (eds) The microscopic analysis of metals. Charles Griffin, London, pp 58–103
91. Guest JJ (1900) On the strength of ductile materials under combined stress. Philos Mag (Ser 5) 50:69–132
92. Landon JW, Quinney H (1923) Experiments with the Hopkinson pressure bar. Proc R Soc Lond A 103:622–643
93. Hopkinson B (1914) The effects of the detonation of gun-cotton. Trans North-East Coast Inst Engineers Shipbuilders 30:199–217
94. Hopkinson B (1906) Explosions of coal-gas and air. Proc R Soc Lond A 77:387–413
95. Hopkinson B (1907) A recording calorimeter for explosions. Proc R Soc Lond A 79:138–154
96. Hopkinson B (1910) On radiation in a gaseous explosion. Proc R Soc Lond A 84:155–172
97. Hopkinson B (1905) The effects of momentary stresses in metals. Proc R Soc Lond 74:498–506
98. Hopkinson J (1872) On the rupture of iron wire by a blow. Proc Manchester Liter Philos Soc 11:40–45
99. Hopkinson J (1872) Further experiments on the rupture of iron wire. Proc Manchester Liter Philos Soc 11:119–121
100. Kolsky H, Shi YY (1958) Fractures produced by stress pulses in glass-like solids. Proc Phys Soc Lond 72:447–453
101. Hopkinson B (1921) The effects of the detonation of gun cotton. In: Ewing JA, Larmor J (eds) The collected scientific papers of Bertram Hopkinson. Cambridge University Press, Cambridge, pp 461–474
102. Robertson R (1921) Some properties of explosives. J Chem Soc, Trans 119:1–29
103. Robertson R (1921) Some war developments of explosives. Nature 107:524–527
104. Davies RM (1948) A critical study of the Hopkinson pressure bar. Phil Trans R Soc Lond A 240:375–457
105. Davies RM, Owen JD, Edwards DH, Thomas DE (1950) Pressure measurements in detonating gases using pressure bar techniques. Proc R Soc Lond A 204:17–19
106. Richard D, Ault S, Qian P, Rutledge G, Leininger L, Baum D, Murphy M (2013) Validation of an impulse measurement technique with high temporal and spatial resolutions. Procedia Eng 58:384–391
107. Clarke SD, Fay SD, Warren JA, Tyas A, Rigby SE, Elgy I (2015) A large scale experimental approach to the measurement of spatially and temporally localised loading from the detonation of shallow-buried explosives. Meas Sci Technol 26:015001
108. Brown V (1966) Gunsight trials at Orfordness in 1916–1917. J Royal Aeronaut Soc 70:247
109. Taylor GI (1946) The testing of materials at high rates of loading. J Inst Civil Engrs 26:486–519
110. Taylor GI, Davies RM (1958) The mechanical properties of cordite during impact stressing (originally written for the ministry of supply of the UK government in (1942) In: Batchelor GK (ed) The scientific papers of Sir Geoffrey Ingram Taylor Vol 1: Mechanics of solids. Cambridge University Press, Cambridge, pp 480–495
111. Volterra E (1948) Alcuni risultati di prove dinamiche sui materiali (Some results on the dynamic testing of materials). Rivista Nuovo Cimento 4:1–28

112. Volterra E (1958) Obituary of Prof. R.M. Davies. Nature 181:879–880
113. Kolsky H (1949) An investigation of the mechanical properties of materials at very high rates of loading. Proc Phys Soc Lond B 62:676–700
114. Hunter SC, Davies EDH (1960) The dynamic compression testing of solids by the methods of the split Hopkinson pressure bar. 1: The theoretical nature of the experiment. Armament Research and Development Establishment, Fort Halstead, Sevenoaks, Kent, Report no. (MX) 7/60
115. Hunter SC, Davies EDH (1960) The dynamic compression testing of solids by the methods of the split Hopkinson pressure bar. 2: Experimental observations of soft metals and polymer materials. Armament Research and Development Establishment, Fort Halstead, Sevenoaks, Kent, Report no. (MX) 8/60
116. Davies EDH, Hunter SC (1963) The dynamic compression testing of solids by the method of the split Hopkinson pressure bar (SHPB). J Mech Phys Solids 11:155–179
117. Krafft JM, Sullivan AM, Tipper CF (1954) The effect of static and dynamic loading and temperature on the yield stress of iron and mild steel in compression. Proc R Soc Lond A 221:114–127
118. Dunn BW (1897) A photographic impact testing machine for measuring the varying intensity of an impulsive force. J Franklin Inst 144:321–348
119. Dunn BW (1909) Measurement of impact stresses. Proc Amer Soc Testing Materials 9:644–651
120. Thurston RH, Johnson JB, Gibbs G, Church IP, Christie J, Dunn BW (1898) Discussion of 'A photographic impact testing machine for measuring the varying intensity of an impulsive force'. J Franklin Inst 145:36–47
121. Anon (ed) (1957) Proceedings of the conference on the properties of materials at high rates of strain. Institution of Mechanical Engineers, London
122. Taylor DBC (1957) Non-uniform yield in a mild steel under dynamic straining. In: (eds) Proceedings of conference on the properties of materials at high rates of strain, Institution of Mechanical Engineers, London, pp 229–238
123. Wright PW, Lyon RJ (1957) A photoelectric method of determining stress and strain in rapidly strained metals. Proceedings of conference on the properties of materials at high rates of strain. Institution of Mechanical Engineers, London, pp 37–50
124. Back PAA, Campbell JD (1957) The behaviour of a reinforced plastic material under dynamic compression. Proceedings of conference on the properties of materials at high rates of strain. Institution of Mechanical Engineers, London, pp 221–228
125. Campbell JD, Duby J (1957) Delayed yield and other dynamic loading phenomena in a medium-carbon steel. Proceedings of conference on the properties of materials at high rates of strain. Institution of Mechanical Engineers, London, pp 214–220
126. Vigness I, Krafft JM, Smith RC (1957) Effect of loading history upon the yield strength of a plain carbon steel. Proceedings of conference on the properties of materials at high rates of strain. Institution of Mechanical Engineers, London, pp 138–146
127. Anon (ed) (1922) Symposium on impact testing of materials. American Society for Testing Materials, Philadelphia, PA
128. Margerum CE (1922) Measurement of pressures caused by impact. Proc Amer Soc Testing Materials 22(2):104–116
129. Elmendorf A (1922) Measuring forces in impact. Proc Amer Soc Testing Materials 22 (2):117–123
130. Whittemore HL (1922) Resumé of impact testing of materials, with bibliography. Proc Amer Soc Testing Materials 22(2):6–36
131. Bell JF (1984) Mechanics of solids. 1: The experimental foundations of solid mechanics. Springer, Berlin
132. Harding J, Wood EO, Campbell JD (1960) Tensile testing of materials at impact rates of strain. J Mech Eng Sci 2:88–96
133. Pope DP, Vreeland T Jr, Wood DS (1964) Machine for producing square torsion pulses of microsecond duration. Rev Sci Instrum 35:1351–1355

134. Duffy J, Campbell JD, Hawley RH (1971) On the use of a torsional split Hopkinson bar to study rate effects in 1100-0 aluminum. Trans ASME J Appl Mech 38:83–91
135. Lewis JL, Campbell JD (1972) The development and use of a torsional Hopkinson-bar apparatus. Exper Mech 12:520–524
136. Lewis JL, Goldsmith W (1973) A biaxial split Hopkinson bar for simultaneous torsion and compression. Rev Sci Instrum 44:811–813
137. Anon (ed) (1985) DYMAT 85: International conference on mechanical and physical behaviour of materials under dynamic loading. Les Editions de Physique, Les Ulis, France
138. Cadoni E, Couque H, Hiermaier S (eds) (2015) DYMAT 2015: 11th international conference on the mechanical and physical behaviour of materials under dynamic loading. EDP Sciences, Les Ulis, France

Chapter 2
Tensile Testing Using the Kolsky-Hopkinson Bar Machine

Arild H. Clausen

2.1 Tension Tests

The uniaxial stress-strain curve provides an important piece of information within several fields of engineering. Design of civil engineering and transport-related structures requires knowledge of stiffness properties, yield strength, and possibly also hardening behaviour and fracture characteristics for the material at hand. People working with machining and forming processes have similar demands. Accurate information about the material properties has become even more important during the last decades because the finite element method now is the primary design tool. Simultaneously, new fields of applications, more severe conditions during use and cost-saving arguments require that the materials and structures have to be exploited beyond the limits that they typically were exposed to in the past.

The tension test has been and is the most important way to explore material properties at quasi-static conditions. The main reasons are:

- A tension test with a dog-bone sample provides well-defined boundary conditions and a gauge part that is convenient for determination of the uniaxial response of the material.
- Until the onset of necking, the stress state in the gauge part is purely uniaxial, i.e. 1D, and the strain field is homogeneous. This facilitates a precise interpretation of the test results, including determination of the true stress-strain curve.

A. H. Clausen (✉)
Structural Impact Laboratory (SIMLab), Department of Structural Engineering,
Norwegian University of Science and Technology (NTNU), NO-7491 Trondheim, Norway
e-mail: arild.clausen@ntnu.no

R. Othman (ed.), *The Kolsky-Hopkinson Bar Machine*,
https://doi.org/10.1007/978-3-319-71919-1_2

- The instrumentation protocol is simple. The basic requirements are measurement of the force, normally provided by the load cell of the test machine, and the elongation of a defined section of the gauge part, which typically is found with an extensometer. Information about transverse strains and Poisson's ratio would in addition call for use of a transverse extensometer.
- The main stages of mechanical response for metallic materials, i.e. elasticity, plasticity and failure, are easily identified through tension tests.
- In a structure or component, failure is often occurring at a point where a tension stress state is dominating. Slender or thin-walled parts subjected to compression are susceptible to buckle, which means that the failure in this case will be dependent on geometry as well as material properties.

For reference later in this chapter, it is useful to review the main aspects of a quasi-static tension test. Figure 2.1a shows a typical experimental set-up for a tension test on a metallic material. The sample is attached to the machine via two pins in order to avoid bending moments. The instrumentation in this photo involves both a longitudinal and transverse extensometer. The transverse one is superfluous in many cases. The initial distance between the knives of the longitudinal extensometer is denoted L_0, and the initial cross-section area is A_0. During the test, the test machine logs the applied force F, while the extensometer measures the elongation ΔL. The nominal, or engineering, stress σ_{nom} and strain ε_{nom} are defined as

$$\sigma_{nom} = \frac{F}{A_0} \tag{2.1}$$

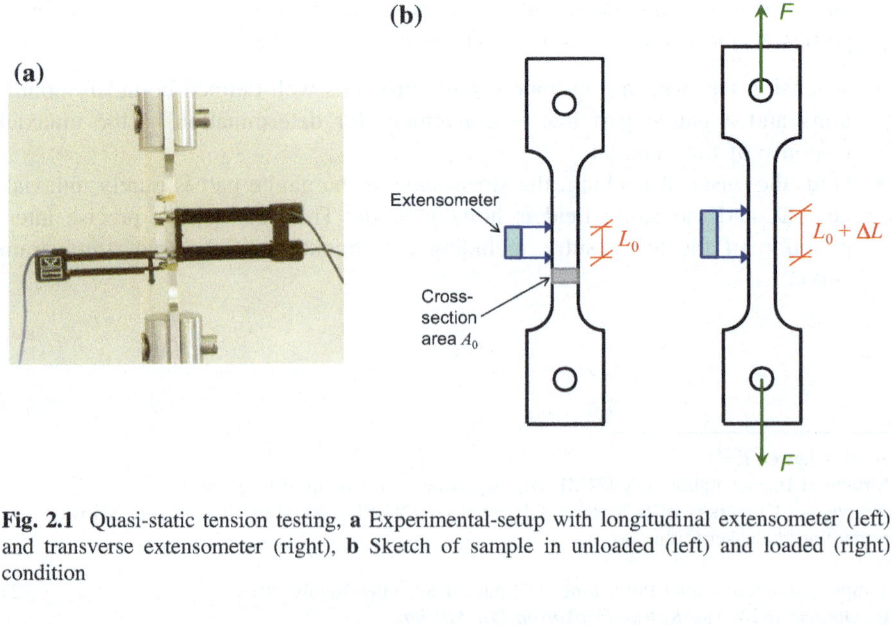

Fig. 2.1 Quasi-static tension testing, **a** Experimental-setup with longitudinal extensometer (left) and transverse extensometer (right), **b** Sketch of sample in unloaded (left) and loaded (right) condition

and

$$\varepsilon_{nom} = \frac{\Delta L}{L_0} \tag{2.2}$$

The true stress σ and logarithmic strain ε are often more relevant, in particular in numerical simulations involving plastic deformations. Alternative terms for these measures are Cauchy stress and true strain, respectively. The incremental logarithmic strain $d\varepsilon$ is defined as an infinitesimal elongation dL relative to a deformed length L, and it is shown in numerous text books (e.g., [1, 2]) that ε is related to the nominal strain through the relation

$$\varepsilon = \ln\left(\frac{L}{L_0}\right) = \ln(1 + \varepsilon_{nom}) \tag{2.3}$$

The true stress σ is defined as the force divided by the current area A of the deformed cross-section. Assuming isochoric plastic deformation, i.e. no change of volume such that $AL = A_0L_0$, the true stress is, again according to standard text books [1, 2], readily calculated as

$$\sigma = \frac{F}{A} = \sigma_{nom}(1 + \varepsilon_{nom}) \tag{2.4}$$

The plastic strain ε^p is an indispensable measure in the theory of plasticity. It stems from the assumption that the total strain ε can be additively decomposed into an elastic part ε^e and a plastic part ε^p. The elastic response is for many materials, including most metals, well described with Hooke's law, which for uniaxial loading reads

$$\sigma = E\varepsilon^e \tag{2.5}$$

where E is Young's modulus. Thus, the plastic strain is found as

$$\varepsilon^p = \varepsilon - \varepsilon^e = \varepsilon - \frac{\sigma}{E} \tag{2.6}$$

Normally, the true and nominal stresses and strains do not differ significantly in the elastic domain.

The calculation of nominal strain in Eq. (2.2), whereon the subsequent relations rely, assumes that the deformation and hence strain field is homogeneous within the measurement length of the extensometer. This is the case until the onset of necking. Necking starts when the tangent of the nominal stress-strain curve is horizontal. Thereafter, the strain field is not homogeneous anymore because the deformation is localized in the neck. Hence, the extensometer data do not have any relevance. Nevertheless, localization starts late in the deformation process for many materials, whereof metals at room temperature is an important case, implying that the measurement protocol described so far works well up to fairly large plastic deformations. On the other hand, there are also situations where necking occurs at a comparatively early stage of deformation. Polymeric materials often experience necking before any large amount of plastic deformation, and so do metals at high temperatures. Figure 2.2 illustrates the completely different situations that usually are encountered for metals and polymers at room temperature.

After onset of necking, local measurements of the deformation in the neck are required to find the true stress and the logarithmic (and plastic) strain. The transverse deformations are now essential because they provide the area A and hence the true stress, see Eq. (2.4). Moreover, the transverse strains can be used to find the longitudinal strain for metallic (isochoric) materials. Several tools, e.g. mechanical or laser-based transducers, see Fig. 2.3, are available for measurement of the diameter, providing a subsequent calculation of the transverse strain in the neck. An alternative strategy is to monitor the test with one or two digital cameras. After the

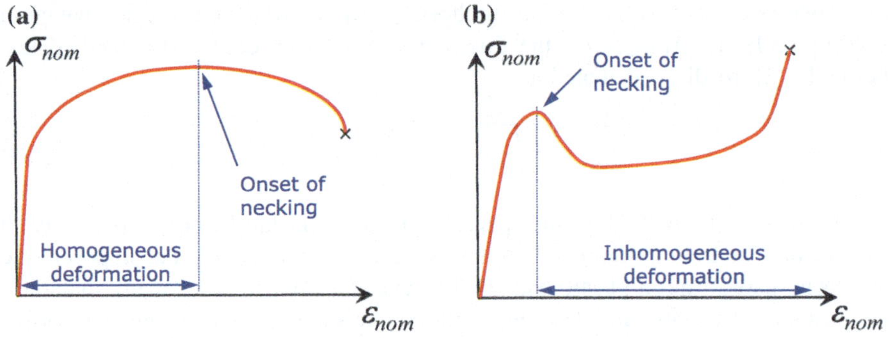

Fig. 2.2 Sketch of typical nominal stress-strain curves, **a** Metals (at room temperature), **b** Ductile polymers

Fig. 2.3 Laser-based system for diameter reduction measurements

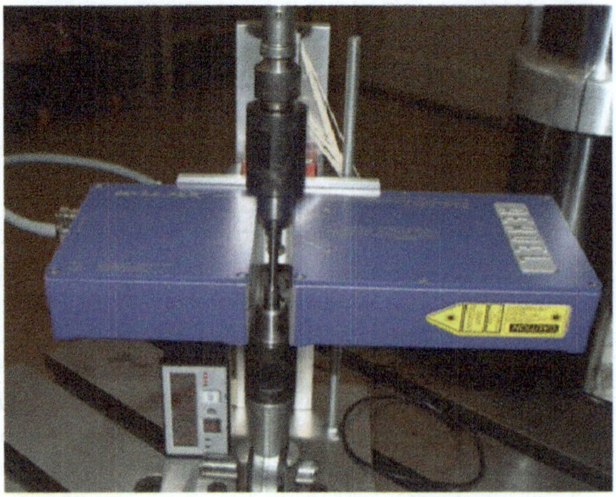

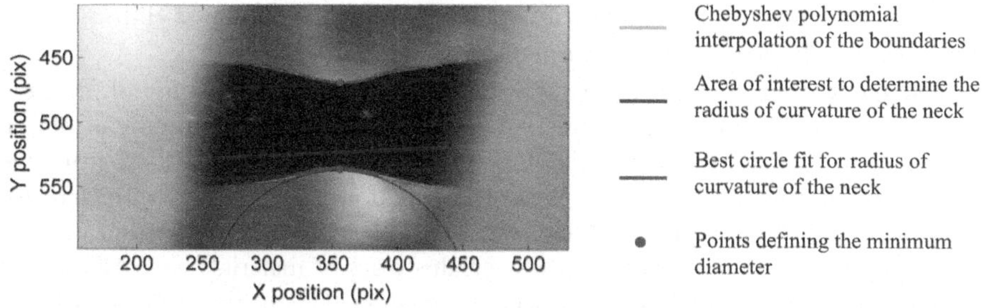

Fig. 2.4 Determination of minimum diameter and radius of curvature in the necked section of a tensile specimen [6]

test, analysis of the pictures with a digital image correlation (DIC) software yields the transverse as well as longitudinal strains. Another powerful application of the digital pictures is to pay attention to the boundaries of the necked specimen, see Fig. 2.4. The number of pixels between the two red points gives the minimum diameter, and hence the area A at the current stage of the deformation process.

Another feature with the curved boundaries of the neck is that they induce a triaxial stress state. The consequence is that the equivalent stress σ_{eq} in the neck, which is a key constituent in the yield criterion used in numerical simulations of plasticity problems, differs from the uniaxial stress σ. There exist geometry-based methods for determination of σ_{eq} [3, 4], whereof a widely acknowledged one for samples with circular cross-section was proposed by Bridgman [5]. It applies the minimum diameter D of the test specimen (i.e., between the two red points in Fig. 2.4) and the radius of curvature R_{neck} of the necked section (i.e., the radius of the red circle in Fig. 2.4). The Bridgman approximation of the equivalent stress reads

$$\sigma_{eq} = \frac{\sigma}{\left(1 + \frac{4R_{neck}}{D}\right) \cdot \ln\left(1 + \frac{D}{4R_{neck}}\right)} \tag{2.7}$$

where σ is the uniaxial stress as calculated with Eq. (2.4). A robust method to find this circle is first to interpolate the boundary with a higher-order Chebyshev polynomial, and thereafter fit a circle to the concave part of the boundary (addressed with blue colour in Fig. 2.4) [6]. Assuming plastic incompressibility, i.e. $AL = A_0L_0$, the local strain in the necked section can also be calculated when the minimum diameter D is known, viz.

$$\varepsilon = \ln\left(\frac{L}{L_0}\right) = 2\,\ln\left(\frac{D_0}{D}\right) \tag{2.8}$$

Equation (2.8) supposes isotropic deformation of the material. In case of anisotropy, it is required to measure two perpendicular diameters D_1 and D_2. Provided that these two diameters coincide with the half-axes of the deformed ellipse-shaped cross-section, the logarithmic strain reads $\varepsilon = \ln\left(D_0^2/D_1D_2\right)$.

The methodology described so far is most relevant for testing of metallic materials. A key feature with metals is that they do not change volume at plastic deformation, which means that the longitudinal strains can be found from the transverse deformation, and vice versa. The situation is particularly attractive for isotropic metals because it is sufficient to deal with one transverse strain component. When it comes to polymers, however, this class of materials is susceptible to change volume during the entire deformation process. This means that longitudinal and transverse strains have to be measured at all stages. Moreover, the early onset of necking, see Fig. 2.2b, implies that extensometers are not useful tools. Instrumentation with digital camera(s) for post-test DIC analysis of the pictures is therefore required during the entire material test of such materials.

It appears from the preceding discussion that optical visibility is favourable in a tension test. Firstly, determination of the true stress-strain response in the post-necking phase requires local measurements of the geometry in the neck. Camera is a convenient tool for this purpose. Moreover, the entire test has to be monitored with a camera in the case of dilating materials, e.g. polymers or foams.

This section has addressed some important assumptions and limitations for quasi-static tension tests. They are relevant in dynamic testing as well. In addition, the dynamic feature enters the scene. The maximum attainable strain rate in conventional uniaxial testing machines is often of order 1 s^{-1}. It might be possible to speed up the machine even more. The machine needs, however, a certain time and stroke to attain the desired speed, and inertia of the moving parts of the machine, e.g. the cross-heads, is another issue to take into account. Stress-strain curves found close to the limiting speeds of conventional machines might therefore have reduced quality.

The remaining sections of this chapter will deal with different aspects of dynamic tension testing with split-Hopkinson (Kolsky) bars. It was emphasised at the start of this section that the tension test has great importance, while Kolsky and other pioneering investigators mainly worked with compression bars. The next section provides a short summary of the one-dimension stress wave propagation theory that is relevant for analysis of split-Hopkinson tension bars (SHTB). Thereafter, Sect. 2.3 presents some of the challenges associated with dynamic tension testing, and different ways for solving them. Thus, Sect. 2.3 includes a number of approaches for dynamic material testing. It has turned out that rather long incoming stress waves are attractive in split-Hopkinson tension bars because this facilitates large strains in the specimen. Section 2.4 outlines such a test rig in detail, and pays particular attention to the response analysis. Improved methods for acquisition of measurement data are discussed in Sect. 2.5. Finally, some considerations on SHTB testing at temperatures differing from room temperature are included in Sect. 2.6 before this chapter on tension testing with Kolsky bars is rounded off with a short summary in Sect. 2.7.

2.2 Review of One-Dimensional Elastic Wave Propagation Theory

All split-Hopkinson systems exploit propagation of elastic waves in the bars, or possibly viscoelastic waves if the bars are made of polymeric materials. Depending on the test rig at hand, these waves are longitudinal, i.e. compression or tension, or they yield shear deformation of the bar. Assuming that the deformation is elastic, there are linear relations between displacement, velocity, stress and strain at any position x along the bar and at any instant t of time. Therefore, the terms elastic wave, stress wave and strain wave are used among each other. The theory of one-dimensional (1D) propagation of elastic waves is treated in a number of text books, whereof Kolsky [7] and Graff [8] are classical references. A short review is provided in this section to make this chapter on split-Hopkinson tension bars self-content.

One-dimensional propagation of waves in an elastic bar is governed by the differential equation

$$\frac{\partial^2 u}{\partial t^2} = c_b^2 \frac{\partial^2 u}{\partial x^2} \tag{2.9}$$

where $u = u(x, t)$ is the displacement component in the longitudinal $x-$ direction, and c_b is the elastic one-dimensional stress wave velocity in the bar material, given as

$$c_b = \sqrt{\frac{E_b}{\rho_b}} \tag{2.10}$$

Here, E_b is Young's modulus, and ρ_b is the density of the bar material. The velocity c_b corresponds also to the speed of sound in the material at hand. It is around 5000 m/s for some common metals (e.g., steel and aluminium), while soft materials (lead, polymers) have a significantly smaller value of c_b due to a comparatively low E_b.

The general solution of Eq. (2.9) reads

$$u(x, t) = f(x - c_b t) + g(x + c_b t) \tag{2.11}$$

where f and g are arbitrary functions of the arguments $x - c_b t$ and $x + c_b t$, with f representing a right-traveling wave (i.e., in the positive x—direction) and g representing a left-traveling wave. The functions f and g can be uniquely solved from the initial and boundary conditions of a specific problem. When the functions $f(x - c_b t)$ and $g(x + c_b t)$ are established, the mathematical expressions for f and g are independent of the time t and the position x along the bar; x and t do only occur as arguments in the functions.

The strain $\varepsilon = \varepsilon(x,t)$ in the bar is found by partial differentiation of $u(x,t)$ with respect to x

$$\varepsilon(x,t) = \frac{\partial u}{\partial x} = f'(x - c_b t) + g'(x + c_b t) \tag{2.12}$$

where the prime f' and g' denotes differentiation with respect to the entire argument of f and g. Assuming linear elastic material behaviour and a one-dimensional stress state, Hooke's law is applicable and the stress $\sigma = \sigma(x,t)$ becomes

$$\sigma(x,t) = E_b\varepsilon = E_b(f'(x - c_b t) + g'(x + c_b t)) \tag{2.13}$$

Partial differentiation of $u(x,t)$ with respect to t gives the particle velocity $v = v(x,t)$ in the bar

$$v(x,t) = \frac{\partial u}{\partial t} = c_b(-f'(x - c_b t) + g'(x + c_b t)) \tag{2.14}$$

The relations (2.13) and (2.14) are useful when analyzing the response of a split-Hopkinson bar system.

2.3 Dynamic Tension Tests with Split-Hopkinson Bars

2.3.1 Introductory Remarks

As outlined in other chapters of this book, it is, at least in principle, rather easy to construct a split-Hopkinson pressure bar (SHPB). The classical approach is to generate a compression wave by firing a striker bar at the end of an input bar. During the test, the contact pressure between the cylinder-shaped sample and the input and output bars ensures that the sample keeps in position. In order to stress the specimen in tension instead of compression, two important issues are:

- How to design the tension sample and attach it to the bars
- How to generate a tension stress wave in the sample

These two issues are related, and the solution of one of them has consequences for the other.

To start with the last one, there are two main strategies for applying tension to a test specimen in a split-Hopkinson bar. The first is, somehow, to generate a tension stress wave. The other approach is to use a SHPB, yet, design the sample in a clever way such that it is loaded in tension instead of compression. This brings the attention over to the first of the issues addressed above; and in particular the connection between the bars and the specimen.

This section will present and to some degree discuss different approaches for how to perform tension tests in a split-Hopkinson bar system. Section 2.3.2 presents practical solutions for how tension tests can be carried out using a split-Hopkinson pressure bar, while Sect. 2.3.3 is devoted to test rigs where tension stress waves in the bars are applied to the specimen. For general reviews of tension testing with a split-Hopkinson system, it is also referred to Nemat-Nasser [9], Gray [10] and Chen and Song [11], although all these references pay their main attention to compression testing.

2.3.2 Tension Testing in a Split-Hopkinson Pressure Bar (SHPB) System

Careful design of the specimen may facilitate tension testing in a SHPB system originally designed for compression tests, which means that the set-up employs the pressure wave propagating through the incident bar. The benefit with such a solution is that the same test rig can be used for split-Hopkinson tests in compression and tension. Only minor changes might be required.

One early approach was suggested by Lindholm and Yeakley [12]. Instead of the cylindrical compression specimen, a hat-formed specimen is introduced between the incident and transmission pressure bars, see Fig. 2.5. The transmission bar is tubular with a slightly larger internal diameter than the diameter of the incident bar. The incident bar can be loaded in the same way as in a conventional SHPB test. An obvious drawback with this design is that direct measurements on the specimen are not feasible, and the machining is complicated.

Another solution was proposed by Nicholas [13]. He utilised the fact that a compression stress wave is reflected as a tension stress wave at a free end. Figure 2.6 shows his experimental set-up. The initial, compression wave is firstly transferred through the ring-formed shoulder without affecting the test specimen to any noteworthy degree. After reflection, however, the tension wave loads the

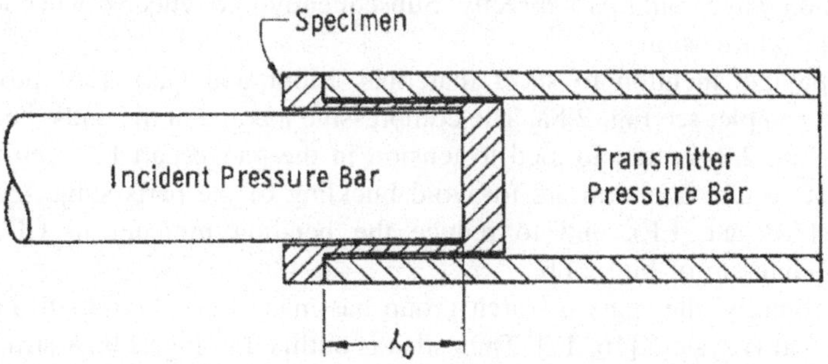

Fig. 2.5 Specimen for tensile testing in a split-Hopkinson bar with compression pulse in the incident bar as proposed by Lindholm and Yeakley [12]

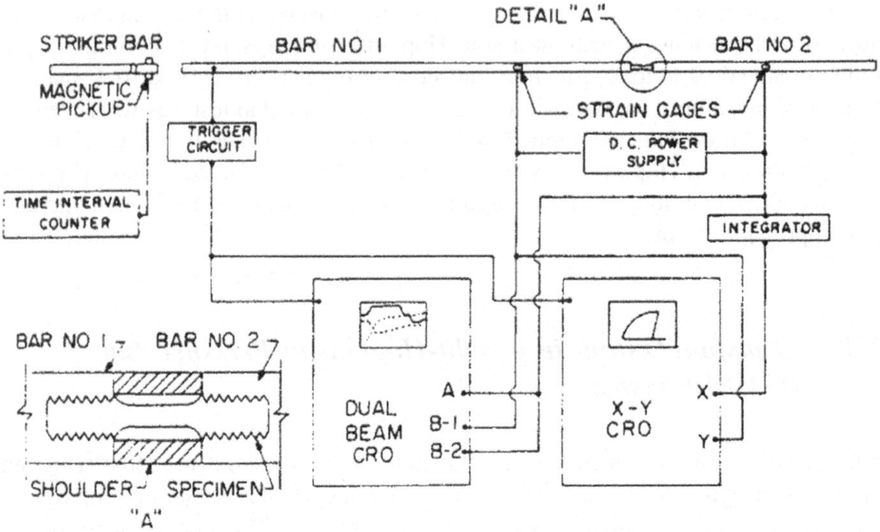

Fig. 2.6 Split-Hopkinson bar for tensile tests as proposed by Nicholas [13]

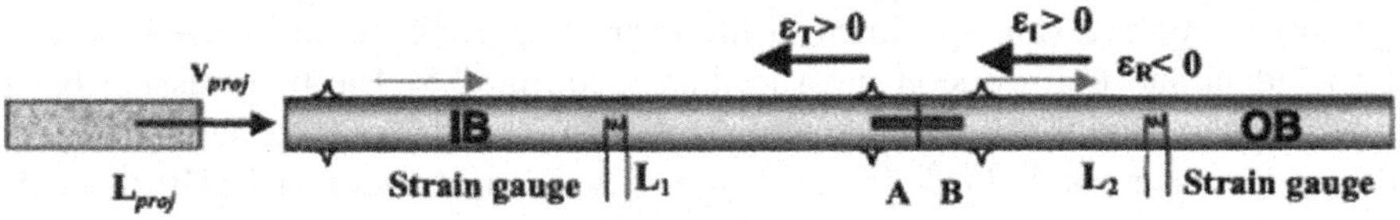

Fig. 2.7 Non-direct tensile loading as proposed by Haugou et al. [14]

threaded specimen, as the shoulder piece is not mechanically joined with the bars [13]. Haugou et al. [14] adopted the same idea in their tests on a sheet material, see Fig. 2.7. Four sheet samples were fixed around the circumferential surface of the bars such that one end of the samples was attached to the input bar, and the other one to the output bar. They placed the bars in contact before the test such that the compression wave could pass directly. Subsequently, the reflected wave loaded the four samples in tension.

Also paying attention to sheet materials, Mohr and Gary [15] proposed an M-shaped sample, see Fig. 2.8a. The compressive loading at the ends, i.e. points A and F in Fig. 2.8, is transformed to tension in the gauge part between C and D. Issues that require attention are to avoid buckling of the parts subjected to compression (AB and EF), and to reduce the bending moment in CD, see the free-body-diagram of Fig. 2.8b.

More recently, the same research group has made new attempts to inverse the load in a SHPB system [16, 17]. Their idea is outlined in Fig. 2.9. A striker ① hits an input bar ② that pushes a load inversion device ③ which ensures that the specimen ④ is subject to tension. Dunand et al. [16] proposed to use two

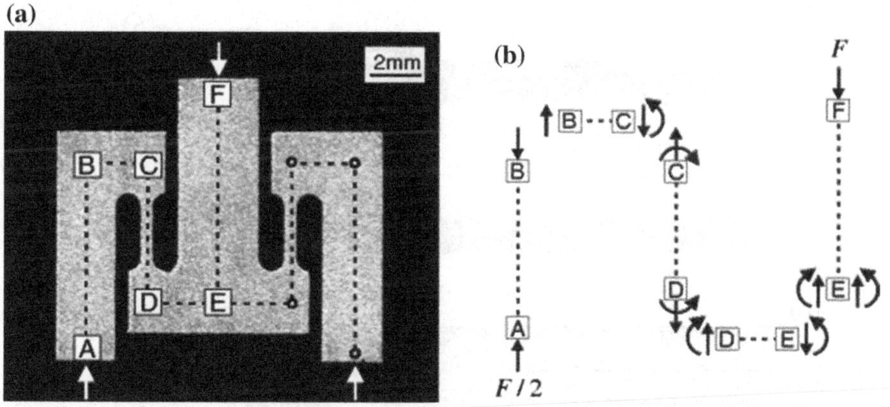

Fig. 2.8 M-shaped specimen as proposed by Mohr and Gary [15]

compression output bars ⑤, see Fig. 2.9a, while Roth et al. [17] presented a set-up with one tension output bar ⑤, Fig. 2.9b. According to Fig. 2.9c, the data acquisition system is up-to-date with a high-speed camera ⑥ in addition to the classical instrumentation with strain gauges ⑦.

The configurations addressed in this section ensure a fast loading of the sample. In a similar way as for compression tests applying a SHPB set-up, the duration of the stress pulse is determined by the length of the striker. Common for the test rigs is that the tension loading is applied to the sample in an indirect way. This might call for complex specimen designs or intermediate parts that somehow serve to pass or invert the compression stress wave. An alternative strategy is to design robust test rigs that provide a more direct tension load to the sample.

2.3.3 Approaches for Direct Tensioning of the Sample

The majority of researchers have looked for a way to apply tension to the incident bar, which subsequently yields a direct tension loading of the test specimen. Harding et al. [18] worked along this line when they in 1960 proposed the very first set-up allowing for dynamic loading in tension. A sketch of their test rig is shown in Fig. 2.10a. They placed their test sample and a so-called inertia bar with a strain gauge in a weighbar tube. A striker bar hit the top of the weighbar, generating a compression stress wave in the tube, and a downward displacement of the yoke. Thus, the sample was deformed in tension. As the instrumentation was limited to strain gauges at the inertia (transmission) bar, Harding et al. [18] had to do an extra test without a test specimen, i.e. with the inertia bar directly attached with the yoke, to characterize the incoming tension stress wave to the sample. An improved set-up,

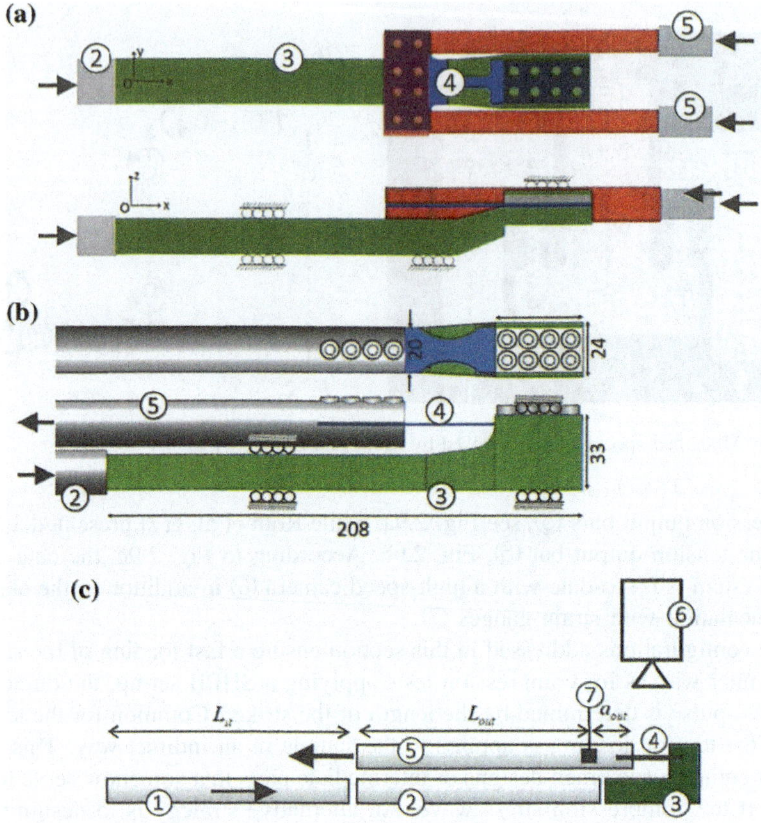

Fig. 2.9 SHPB system with load inversion as proposed by Roth et al. [17], **a** Two compressive output bars, **b** One tensile output bar, **c** Elevation sketch of entire set-up

shown in Fig. 2.10b, was applied by Harding and Welsh [19], who studied the dynamic mechanical properties of fibre-reinforced materials. They introduced an input bar instrumented with gauges between the yoke and specimen, and this bar transferred a tension stress wave. The extra test without any sample was avoided with this improved set-up. According to Nicholas [13], the rise time of the stress wave pulse is a problem in the SHTB proposed by Harding et al. [18]. It is obviously also a drawback that the sample is hidden within the tube.

Yet, the perhaps most common type of split-Hopkinson tension bars is inspired by the arrangement shown in Fig. 2.10b, i.e., the impactor hits the end of an input bar. An important difference, however, is that the yoke is replaced with a flange that is hit by a striker. The striker is typically a tube surrounding the input bar, see Fig. 2.11. It appears from the figure that this arrangement facilitates that more or

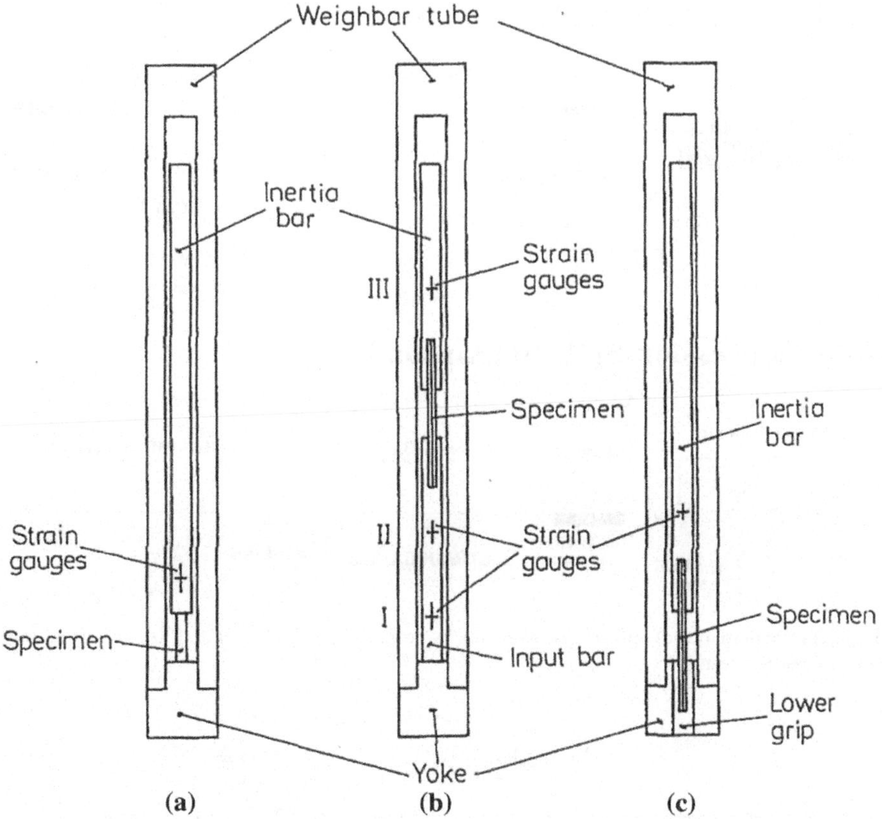

Fig. 2.10 Schematic arrangement of tensile Hopkinson bar [19], **a** Original design by Harding et al. [18], **b** Modified design for CFRP sheets by Harding and Welsh [19], **c** Modified design for GFRP sheets by Harding and Welsh [19]

less the same test rig can be applied in dynamic tension and compression tests. Moreover, it ensures optical visibility of the specimen provided that the incident tube is rather short. Although Fig. 2.11 stems from a more recent study by Thakur et al. [20], Ogawa [21] was back in 1984 among the first ones to present this concept for a SHTB.

It might be a practical problem for the strain gauge instrumentation that the input bar is surrounded by the striker, and there are also challenges with obtaining an appropriate pulse shaping. One way to avoid this is to locate the striker inside a hollow input bar [22]. Another issue is where to place the firing mechanism for the striker. Van Slycken et al. [23] put the accelerator in an eccentric position relative to the input bar, see Fig. 2.12.

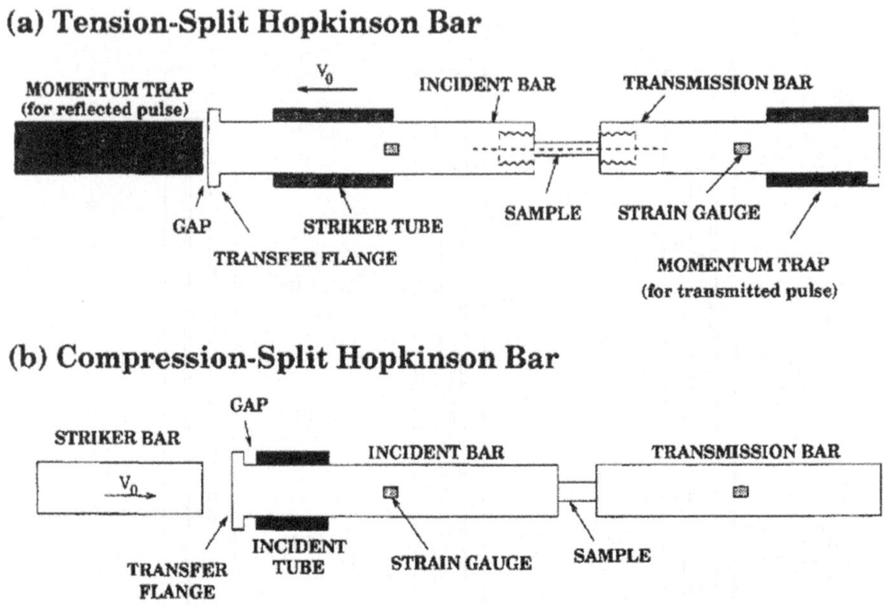

Fig. 2.11 Schematics of split-Hopkinson bars as proposed by Thakur et al. [20], **a** Tension tests, **b** Compression tests

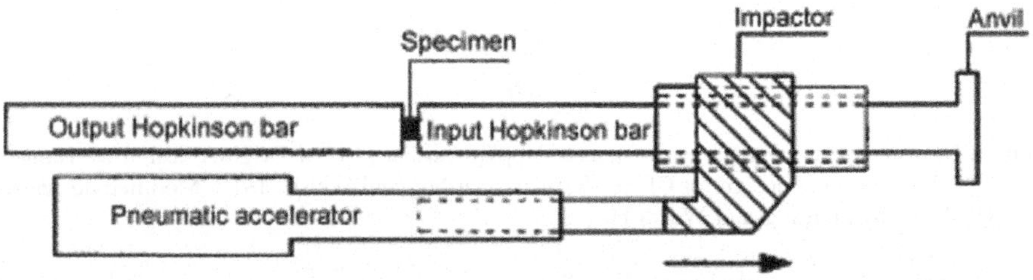

Fig. 2.12 Schematic view of SHTB with concentric accelerator as proposed by Van Slycken et al. [23]

According to Gerlach et al. [24], there are limitations on the maximum length of the striker tube because the input bar needs supports to avoid elastic sagging deformations. A long striker implies an increased duration of the pulse, which again allows for testing of ductile materials or at lower strain rates. The U-shaped striker shown in Fig. 2.13 avoids many of the difficulties related to pulse shaping, instrumentation and support of the test rig.

All examples shown so far in Sect. 2.3 involve an impacting striker. The pressure generated at the contact surface between the striker and the input bar is transformed to tension in the input bar by means of stress wave reflection at a flange or free end, alternatively by employing a clever design of the specimen geometry

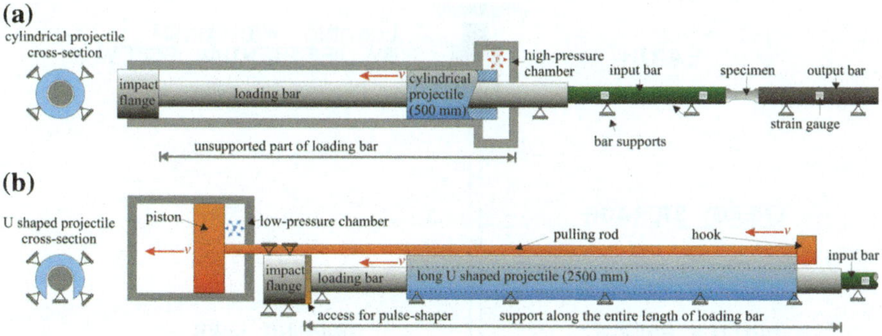

Fig. 2.13 Classic (**a**) and modified (**b**) design for SHTB as proposed by Gerlach et al. [24]

such that the compression wave in the input bar stresses the sample in tension. A more direct approach was proposed by Albertini and Montagnani [25, 26], who loaded a part of the incident bar with a tension force, see Fig. 2.14. Before the test, pressure is applied to the gripping jaws such that the input bar (denoted BAR 1 in the figure) is clamped at this location. Thereafter, the upper part of the input bar was stressed in tension by means of a differential screw. The stored mechanical energy is released by breaking the brittle intermediate piece in the clamp (denoted BRITTLE BRIDGE in Fig. 2.14). Albertini and Montagnani [25] reported that the rise time of the tension stress wave is around 25 μs.

A similar SHTB concept, see Fig. 2.15, was proposed some years later by Staab and Gilat [27]. The arrangement of their clamping mechanism appears from the lower right-hand part of Fig. 2.15, and the typical rise time is 30–40 μs. Pre-stressed split-Hopkinson bars have also been applied by e.g. Cadoni et al. [28] and Chen et al. [29]. In the latter case, the length of the pre-stressed bar is around 6 m, facilitating that the incoming tension stress wave has a duration of more than 2 ms. This comparatively long duration is beneficial because rather ductile materials can be tested up to failure during the first pass-through of the stress wave, and it is also, depending on the geometrical measures of the sample, feasible to test materials at strain rates of an order of only $10^2\ s^{-1}$.

The next section will pay particular attention to the design and analysis of a pre-stressed tension bar of Albertini-type. This type of SHTBs differs more from the conventional pressure bars, treated in several other chapters of this book, than the bars with reversed direction of the striker impact do. Pre-stressed bars can be applied for axisymmetric as well as sheet specimens, and they are also well suited for exploring material response at high temperatures because of the long duration of the incoming stress wave. A final and important benefit with such designs of SHTBs is that there are no challenges associated with optical visibility of the test sample.

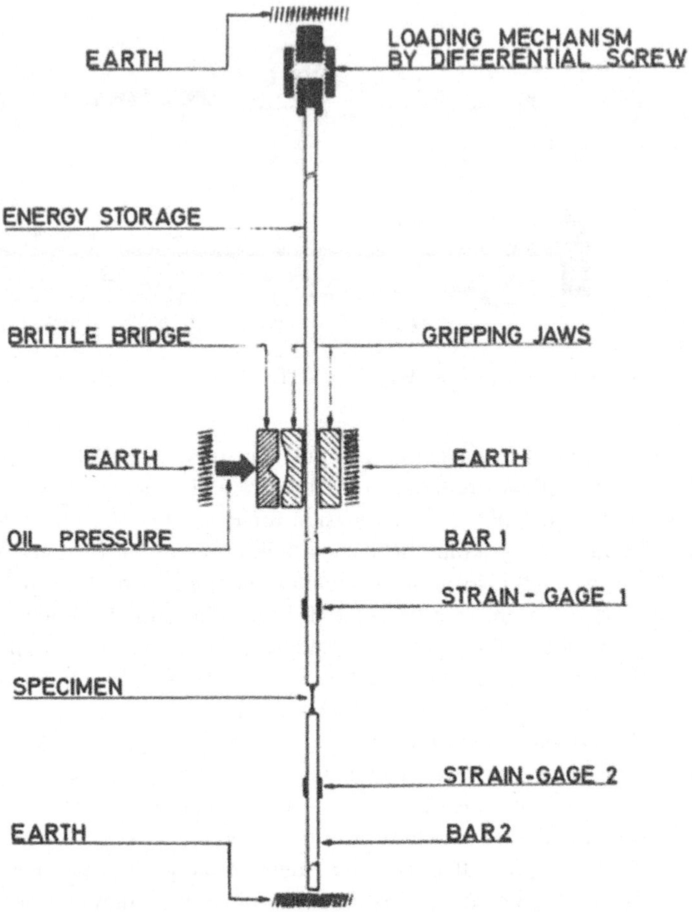

Fig. 2.14 Split-Hopkinson bar with pre-stressing as proposed by Albertini and Montagnani [26]

2.4 Split-Hopkinson Tension Bar Loaded by Direct Tension

2.4.1 Bar, Mechanical Devices and Instrumentation

As outlined in the previous section, it seems that Albertini and Montagnani [25, 26] were the first ones to build a proper split-Hopkinson test rig employing pre-tension of the input bar. Their concept will now be discussed in more detail. Sects. 2.4.1–2.4.3 look closer at how such a pre-stressed bar works. The comparatively long stress wave and the geometry of the tension test sample call for some precautions in the data treatment that are not occurring for compression bars. The conventional

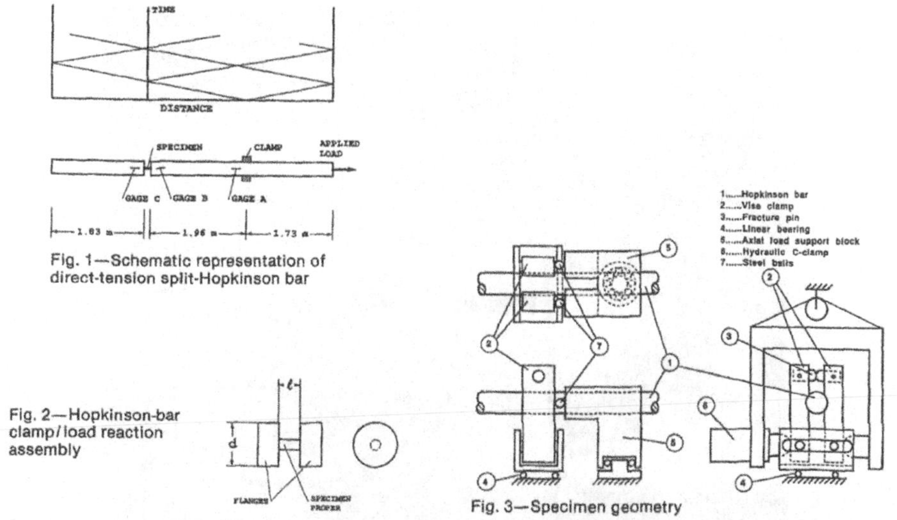

Fig. 2.15 Split-Hopkinson bar with pre-stressing as proposed by Staab and Gilat [27]

response analysis based on strain gauge measurements is presented in Sects. 2.4.4–2.4.6, while a more modern instrumentation system involving high-speed camera is outlined in Sect. 2.5.

A split-Hopkinson tension bar presented by Vilamosa et al. [6] and Chen et al. [29], see photo in Fig. 2.16, is selected as an example of a pre-stressed SHTB. A sketch of the test rig is shown in Fig. 2.17. The SHTB consists of two steel bars, whereof the input bar ABC has a total length of 8140 mm, and the output bar DE is 7100 mm long. Both bars have cylindrical cross-section with diameter 10 mm. The cross-section area is thus $A_b = 78.5\ \text{mm}^2$. The bars are made of high-strength steel Tibnor 52SiCrNi5 with yield stress of approximately 900 MPa. A rigid steel frame supports the bars. The bearings between the bar and the frame are made of 60 mm long PVC tubes with minimal friction and large electrical resistivity. The centre distance between each of these tubes is 625 mm. Except at points A and B, these PVC bearings are the only connection between the frame and the bars.

The test specimen is connected to the bars between points C and D. An example of test specimen geometry is shown in Fig. 2.18. The threaded ends of the axi-symmetric specimen have diameter 5 mm in order to fit into the threaded holes M5 in the bars at C and D. It is feasible to reduce the thickness of the sample by introducing an intermediate piece where the diameter of the threaded section is 3 mm. Sheet materials, however, are typically 1–2 mm thick, and call therefore for a different design of the connection between the bars and specimen. Tarigopula et al. [30] developed a fixture for steel sheets with thickness 1.5 mm, see Fig. 2.19. The sample was glued into the slit of the fixture, and thereafter the M5 threads were machined on the extensions of the fixture containing the specimen in the slit. This arrangement ensured a smooth transition of the stress wave from the input bar to the

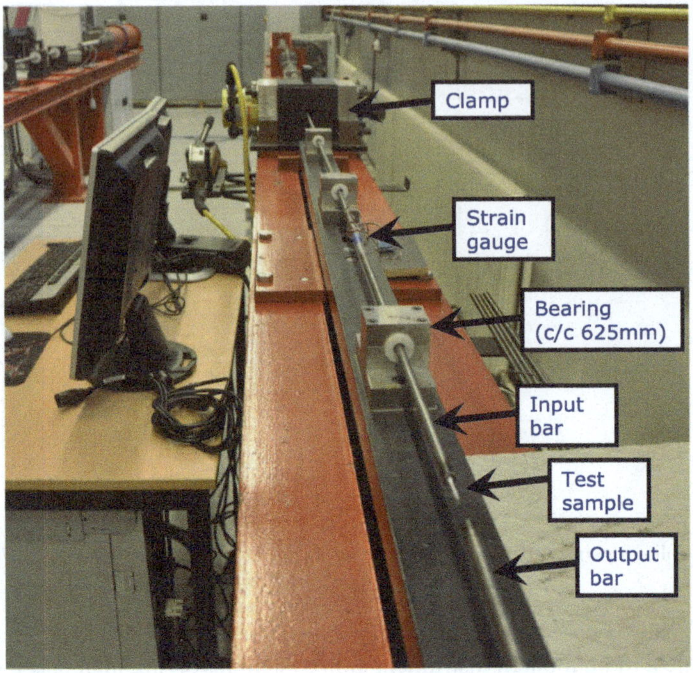

Fig. 2.16 Photo of split-Hopkinson tension bar with pre-stressing arrangement

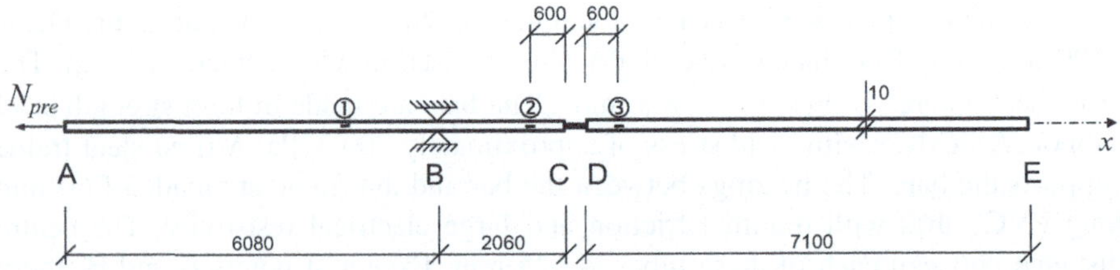

Fig. 2.17 Sketch of pre-stressed split-Hopkinson tension bar system. Measures in mm

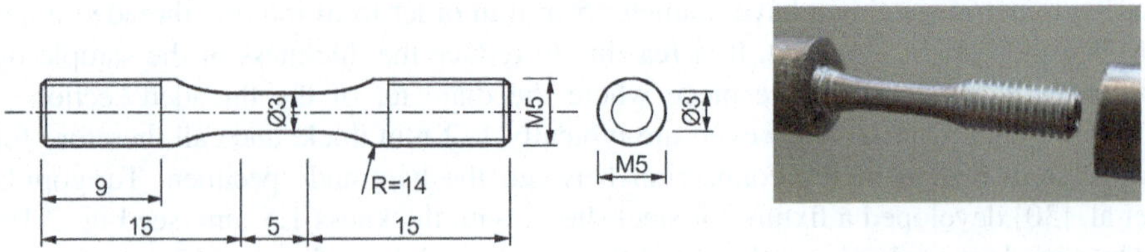

Fig. 2.18 SHTB test specimen with circular cross section. Measures in mm

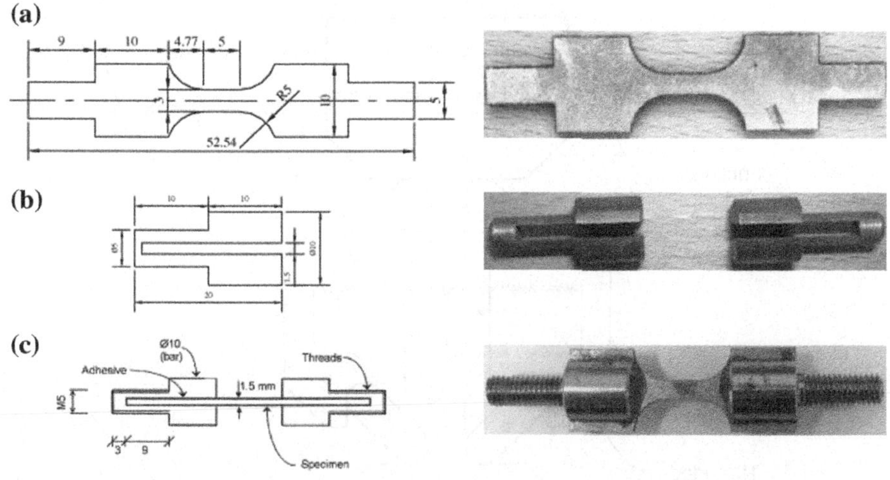

Fig. 2.19 SHTB test specimen made of sheet material [30]. Measures in mm, **a** Geometry of the specimen, **b** Geometry of the fixture, **c** Assembly of the specimen and fixture

sample. A similar solution, where the sample was bolted directly to slits in the bars, was proposed by Huh et al. [31], while Van Slycken et al. [23] glued their sheet sample directly in the slits of their Hopkinson bars.

Before the test, part AB of the input bar in Fig. 2.17 is pre-stressed by applying a force N_{pre} with a hydraulic jack at end A of the bar. Point B is prevented from movements by means of a locking mechanism. The lock is shown in detail in Fig. 2.20, where the bar diameter of 10 mm serves as a guideline of the geometrical dimensions of the other components in the lock. The lock works by applying pressure with a hydraulic jack at component (a). Via the components (b)–(h), constituting a truss structure, the pressure is transferred to component (i). The fixed support in the longitudinal direction of the bar at point B is now established by means of friction forces. The key component in the lock is the high-strength notched bolt (f). The diameter of the threaded part of the bolt is 12 mm, whilst the notch diameter is 6 mm. After machining, the bolts are heat treated to a hardness of approximately HRC 50. An important characteristic is that the bolt ruptures very suddenly, and the notch amplifies the brittleness. After fracture, the friction force between (i) and (j) and the bar is abruptly removed, and the stress caused by the force N_{pre} in part AB of the bar is released as a stress wave which is propagating towards the specimen. In order to have a short rise time of the stress wave, it is important that bolt (f) has low ductility. On the other hand, the behaviour of the bolt is very unpredictable if the brittleness is too high, because the pressure at which the bolt fractures then is almost random and often much lower than 300 kN. A medium high hardness of HRC 50 was found to be suitable.

When the notched bolt in the lock at point B has fractured, the tension stress wave propagates towards end C of the input bar, where the wave is partly reflected

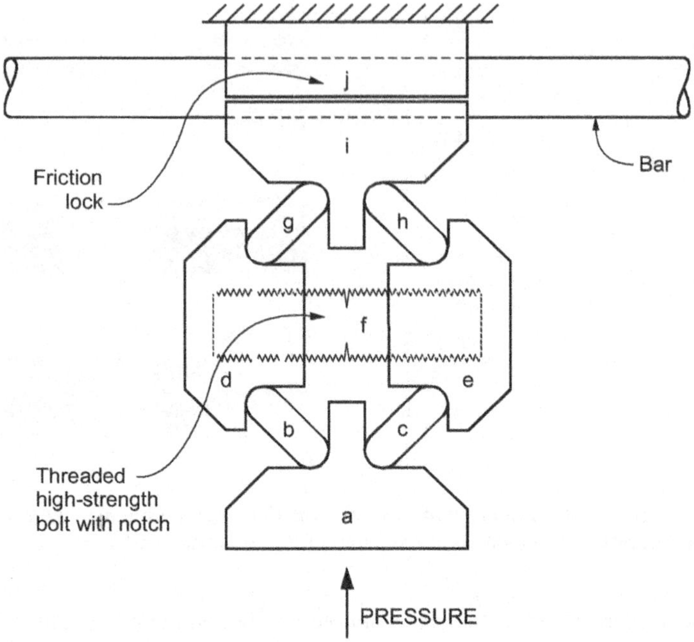

Fig. 2.20 Sketch of friction lock mechanism

and partly transmitted into the specimen and further into the output bar. The sample is loaded in tension during this process. A detailed analysis follows in Sect. 2.4.4.

The classical instrumentation protocol of this split-Hopkinson tension bar system involves three pairs of strain gauges glued at the locations ①, ② and ③ in Fig. 2.17. Strain gauge ① is simply used to monitor the pre-tension force N_{pre}. The two other gauges ② and ③, which both are placed 600 mm from the sample, measure the strain-time history after release of the stress wave. The sampling frequency of the gauges has to be of order 1 MHz in order to have a sufficient number of data points during a test which typically has duration around 1 ms. Subsequently, the stress, strain and strain rate experienced by the sample are determined from the gauge recordings and one-dimensional stress wave theory.

The strain gauges are sufficient to find the response of a metallic sample until the onset of necking. In addition, it is common, in particular in recent years, to monitor split-Hopkinson tests with a high-speed camera. The pictures facilitate determination of the true stress and logarithmic (true) strain also after necking. Such instrumentation and the subsequent treatment of data will be discussed in Sect. 2.5. For some applications, it is also of interest to investigate material response at elevated temperatures or cold conditions. Heating or cooling equipment is then needed. Such tests call also for a temperature measurement device in addition to the data acquisition system used to determine stresses and strains. Section 2.6 pays more attention to tests at different temperatures.

The next two sub-sections, however, take a closer look at the release of the stress wave and possible dispersion. Thereafter, the classical application of stress wave theory in the calculation of the stress and strain in the sample is outlined in Sect. 2.4.4.

2.4.2 Pre-stretching of Input Bar and Release of Stress Wave

The first stage in a SHTB test is to pre-stress part AB of the input bar. The resulting stress $\sigma_{pre} = N_{pre}/A_b$ is constant in AB, see the upper part of Fig. 2.21. When the friction lock at point B is broken, a tensile wave with amplitude $\sigma_{pre}/2$ propagates towards the sample, while a release wave of the same amplitude propagates towards the free end A [27]. The release wave is reflected at A. The total length of the tension stress wave is thus $2L_{AB}$. The lower part of Fig. 2.21 illustrates the stress state in the input bar at time $t = L_{BC}/c_b$ after the release of the clamp at B, where it is recalled that c_b is the uniaxial wave propagation velocity of the bar.

A convenient way to investigate the wave propagation in the entire SHTB system is to connect the two bars with a M5 set screw, see Fig. 2.22. The input and output bars perform now as a continuous system. This arrangement with a set screw also serves to evaluate if there are any disturbances associated with the threaded fixtures.

Figure 2.23 shows the signals recorded by the strain gauges ①, ② and ③ after pre-stretching the input bar to a force of $N_{pre} = 13.2$ kN and thereafter releasing the stress wave. The clamp is removed at time $t = 0$. Strain gauge ① is located in the pre-tensioned part of the bar, and is therefore strained in advance, i.e. before $t = 0$. According to Hooke's law, the initial strain of approx. 800×10^{-6} corresponds to a

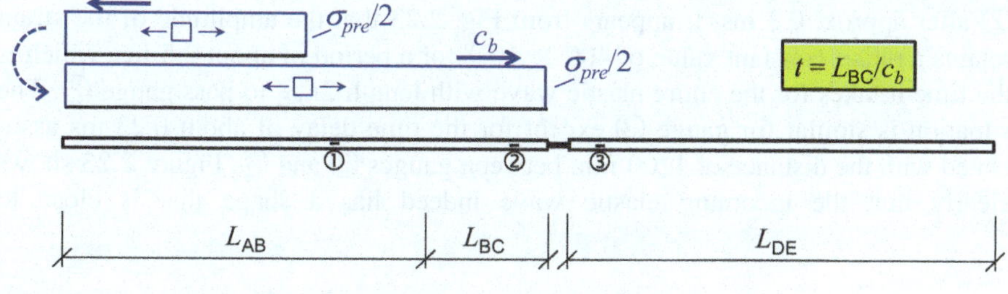

Fig. 2.21 Release of stress wave by removal of clamp at B

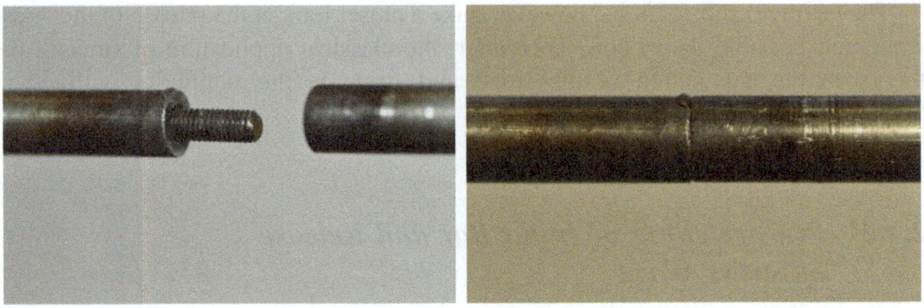

Fig. 2.22 Connection of input and output bar with a set screw

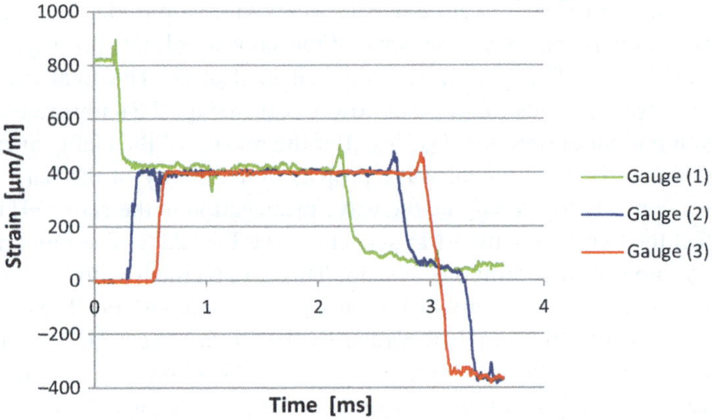

Fig. 2.23 Strain gauge recording in test with a set screw, see Fig. 2.22. Sampling frequency is 1 MHz

stress σ_{pre} = 170 N/mm^2. Figure 2.23 illustrates clearly that the initial stress σ_{pre} is split in two parts with amplitude $\sigma_{pre}/2$ and an associated strain of 400 $\times$ 10^{-6} when the stress wave starts to propagate, i.e. the wave propagation sketch of the lower part of Fig. 2.21 is confirmed. Further, Fig. 2.23 shows that the strain in gauge ① is halved after 0.2 ms, which is the time it takes for the tail of the unloading elastic wave to propagate from B to the position of gauge ①. The front of the elastic wave propagating in the right-hand direction approaches strain gauge ② after approx. 0.3 ms. It appears from Fig. 2.23 that the amplitude of the strain retains a rather constant value of 400 $\times$ 10^{-6} for a period of about 2.5 ms, which is the time it takes for the entire elastic wave with length $2L_{\mathsf{AB}}$ to pass gauge ②. The situation is similar for gauge ③ except for the time delay of about 0.23 ms associated with the distance of 1200 mm between gauges ② and ③. Figure 2.23 shows clearly that the incoming elastic wave indeed has a shape that is close to

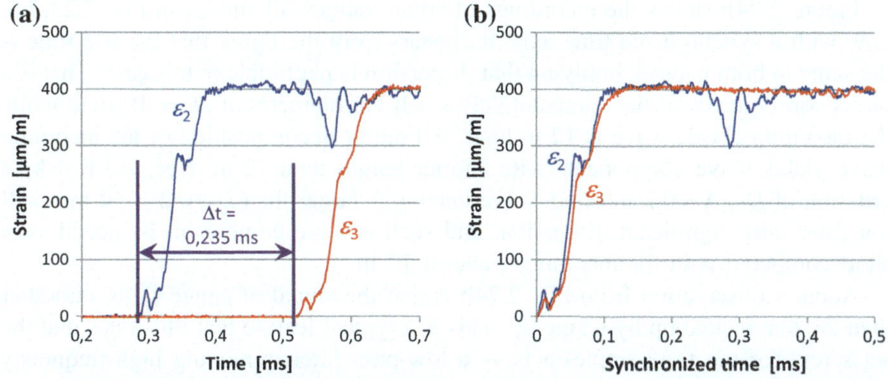

Fig. 2.24 **a** Excerpt of Fig. 2.23, addressing the front of the elastic waves, **b** Excerpt with synchronized time axis for gauges ② and ③

rectangular, and that the threads of the set screw do not impose any major disturbance to the wave.

Figure 2.24a presents an excerpt of Fig. 2.23, paying attention to the rise of the signal registered by gauges ② and ③ when the front of the elastic wave passes these two positions. The rise time of the signal is around 60 µs in this particular case. It is related to the fracture characteristics of the notched bolt at position B, see Fig. 2.20, and differs therefore slightly from test to test. Normally, the rise time is somewhat lower than 60 µs. It appears also from Fig. 2.24a that the time lag between the arrival of the front of the elastic wave at gauges ② and ③ is 0.235 ms. The distance between these two gauges is 1200 mm. Therefore, the wave propagation velocity is c_b = 1200 mm / 0.235 ms = 5106 m/s, which agrees well with the value obtained from Eq. (2.10). Thus, c_b = 5100 m/s is adopted in all subsequent calculations.

2.4.3 Dispersion

The test with a set screw also serves to evaluate the presence of dispersion of the propagating elastic wave in this test rig. Dispersion is a phenomenon observed in many fields of wave physics, and means that waves with different wave lengths have different propagation velocities. More specific, long waves move faster than short waves. In split-Hopkinson bars, dispersion tends to smear out the original pulse, which is close to rectangular in Fig. 2.23. An initially sharp pulse becomes therefore less steep [32]. Analytical treatments of the dispersion phenomenon demonstrate that dispersion becomes important for small ratios D_b/Λ, where D_b is the diameter of the bar, and Λ is the wave length [7, 8, 32]. Kolsky [7] suggests that dispersion can be neglected when $D_b/\Lambda < 0.2$.

Figure 2.24b shows the recordings of strain gauges ② and ③ in Fig. 2.23, but now with a synchronized time axis. It appears from the figure that the rise time is the same in both gauges, implying that dispersion is negligible in this case. This is a direct consequence of the comparatively small bar diameter of D_b = 10 mm, while the incoming elastic wave is 12 m long. A Fourier decomposition of the incoming wave yields wave components with shorter length than 12 m. Keeping Kolsky's criterion of $D_b/\Lambda < 0.2$ in mind, wave lengths Λ larger than $D_b/0.2$ = 50 mm will not cause any significant dispersion, and such a wave component is indeed very short compared with the incoming wave of 12 m.

Another observation from Fig. 2.24b is that the signal of gauge ③ is smoother than the one picked up by gauge ②. This is a typical feature that illustrates that the set screw (or the test sample) acts as a low-pass filter, removing high-frequency components from the incoming stress wave.

2.4.4 Conventional Response Analysis

The previous two sections, addressing a test with a set screw, contained two important results, see Fig. 2.23:

- The incoming elastic wave has long duration and constant amplitude. It can be idealized as a wave with a rectangular shape.
- The incoming elastic wave does not change shape between strain gauges ② and ③, i.e. the dispersion is negligible. With reference back to Fig. 2.17, this implies that with the exception of a time shift, the elastic wave at point C is identical to the wave recorded by gauge ②. Similarly, gauge ③ gives the response at point D.

These two observations are essential in the subsequent analysis of the stress and strain in the test sample. This section outlines the conventional protocol for determination of the response in the test specimen, applying strain gauge signals and one-dimensional wave propagation theory. More modern techniques involving use of high-speed camera will be presented in Sect. 2.5.

Figure 2.25 shows the strains recorded by gauges ② and ③ in a test on a steel alloy. These strains are denoted ε_2 and ε_3, respectively. Similar to the test with the set screw, see Fig. 2.23, the origin of the time axis, i.e. $t = 0$, is defined when the elastic wave is released by breaking the lock at B. The signal in gauge ② suddenly rises when the front of the incoming wave arrives around t = 0.32 ms. Thereafter, the strain ε_2 is rather constant at a level around 580×10^{-6} for a time of slightly more than 0.2 ms. At t = 0.56 ms, strain ε_2 is reduced while strain ε_3 starts to increase. The time difference between the arrival of the elastic wave at gauges ② and ③ is 0.235 ms. This corresponds to the time it takes for the stress wave to propagate from gauge ② to C and back to ②, alternatively from gauge ② to C,

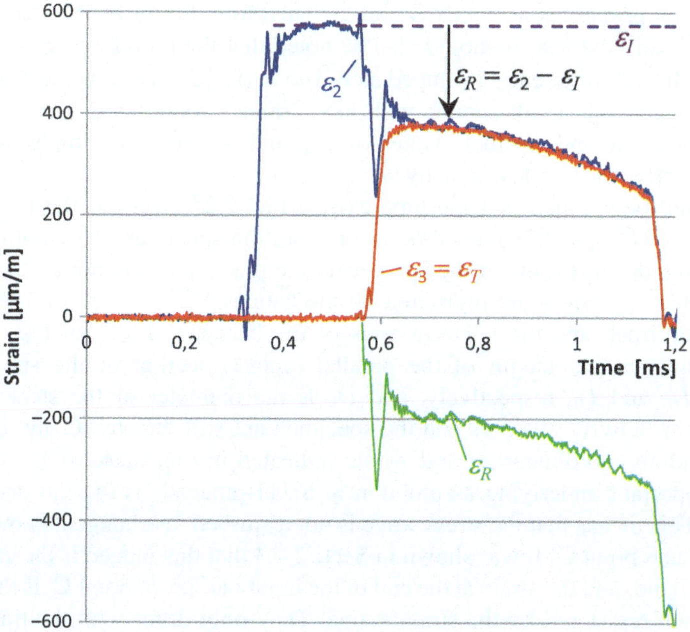

Fig. 2.25 Strains recorded in a test on a steel alloy. Sampling frequency is 1 MHz

through the sample CD and further from D to gauge ③. In both cases, the distance is 1.2 m, and the required time is 1.2 m / 5100 m/s = 0.235 ms, where $c_b = 5100$ m/s is the wave propagation velocity in the steel bars.

The strain in gauge ② is reduced from 580×10^{-6} to approximately 380×10^{-6} when $t = 0.58$ ms. Recalling that the incoming stress wave is a rectangular pulse with length $2L_{AB} = 12.16$ m and duration 2.4×10^{-3} s, the recorded signal ε_2 is from now on the sum of the original stress wave still propagating from part AB of the bar and the reflected signal moving in the opposite direction. For convenience, these two strain components are denoted respectively the incoming strain ε_I and reflected strain ε_R, see Fig. 2.25, thus $\varepsilon_2 = \varepsilon_I + \varepsilon_R$. The reflected wave is in compression because $\varepsilon_2 < \varepsilon_I$ when $t > 0.58$ ms. In a similar way, it is common to denote the strain in gauge ③ as the transmitted strain ε_T, i.e. $\varepsilon_3 = \varepsilon_T$. No superposition occurs here because the test is finished, i.e. the sample has ruptured, before the head of the transmitted wave has propagated to end E of the bar and back again to gauge ③.

As the reflected and transmitted signals call for some interaction between the elastic wave and the test specimen, the part of the signals referring to the test starts around $t = 0.58$ ms in Fig. 2.25. In the subsequent time period until $t = 1.17$ ms, the specimen is strained to fracture. This deformation process, which involves strain hardening and eventually necking of the steel sample, influences the reflection and transmission properties at points C and D. Therefore, the recorded signals are not constant during the test. It should also be noted that the ratio between ε_R and ε_T is governed by the difference in impedance $A\rho c = A\sqrt{\rho E}$ between the bars and the sample. In general, a soft sample with low Young's modulus E yields a comparatively large ε_R and a small value of ε_T, and so does a sample with small cross-section area A or low density ρ.

It should be recalled that the time axis in Fig. 2.25 refers to the recordings at strain gauges ② and ③ situated 600 mm from the specimen. This implies a time lag, and the actual events in the specimen take place 0.12 ms before the reflected and transmitted signals are registered by the gauges.

The specimen and the adjacent ends of the bars are shown in Fig. 2.26. The initial diameter and length of the parallel (gauge) section of the specimen are denoted D_0 and L_0, respectively, and D_b is the diameter of the steel bars. The connections between the bars and the specimen are still denoted C and D, and the corresponding coordinates x_C and x_D are indicated in Fig. 2.26.

An important underlying assumption in SHTB analysis is that no dispersion is present. This means that the stress wave is not disturbed, or changed, between strain gauge ② and point C. It was shown in Sect. 2.4.3 that this indeed is the case for the test rig at hand, and the strain at the end of the input bar, i.e. at point C, is thus similar to the strain registered by the strain gauge. They only differ with the time shift of approx. 0.12 ms because the elastic wave approaches gauge ② before it reaches point C. Similarly, the strain at point D corresponds to the strain in gauge ③.

According to Fig. 2.25, the signal in gauge ② contains a component of both the incoming and reflected wave during the straining process of the specimen at $t > 0.58$ ms. Assuming that the strain $\varepsilon(x_C, t)$ at C is equal to the strains in strain gauge ② (except the time lag), the strain at C is also the sum of the incoming and reflected strains ε_I and ε_R, respectively. Applying Eq. (2.12), $\varepsilon(x_C, t)$ can be expressed by differentiation of the functions $f(x, t)$ and $g(x, t)$

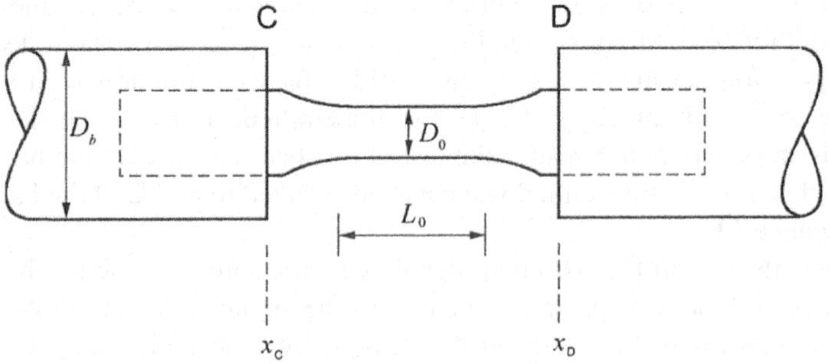

Fig. 2.26 Transition between bars and specimen

$$\varepsilon(x_{\mathrm{C}},t)=\varepsilon_I+\varepsilon_R=f'(x_{\mathrm{C}}-c_bt)+g'(x_{\mathrm{C}}+c_bt) \tag{2.15}$$

where it is assumed that the incoming strain wave ε_I propagates in the positive x–direction and hence is affiliated with the function $f'(x_{\mathrm{C}}-c_bt)$, while ε_R corresponds to $g'(x_{\mathrm{C}}+c_bt)$. Further, the velocity $v(x_{\mathrm{C}},t)$ at C can be determined with Eq. (2.14)

$$v(x_{\mathrm{C}},t)=c_b(-f'(x_{\mathrm{C}}-c_bt)+g'(x_{\mathrm{C}}+c_bt))=c_b(-\varepsilon_I+\varepsilon_R) \tag{2.16}$$

where Eq. (2.15) was employed in the last equality. The displacement $u(x_{\mathrm{C}},t)$ of point C is found by integration of Eq. (2.16)

$$u(x_{\mathrm{C}},t)=\int_0^t v(x_{\mathrm{C}},\tau)d\tau=c_b\int_0^t(-\varepsilon_I+\varepsilon_R)d\tau \tag{2.17}$$

For practical purposes, the signals ε_I and ε_R are integrated numerically. Finally, the force $F(x_{\mathrm{C}},t)$ in the bar at point C is determined by multiplying the strain $\varepsilon(x_{\mathrm{C}},t)$ with the elastic stiffness E_bA_b of the bar:

$$F(x_{\mathrm{C}},t)=E_bA_b\varepsilon(x_{\mathrm{C}},t)=E_bA_b(\varepsilon_I+\varepsilon_R) \tag{2.18}$$

Eqs. (2.15)–(2.18) demonstrate how the strain signals ε_I and ε_R in strain gauge ② can be used to find the response at end C of the incident bar. The strain, velocity, displacement and force at point D can be determined in a similar way from the transmitted strain ε_T in gauge ③. This case is somewhat simpler because no stress wave is reflected from the end of the output bar at point E before the test specimen has fractured. Thus, only the function f' has to be taken into account. The velocity, displacement and force at D are respectively

$$v(x_{\mathrm{D}},t)=-c_b\varepsilon_T \tag{2.19}$$

$$u(x_{\mathrm{D}},t)=-c_b\int_0^t\varepsilon_Td\tau \tag{2.20}$$

$$F(x_{\mathrm{D}},t)=E_bA_b\varepsilon_T \tag{2.21}$$

Equilibrium of the specimen demands that the forces $F(x_{\mathrm{C}},t)$ and $F(x_{\mathrm{D}},t)$ at points C and D have to be equal. Otherwise, inertia forces or large transient effects are present. Supposing that $F(x_{\mathrm{C}},t)=F(x_{\mathrm{D}},t)$, Eqs. (2.18) and (2.21) give a relation between the strains ε_I,ε_R and ε_T

$$\varepsilon_I+\varepsilon_R=\varepsilon_T \tag{2.22}$$

During the processing of results from a SHTB test, the equilibrium requirement can be checked by ensuring that the signals $\varepsilon_I + \varepsilon_R$ and ε_T really are coincident. It is recalled that ε_2 is the sum of ε_I and ε_R, and, indeed, the curves ε_2 and ε_3 in Fig. 2.25 show that this criterion is fulfilled after some introductory oscillations. On the other hand, equilibrium is strictly speaking not established when the curves are not coincident, and the accuracy of the results obtained in a split-Hopkinson machine can therefore be questioned at small strains, typically in the elastic domain of the deformation process of the specimen.

Assuming that $F(x_\mathsf{C}, t)$ and $F(x_\mathsf{D}, t)$ are equal, the nominal stress σ_{nom} in the specimen can be found by dividing one of these forces with the initial cross section area A_0 of the sample. Usually, the signal in gauge ③ is smoother than the one in gauge ②, and $F(x_\mathsf{D}, t)$ is therefore selected

$$\sigma_{nom} = \frac{F(x_\mathsf{D}, t)}{A_0} = \frac{E_b A_b}{A_0} \varepsilon_\mathrm{T} \tag{2.23}$$

The strain in the sample can be found from the displacements of points C and D. It is clear from Eqs. (2.17) and (2.20) that the displacements $u(x_\mathsf{C}, t)$ and $u(x_\mathsf{D}, t)$ are negative, implying that both ends of the specimen in Fig. 2.26 move in the left-hand direction during the test. Moreover, Eq. (2.22) and the fact that ε_R is negative yield that the displacement of C is larger than the displacement of D. It follows that the specimen experiences a net elongation. Recalling that x increases from point C to D and that the positive direction of the displacement u (and the velocity v) thus is in the right-hand direction in Fig. 2.17, the elongation of the specimen, expressed as the increase of distance between C and D, is $\Delta L = u(x_\mathsf{D}, t) - u(x_\mathsf{C}, t)$. It is emphasised that this elongation is distributed over the entire length of the specimen, including also the transition part (shoulders) where the diameter increases from 3 mm to 5 mm. Nevertheless, in the inelastic regime of the stress-strain curve, it can be assumed that most deformation of the specimen takes place in the parallel section L_0 until the onset of necking. Applying Eqs. (2.20) and (2.17), the nominal strain ε_{nom} in the specimen is then

$$\varepsilon_{nom} = \frac{\Delta L}{L_0} = \frac{u(x_\mathsf{D}, t) - u(x_\mathsf{C}, t)}{L_0} = \frac{c_b}{L_0} \int_0^t (-\varepsilon_T - (-\varepsilon_I + \varepsilon_R)) d\tau \tag{2.24}$$

Introducing $\varepsilon_I + \varepsilon_R = \varepsilon_T$, see Eq. (2.22), the final expression for ε_{nom} involves only ε_R

$$\varepsilon_{nom} = -2 \frac{c_b}{L_0} \int_0^t \varepsilon_R d\tau \tag{2.25}$$

Finally, the nominal strain rate $\dot{\varepsilon}_{nom}$ in the specimen is obtained by differentiating the strain ε_{nom} in Eq. (2.25) with respect to t

$$\dot{\varepsilon}_{nom} = -2\frac{c_b}{L_0}\varepsilon_R \tag{2.26}$$

To summarize, the reflected stress wave registrations in strain gauge ② is used to determine the strain and strain rate in the specimen. It is assumed that all deformation occurs in the gauge part of the sample, i.e. the deformation of the shoulders outside L_0 is neglected. The stress in the specimen is calculated from the transmitted stress wave registrations in strain gauge ③.

The nominal stress-strain curve for the steel sample addressed in the strain recordings of Fig. 2.25 is shown in Fig. 2.27. The figure also includes references to a series of pictures taken with a high-speed camera during the test, see Fig. 2.28. Necking starts at the ultimate value of the nominal strain, which corresponds to picture II. The series of pictures also demonstrate that both bars move in the left-hand direction, and the sample is strained in tension because the displacement of the incident bar (to the left) is larger than for the transmission bar. The strain rate in this test was $\dot{\varepsilon}_{nom} \approx 500\ \text{s}^{-1}$.

A rather obvious shortcoming of the nominal stress-strain curve in Fig. 2.27 is that the elastic stiffness is too low. The slope of the linear part of the curve corresponds to a Young's modulus of only 50×10^3 N/mm^2. The next section will discuss the cause of this unexpectedly low value, and also demonstrate a way to correct the stress-strain curve.

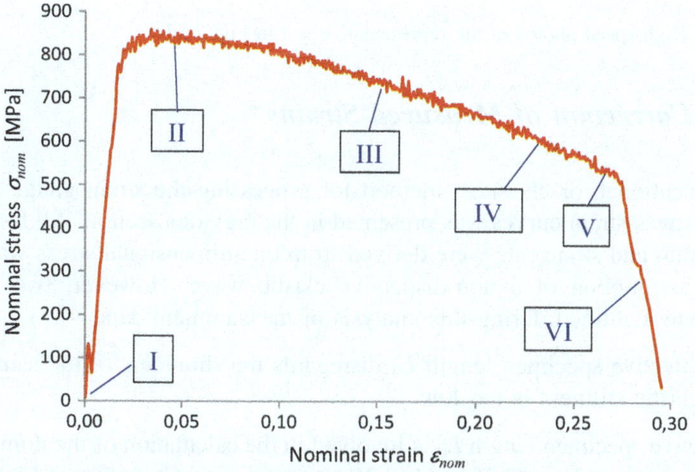

Fig. 2.27 Nominal stress-strain curve for the steel alloy including references to the pictures in Fig. 2.28. Young's modulus is not corrected

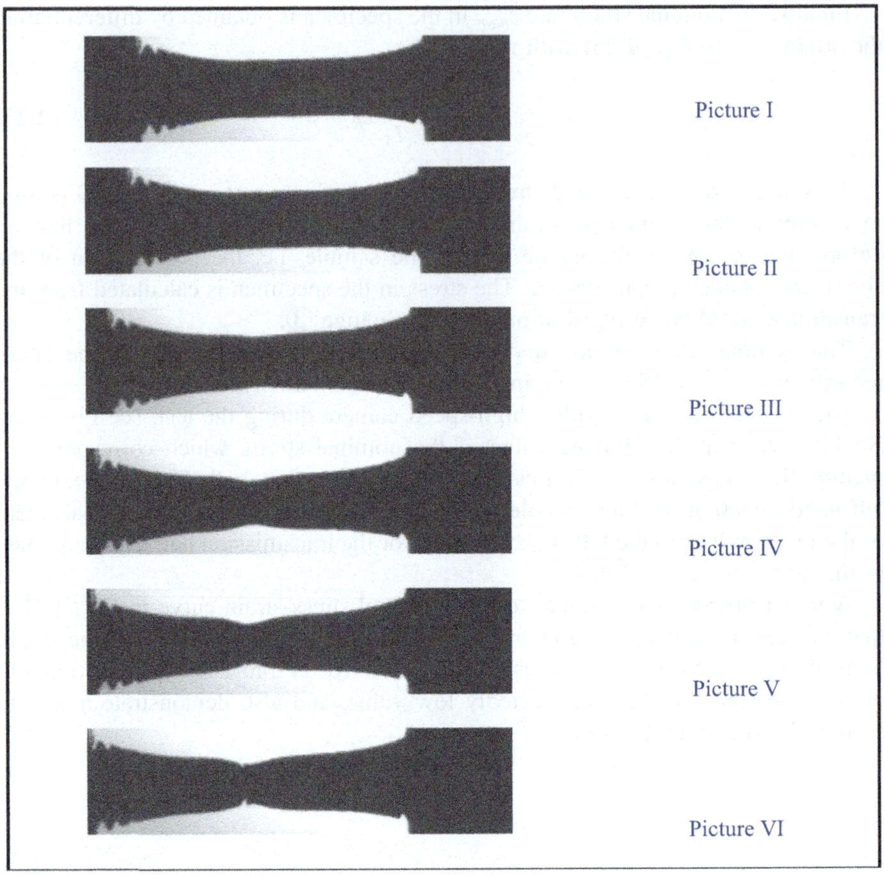

Fig. 2.28 High-speed photos of the representative test in Fig. 2.27

2.4.5 Correction of Measured Strains

The conventional, or classical, method for processing the strain gauge signals to nominal stress-strain curves was presented in the previous section. All formulas for stress, strain and strain rate were derived from one-dimensional stress wave theory and the assumption of a non-dispersive elastic wave. However, two important issues were identified during this analysis of measurement data:

- The effective specimen length L_0 disregards the shoulders of the sample
- The elastic stiffness is too low

The effective specimen length L_0 is involved in the calculation of the nominal strain and strain rate, see Eqs. (2.25) and (2.26), respectively. According to Fig. 2.18, the gauge section of the specimen has length 5 mm. Equations (2.25) and (2.26) assume that all deformation in the specimen is located in this parallel part. During elastic deformation, however, some strains must also be expected in the transition part, where the specimen diameter increases from 3 to 5 mm. Another feature is that some relative displacement is also likely to occur in the threaded connections between the sample and the two bars. On the other hand, the plastic deformation will take place in the gauge part defined by L_0 after the onset of yielding. The general observation is therefore that the elastic strains predicted with Eq. (2.25) are too large. On the other hand, also the stresses might be inaccurate in the first phase of a test, i.e. in the elastic domain, due to lack of equilibrium, see the discussion preceding Eq. (2.22). Nevertheless, it is of interest to correct the elastic strains, and a convenient way to do this was proposed by Albertini and Montagnani [33].

Figure 2.29 shows two generic, nominal stress-strain curves. The curve labelled "Dynamic" is the measured one from a test in a split-Hopkinson tension bar, applying Eq. (2.25) to determine the nominal strain. According to Eq. (2.24), the strain ε_{nom}^{meas} actually refers to the entire relative displacement $u(x_\mathsf{D}, t) - u(x_\mathsf{C}, t)$ between points C and D in Fig. 2.26. The correct stress-strain curve exhibits a stiffer behaviour because the strain ε_{nom}^{corr} only addresses the elongation of the gauge part L_0 of the sample. The difference $\Delta\varepsilon$ between the measured and correct strain is estimated with use of Hooke's law

$$\Delta\varepsilon = \frac{\sigma_{nom}}{E_{dyn}} - \frac{\sigma_{nom}}{E} = \sigma_{nom} \cdot \frac{E - E_{dyn}}{E_{dyn} \cdot E} \tag{2.27}$$

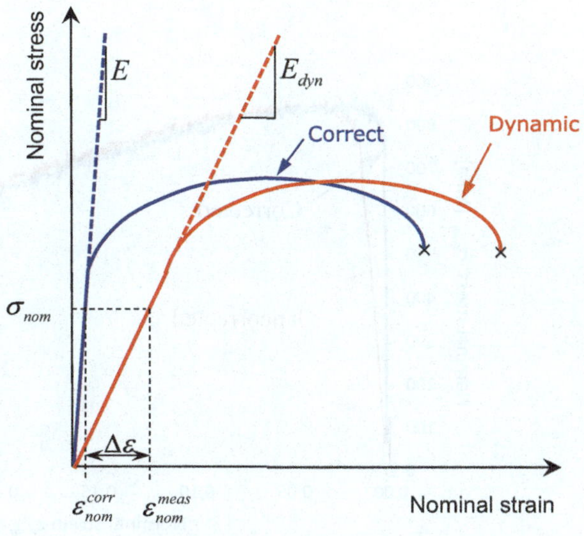

Fig. 2.29 Correction of strains

where E_{dyn} is the dynamic Young's modulus as determined from the test data, and E is the correct stiffness. It follows that a corrected nominal strain is found as [33]

$$\varepsilon_{nom}^{corr} = \varepsilon_{nom}^{meas} - \Delta\varepsilon = \varepsilon_{nom}^{meas} - \sigma_{nom} \cdot \frac{E - E_{dyn}}{E_{dyn} \cdot E} \tag{2.28}$$

where ε_{nom}^{meas} and σ_{nom} respectively are given by Eqs. (2.25) and (2.23). Although Young's modulus has a slight dependency on strain rate, it is common practise to apply the quasi-static value for E when the nominal strains are corrected with use of Eq. (2.28). Once again using the test on a steel alloy as an illustration, Fig. 2.30 compares the uncorrected stress-strain curve (the same curve as already shown in Fig. 2.27) with a curve where the strains have been corrected. It is obvious that Eq. (2.28) imposes a major adjustment of the nominal strains.

As a further exploration of the validity of the correction given in Eq. (2.28), some tests were carried out on specimens equipped with strain gauges. The specimens were made of the same steel material as the specimen already discussed in the previous section. However, three different geometrical shapes were considered in order to investigate the importance of the shoulder between the gauge section and the threaded part. In addition to the design shown in Fig. 2.18, specimens with shorter transition zones were also prepared for this particular study. Thus, the radius R = 14 mm was exchanged with R = 10 mm or R = 6 mm for some samples. The length L_0 = 5 mm of the gauge part was kept in all cases, so the total length of the specimen was analogously reduced with decreasing radius. Before testing in the split-Hopkinson tension bar, two strain gauges were glued diametrically opposed to each other in the gauge part of the specimen. Unfortunately, only three of the tests

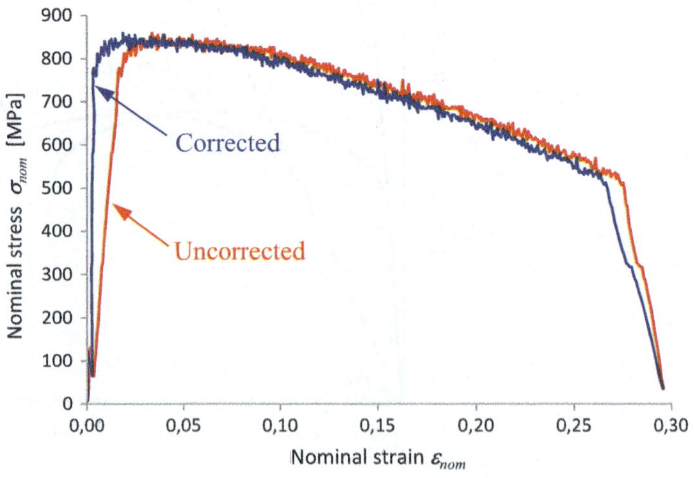

Fig. 2.30 Uncorrected and corrected nominal stress-strain curves for a steel alloy. The uncorrected curve is the same one as shown in Fig. 2.27

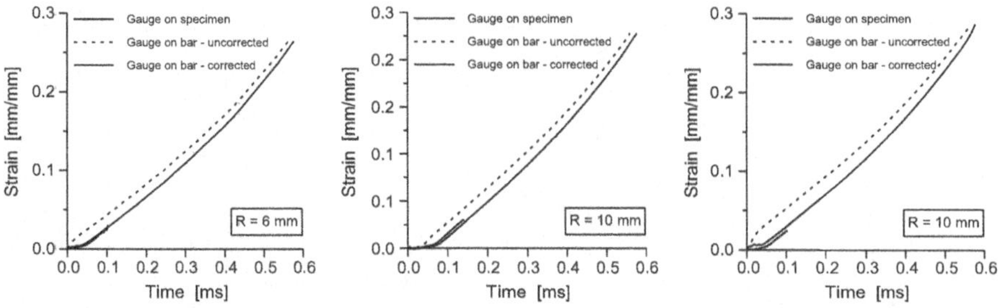

Fig. 2.31 Strain vs. time in tests with strain-gauge instrumented samples with shoulder radius of 6 and 10 mm. Recordings from gauge directly on the specimen, and strains, without and with correction, determined from the signals in gauge ② on the bar

with strain gauge instrumentation were successful; two specimens with radius R = 10 mm and one specimen with radius R = 6 mm. Thus, none of the test specimens with the default radius R = 14 mm were useable for the desired purpose.

The strain registrations as function of time are shown in Fig. 2.31. The three individual diagrams in this figure correspond to the three successful tests. Moreover, three curves are shown in each diagram. Firstly, the strain measured with the gauges mounted directly on the specimen is depicted with the rather thick solid line. These curves are abrupt already at approximately 0.1 ms, and it is likely that the cause is loosening of gauges. Nevertheless, the largest strain picked up by the gauge is around 0.05, which is well into the plastic domain. Moreover, the strains determined with use of Eq. (2.25) are shown with the dashed lines. These strains are found from the registrations in gauge ② on the bar, and cover thus the entire test. They are, as already addressed for another sample in Fig. 2.30, too large. It is, however, interesting to note that the entire difference seems to arise in the start of the test, i.e. before 0.5 ms. Thereafter, the thick and dashed curves are rather parallel. In the last curve, depicted with a thin solid line, Eq. (2.28) is applied to the uncorrected strains. It appears that the method suggested by Albertini and Montagnani [33] predicts nominal strains that are very close to the ones measured directly on the sample with strain gauges. Moreover, there is no clear difference between the two cases of R = 6 mm and R = 10 mm. Thus, it is likely that the strain correction method based on Eq. (2.28) also works well for the specimen with R = 14 mm.

Gilat et al. [36] also compared the strain determined with the conventional method with recordings from strain gauges glued directly on the sample. Moreover, they monitored their tests with a high-speed camera, and found the strains with use of the digital image correlation (DIC) software ARAMIS. They also experienced that the strain gauge only worked for comparatively small strains. Their results show that the strain gauge recording compares reasonably well with the strains determined from the propagating waves in the bars, i.e. the conventional method,

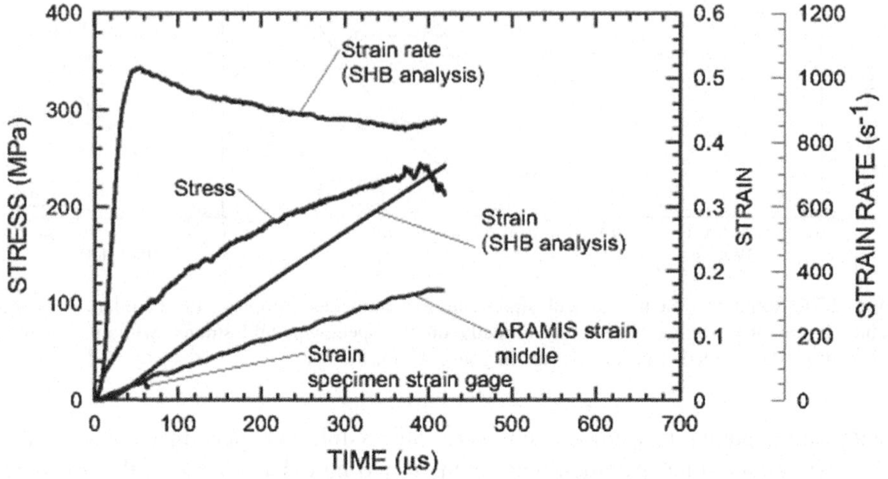

Fig. 2.32 Stress, strain and strain rate in a SHTB test on annealed copper as presented by Gilat et al. [36]. The strain was determined with the conventional method ("SHB analysis"), from strain gauge glued on the specimen, and with DIC ("ARAMIS")

yet without the correction in Eq. (2.28). On the other hand, their general finding is that the conventional method predicts larger strains than what DIC does, in particular for a material with considerable strain hardening (annealed copper), see Fig. 2.32. Again, a plausible explanation is that significant deformation takes place in the shoulders of the sample, and this occurs also in the plastic domain for a strain hardening material.

This section has demonstrated that the conventional method for determination of the strain in SHTB tests, see Eq. (2.25), might be inaccurate because the deformation of the specimen is not limited to the gauge section. The shoulders are also deformed, and there might be minor movements in the threaded fixtures. It is therefore desirable with a more direct way to measure the strains in the sample, in particular when the behaviour at large plastic strains up to failure is requested. Section 2.5 will take a closer look at this issue.

2.4.6 True Stress and Plastic Strain

The purpose with material tests in a split-Hopkinson apparatus is often to gain information for calibration of the constitutive model applied in numerical simulations of a dynamic problem. For metals, the strain-rate effect is normally associated with the viscoplastic part of the material model. Hence, identification of the relevant coefficients in the model calls for test data expressed as true stress vs. plastic strain.

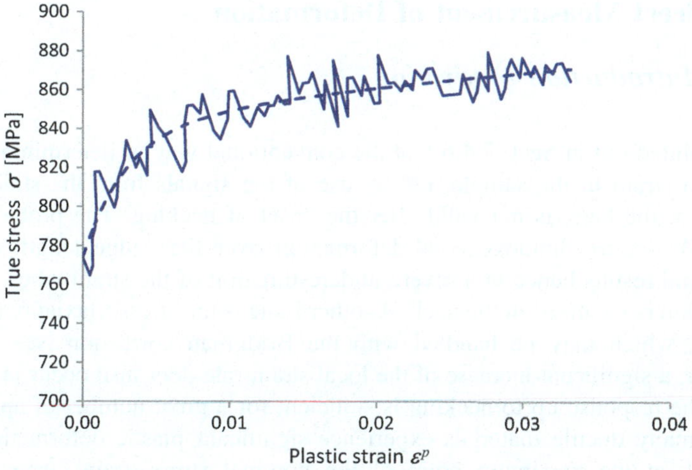

Fig. 2.33 True stress versus plastic strain curve for steel alloy (solid line) and fit of data to a smooth Voce function (dashed line)

Still paying attention to the test on a steel alloy, see the previous sections, the true stress—plastic strain curve is shown in Fig. 2.33. The true stress σ is calculated with Eq. (2.4), while Eq. (2.6) is employed to find the plastic strain ε^p. It should be noted that the ordinate axis of Fig. 2.33 is limited to the range from 700 to 900 MPa, thus, the noise of the experimental data is exaggerated. Another important observation is that the curve ceases at a plastic strain of only 0.033. The reason is that this strain corresponds to the onset of necking. It was emphasised in Sect. 2.1 that the strain state is not homogeneous any more after the global maximum of the nominal stress-strain curve. Hence, local measurements of the deformation are required for determination of the plastic strains after necking, and such data are not available with use of the conventional technique for processing of SHTB results. A more thorough discussion on this issue is provided in Sect. 2.5.

It is often convenient to operate with smooth curves during identification of strain-rate coefficients in constitutive models. Figure 2.33 also includes a fit of the experimental data to the Voce hardening function defined as

$$\sigma = Y + Q_1(1 - \exp(-C_1\varepsilon^p)) + Q_2(1 - \exp(-C_2\varepsilon^p)) \qquad (2.29)$$

where a least squares' algorithm is adopted to optimize the values of the five coefficients Y, Q_1, C_1, Q_2 and C_2. The smooth curve should be applied if, for instance, the true stress at a plastic strain of 0.02 is requested in a calibration procedure.

2.5 Direct Measurement of Deformation

2.5.1 Introduction Motivation

It was pointed out in Sect. 2.4.6 that the conventional way of determining the true stress and strain in the sample, i.e. by use of the signals from the strain gauges mounted at the bars, is not valid after the onset of necking. The problem is that Eq. (2.25) assumes homogeneous deformation over the gauge length L_0 of the sample, and results hence in a severe underestimation of the strain when the entire deformation is localized in the neck. Another issue is the stress triaxiality caused by the neck, which may be handled with the Bridgman correction, see Eq. (2.7). Moreover, a significant increase of the local strain rate does also occur in the neck. Indeed, the response up to necking is sufficient for a great number of applications because many ductile materials experience significant plastic deformation before they neck at the maximum point of the nominal stress-strain curve. Yet, the equivalent stress vs. plastic strain curve after necking is of utmost importance for the modelling of ductile failure.

The onset of necking does not always occur at a late stage of deformation. One example of a case where necking starts at comparatively small strains is the steel material which was addressed in the previous sections (e.g., Figs. 2.30 and 2.33). It is also a trend for many materials that the strain at onset of necking decreases with increasing temperature. Figure 2.34 shows the nominal and true stress-strain curves as found with the conventional method at room temperature and 300 °C for an aluminium alloy. The fracture strain was measured as 0.70 and 3.1, respectively, for the two samples [6]. The material is in other words much more ductile at 300 °C, yet the onset of necking occurs at a smaller strain. Indeed, the true stress-strain curve at 300 °C in Fig. 2.35b is abrupt at a strain less than 0.15, which is only 5% of the strain at failure.

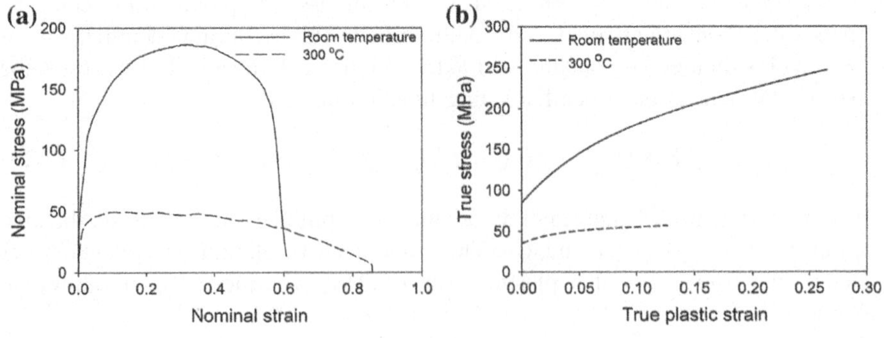

Fig. 2.34 Test results for an aluminium alloy AA6060 at a nominal strain rate of 400 s^{-1} and two different temperatures as found with the conventional technique for analysis of SHTB results: **a** Nominal stress-strain curves until fracture and **b** true stress versus true plastic strain curves until necking [6]

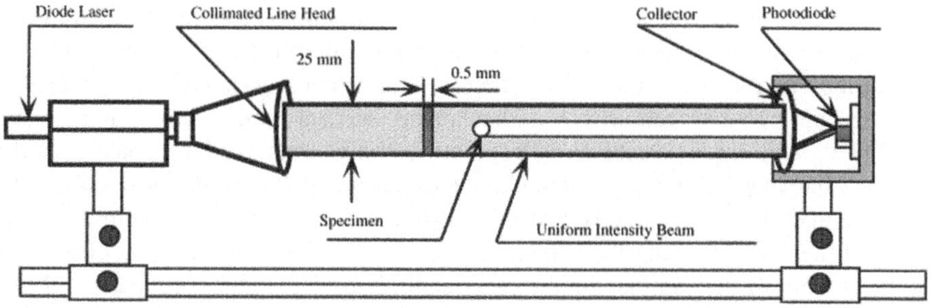

Fig. 2.35 Schematic of the laser occlusive radius detector (LORD) system as proposed by Li and Ramesh [35]. The specimen axis is normal to the plane of the paper

Another drawback associated with the use of the conventional method is that the initial gauge length L_0 in Eq. (2.25) neither is well defined before the onset of necking. Nevertheless, Sect. 2.4.5 devised a correction formula, see Eq. (2.28), which was shown to work well, at least for moderate strain levels.

More direct ways to measure the deformation of the specimen are therefore of interest. Strain gauges are not a convenient tool because they might loosen before large strains are achieved. Moreover, it is challenging to place the strain gauge exactly where the neck occurs without introducing any initial geometrical imperfection prior to the test, and the double-curved surface at the neck is also difficult to handle for strain gauges. Alternatives to strain gauges are therefore required.

Verleysen and Degrieck [34] applied a rotating drum camera, which is an analogue system, and a sheet specimen with grid lines. They used the camera recording to get the relative displacement between the grid lines, and subsequently the longitudinal strains. This method served to find the strain in the neck, but is less suited for determination of the transverse deformation. Another approach was proposed by Li and Ramesh [35], who used a laser system to measure the diameter of an axisymmetric sample during deformation. The width of their laser beam was 25 mm, see Fig. 2.35, and this was sufficient to capture the diameter along the entire gauge length of the specimen. By assumption of isochoric deformation, it is then straightforward to calculate the longitudinal strain with Eq. (2.8).

High-speed cameras have become common in many research labs during the last 10–20 years, and are well suited for the purpose of local measurements of deformations. The time scale of a split-Hopkinson test is of order 10^{-3} s. Modern high-speed cameras may provide a frame rate of more than 1 MHz, which is a typical order of the data acquisition frequency from strain gauges, but this figure is in general dependent on the frame size (i.e., number of pixels) and illumination of the specimen. Vilamosa et al. [6] and Gilat et al. [36] applied a frame rate of order 10^5 Hz when monitoring split-Hopkinson tension bar tests with high-speed cameras. It is feasible to get a sufficient large series of pictures from a SHTB test with such an acquisition frequency. If the purpose with the series is to determine the strains up to failure, it should consist of, say, minimum 100 pictures in order to get a good representation of the stress-strain curve.

A series of digital pictures obtained in a split-Hopkinson test can be used in two different ways. One option is to find the longitudinal and transverse strain field at the surface of the specimen during the test with use of digital image correlation (DIC). This method requires a rather plane surface unless two cameras are employed. The other possibility is to quantify the geometry of the contour of the sample (diameter, radius of curvature at neck) during all stages of deformation. The two options might be mutually exclusive if one camera is applied because a good focus of the surface facing the camera is likely to be of the expense of a clear view of the contour of the sample. On the other hand, both approaches are feasible if two cameras are employed.

2.5.2 Digital Image Correlation (DIC)

Digital image correlation (DIC) is a widely applied technique for determination of the in-plane strains at the surface of a material test sample or a component during mechanical testing. The key idea is that the deformation gradient is found by correlating the positions of a set of pixels in a sequence of pictures obtained with a digital camera. Subsequently, the in-plane strain field may be calculated. Determination of out-of-plane deformations calls for use of another camera and stereo DIC. It is required that the surface under consideration exhibits a spectre of grey-scale values, and it is therefore common to apply a speckled pattern with spray paint to the specimen prior to the test. Digital cameras and DIC have been applied as a measurement tool at quasi-static test conditions for around 20 years, and a number of DIC software packages are available at the commercial market. It should also be mentioned that other methods for calculation of deformations from pictures exist. One example is digital speckle photography (DSP), which was successfully applied by Kajberg and Wikman [37] in analysis of dynamic compression tests on steel.

Determination of strains with DIC in split-Hopkinson tension bar tests requires instrumentation with at least one high-speed camera. Studying sheet samples of high-strength steel and copper, respectively, Tarigopula et al. [30] and Gilat et al. [36] were rather early to present application of DIC in conjunction with SHTB testing. As of 2018, many laboratories apply high-speed cameras and DIC software in the analysis of results from SHTB testing, and there are published a lot of studies on the dynamic behaviour of different materials where DIC is a part of the instrumentation. Indeed, it seems that DIC rather soon will be a part of the state-of-the-art data acquisition system for dynamic material tests.

2.5.3 Contour Tracing

While DIC is well suited for sheet samples, it is more challenging to apply this technique for specimens with circular cross-section due to the curved surface.

An additional challenge arises when curvature about two perpendicular axes occur after the onset of necking. In this case, two cameras and stereo DIC are a typical choice for instrumentation. On the other hand, it is sufficient to measure the minimum diameter D in the neck and the radius of curvature R_{neck} of the necked contour for determination of the equivalent stress and logarithmic strain for circular samples made of metals, see Eqs. (2.7) and (2.8). This means that sufficient information is provided by a series of digital pictures with good contrast between the specimen and the background such that the contour is easily detectable.

Hopperstad et al. [38] were among the first ones to apply contour tracing in a tension test carried out with a split-Hopkinson apparatus. They studied notched samples made of structural steel, but limited their analysis to determination of the minimum diameter. The methodology was refined by Vilamosa et al. [6], see Fig. 2.4. In addition to the minimum diameter, identified with to red spots in the figure, they interpolated the shape of the boundaries with Chebyshev polynomials (green and blue curve). The radius of curvature, addressed with a red circle, was estimated from the blue, concave part of the polynomial. Figure 2.36 shows a photo of the experimental set-up applied by Vilamosa et al. [6], involving a split-Hopkinson tension bar, a high-speed camera and lights for sufficient illumination of the sample. This set-up was also augmented with an induction heater for high-temperature testing, see the next section.

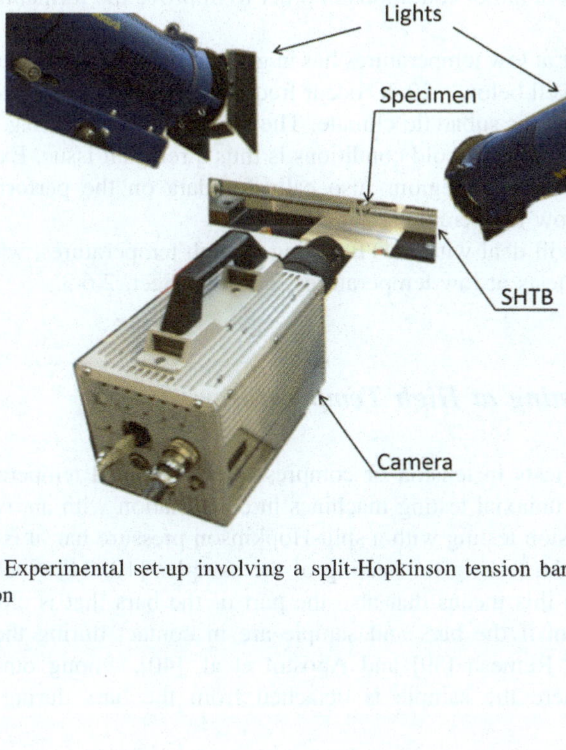

Fig. 2.36 Experimental set-up involving a split-Hopkinson tension bar, high-speed camera and illumination

2.6 SHTB Testing at Different Temperatures

2.6.1 Motivation

Most materials have mechanical properties that are dependent on strain rate. The temperature dependency is in general even more pronounced. Thus, advanced constitutive models take strain rate $\dot{\varepsilon}$ and temperature T into account, and they often treat $\dot{\varepsilon}$ and T as independent variables. However, the combined effect of strain rate and temperature may also be relevant in a number of engineering applications.

The perhaps most important example of coupling between $\dot{\varepsilon}$ and T is adiabatic heating during rapid plastic processes. Plastic work will always generate heat in the material at hand. At quasi-static conditions, the comparatively long time scale ensures that the heat is transferred to the surroundings through conduction, convection and/or radiation without giving any significant local increase of temperature in the plastically deformed zone. This is not the case for dynamic loading, where high strain rates may cause extensive local plastic deformation and subsequent rise of T and temperature softening because the heat does not have time to transfer away. Possible adiabatic heating should be taken into account in numerical modelling of impact and penetration problems.

Another field where increased temperatures occur simultaneously with high strain rates is plastic forming operations like extrusion, rolling and forging. Such processes are carried out at hot conditions in order to improve the formability of the material.

Material behaviour at low temperatures has in general gained less attraction. Yet, winter temperatures well below −20 °C occur frequently in many habituated places with continental, alpine or subarctic climate. The crashworthiness of e.g. cars and various protective structures at cold conditions is thus a relevant issue. Exploration of natural resources in polar regions also calls for data on the performance of various materials at low temperatures.

The next section will deal with SHTB testing at high temperatures, while some considerations about tests at low temperatures follow in Sect. 2.6.3.

2.6.2 SHTB Testing at High Temperatures

Quasi-static material tests in tension or compression at elevated temperatures are usually performed in uniaxial testing machines in combination with an oven. Also for dynamic compression testing with a split-Hopkinson pressure bar, it is common to attach an oven to the test rig for heating of the samples [10, 11]. Such heating takes some time, and this means that also the part of the bars that is close to the sample may get warm if the bars and sample are in contact during the heating process. Lennon and Ramesh [39] and Apostol et al. [40], among others, have devised methods where the sample is detached from the bars during heating, and put in contact with the bars immediately before testing. Such a technique is most easy to implement for SHPB systems where the specimen does not have to be screwed into the bar.

Although heating with an oven also is applied for split-Hopkinson tension bars, it is normally more attractive to choose a heating device that is faster in order to reduce the temperature increase in the bars. Four possibilities are infra-red spot heating [39], heat gun [41], direct resistance heating [42] and induction heating [43]. In addition to the reduced heating time, an important feature facilitating these options rather than an oven is that they usually provide optical access to the sample, implying that the test can be instrumented with one or two cameras for determination of the strain field, see Sect. 2.5.

Lennon and Ramesh [39] applied an infra-red spot heating device in a split-Hopkinson compression bar set-up, see Fig. 2.37. The technique should be feasible in a tension bar as well. They report that the infra-red radiation is concentrated to an ellipsoidal spot with axes of approx. 5 × 5 × 8 mm. The heating time is of order 1 min.

Figure 2.38 shows the heating device applied by Lemanski et al. [41]. They used a heat gun, and a surrounding heat shield ensured that the hot air was directed over the specimen. This heating process takes some minutes. Optical visibility can be obtained by removing the shield immediately before the test.

Clausen et al. [42] presented high-temperature tests on aluminium in a split-Hopkinson tension bar with direct resistance (ohmic) heating of the specimen. The set-up is shown in Fig. 2.39. A transformer with adjustable input current was

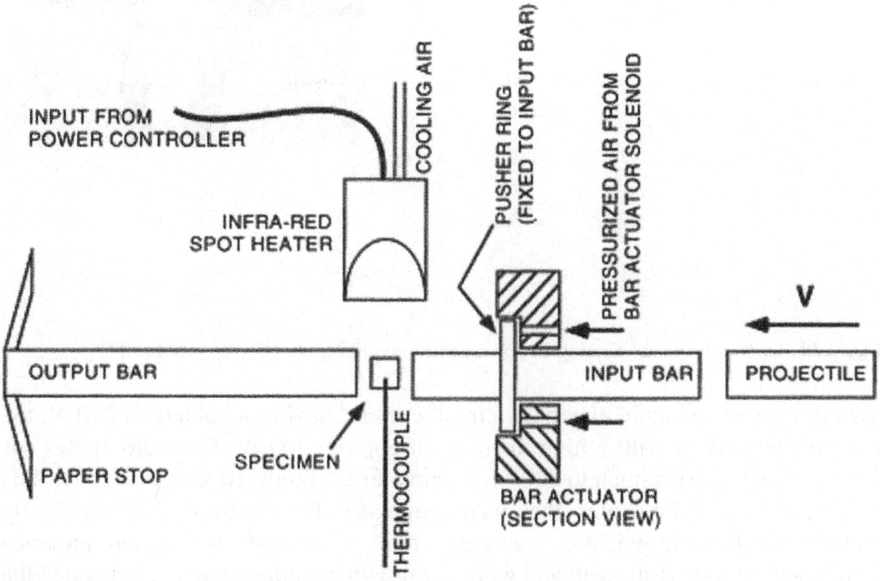

Fig. 2.37 Schematic of set-up for infra-red radiation heating of sample in a split-Hopkinson pressure bar test as proposed by Lennon and Ramesh [39]

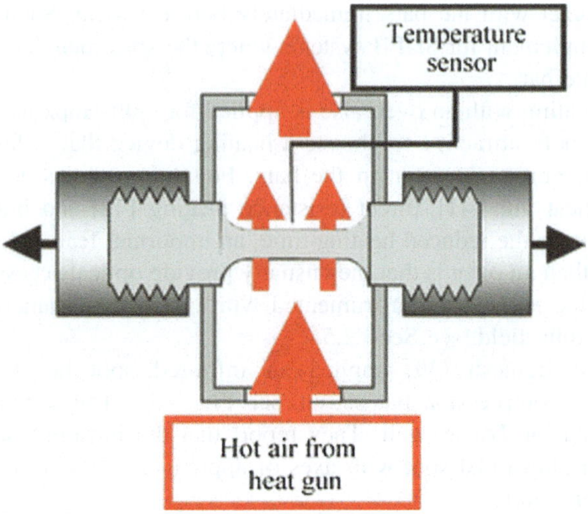

Fig. 2.38 Schematic of SHTB specimen subjected to direct heating by hot air as proposed by Lemanski et al. [41]. The specimen is surrounded by a heat shield

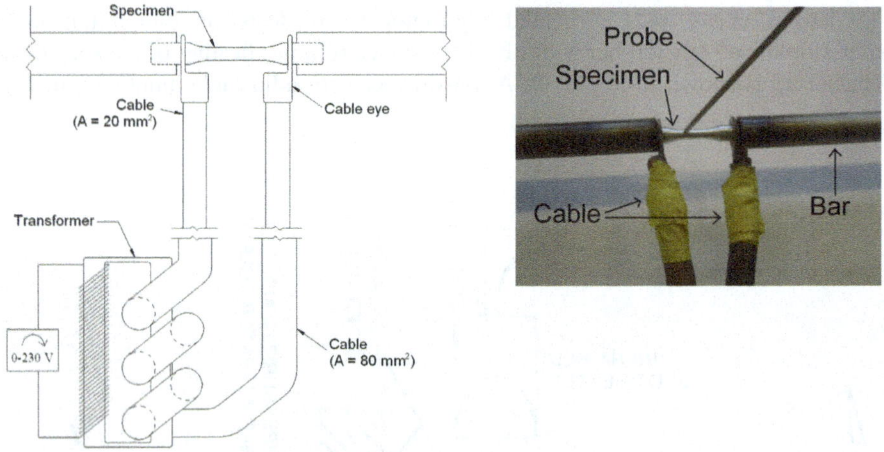

Fig. 2.39 Direct resistance heating of specimen in a split-Hopkinson tension bar [42]

used to convert the main electricity circuit current, having a voltage of 220 V, to a low-voltage current with a high intensity of approx. 200 A. This current was sufficient to heat the specimen to 300 °C within a time of only 10 s. In order to avoid a similar heating of the cables, they were required to have considerably larger cross section area than the specimen's area of 7 mm^2. Experimental tests and numerical simulations of the set-up with and without this cable arrangement demonstrated that there were not introduced any significant disturbances of the stress waves and the performance of the SHTB test [42].

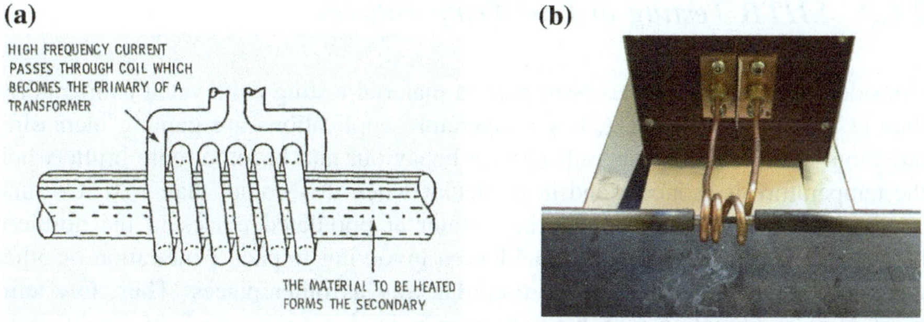

Fig. 2.40 Induction heating of split-Hopkinson tension bar sample, **a** Sketch of induction-coil heater as presented by Rosenberg et al. [43], **b** Induction apparatus used in conjunction with the set-up shown in Fig. 2.36

Induction heating provides a non-contact alternative to direct resistance heating. Such a set-up was devised by Rosenberg et al. [43] as early as in 1986. The principle is outlined in Fig. 2.40a. An induction coil carrying high-frequency current surrounds the metallic specimen. The alternating current induces a current in the sample, and it is in a similar fashion as for the previous method (Fig. 2.39) heated as a result of internal loss of energy due to the ohmic resistance. The heating rate is highest for magnetic materials. Rosenberg et al. [43] heated a steel specimen to 700 °C in about one minute, and a similar heating rate of 10 °C/s was reported in a more recent study on aluminium alloys [6]. Figure 2.40b shows the set-up of Vilamosa et al. [6]. This was applied in conjunction with the high-speed camera instrumentation depicted in Fig. 2.36.

A relevant concern in high-temperature tests in a SHTB apparatus is whether the hot specimen causes any significant heating of the bars. As Young's modulus E_b and hence the wave propagation velocity c_b of the bars decrease with increasing temperatures, both the stress and strain calculations are affected, see Eqs. (2.23) and (2.25), respectively. On the other hand, Young's modulus for steel, which is the usual bar material in SHTB systems, is not reduced much at moderate temperatures up to 300 °C. Provided that the heating device works fast, this implies that only the part of the bars that are close to the specimen will experience any significant temperature increase. Vilamosa et al. [6] investigated this issue both experimentally and numerically. They measured the temperature in the bars with thermocouples, and found that a process where an aluminium sample was heated to 300 °C gave bar temperatures larger than 100 °C only in the closest vicinity (around 50 mm) to the specimen. A subsequent thermo-mechanical numerical simulation, applying a temperature-dependent Young's modulus for the steel bars, revealed that the strain gauge response and hence the stress-strain curves were not affected by this heating of the bars. Yet, slow heating and/or higher temperatures in the sample might call for precautions, for instance cooling of the bars.

2.6.3 SHTB Testing at Low Temperatures

Considerably more effort has been paid to material testing at elevated temperatures than at cold conditions. Yet, low-temperature applications are gaining increasing attention. Many ductile materials change behaviour and become more brittle when the temperature is reduced, and in particular when it is colder than −20 °C. Such conditions are quite common during winter at populated places in the northern hemisphere or at high altitudes. Incidences involving impact, penetration or other rapid loading may occur here in a similar way as other places. Thus, low temperature in combination with high strain rates is also a relevant field for research.

Børvik et al. [44] did dynamic tension tests at −10 °C using a temperature chamber made of plexiglass. These tests were performed in a hydro-pneumatic machine at a rate of about 50 s^{-1} rather than a split-Hopkinson bar. Yet, the specimen is connected to two bars also in such a machine. Therefore, the methodology is transferable to SHTB testing. The chamber shown in Fig. 2.41 surrounded the specimen and the adjacent parts of the bars without being in direct contact with any moving parts. The temperature was reduced by flushing liquid nitrogen into the chamber, and a regulator together with two thermocouples controlled the temperature. The plexiglass chamber ensures optical visibility to the sample. Applying a similar chamber, although with plane walls, Johnsen et al. [45] have recently demonstrated that the transparent walls of the chamber do not disturb the pictures taken with a digital camera nor the subsequent DIC analysis.

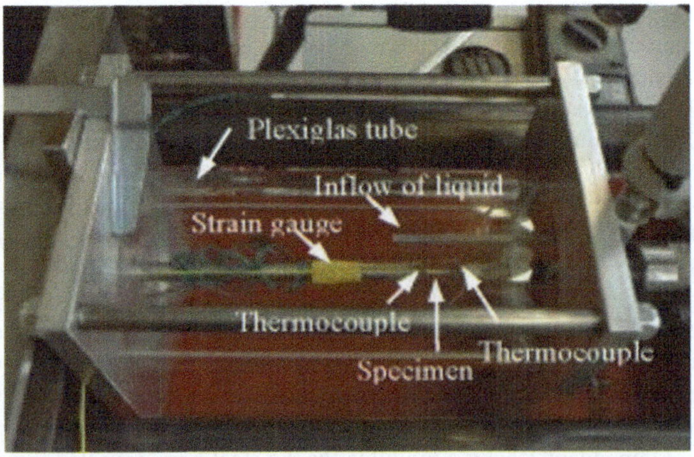

Fig. 2.41 Temperature chamber used for dynamic tension tests at low temperatures [44]

2.6.4 Temperature Measurement

Tests at hot as well as cold conditions call for measurements of the temperature in the sample. Normally, the temperature immediately before the test is of main concern. When an oven is used to heat the specimen, the temperature in the oven defines the testing temperature of the sample provided that the preconditioning time is sufficiently long. The same applies for tests at low temperatures with a cooling chamber.

Direct measurements of the temperature of the sample are required when they are heated in other ways, for instance by induction or with a heat gun. A classical way to measure temperatures, at least for metals, is to apply thermocouples to the surface of the specimen. Other alternatives are measurement probes, laser-based infrared pyrometers or infrared cameras. Common for these devices is that they give the surface temperature of the specimen. To ensure that the entire volume of the specimen gets the desired temperature before a test, a certain holding time is required. This time depends on the thermal properties of the material at hand.

Fast plastic deformation in a dynamic test causes adiabatic heating of the material. Provided that the sampling rate of the temperature measurement system is high enough, it is possible to measure this temperature increase. Recalling that the duration of a split-Hopkinson test is of order 1 ms, it follows that an acquisition frequency of order 10^4–10^5 Hz is required. It is, however, possible to get an estimate of the temperature increase ΔT if the deformation process is assumed to be adiabatic, viz.

$$\Delta T = \int_0^{\varepsilon^p} \chi \frac{\sigma_{eq} d\varepsilon^p}{\rho C_p} \tag{2.30}$$

where ρ is the density of the material, C_p is the specific heat capacity, and χ is the Taylor-Quinney coefficient that defines the proportion of plastic work converted into heat, normally assumed to have a value of 0.9. The nominator $\sigma_{eq} d\varepsilon^p$ of Eq. (2.30) represents the incremental plastic work, which is integrated over the entire plastic deformation to obtain the increase of temperature.

2.7 Summary

Tension tests at quasi-static and dynamic conditions provide very useful information for a wide range of engineering applications. This chapter has outlined various set-ups and methods for dynamic tension tests, applying split-Hopkinson bars. Particular attention has been paid to a pre-tension set-up that allows for comparatively long incoming stress wave, which again facilitates large straining of the sample and/or tests at strain rates down to $10^2\ s^{-1}$. Experimental techniques for tests at hot or cold conditions have also been reviewed.

The chapter has put some emphasise on measurement techniques and the subsequent data treatment. Normally, the outcome from a split-Hopkinson tension test is to be used in a numerical model, which calls for an accurate determination of the true stress and strain in the sample during deformation, preferably also after the onset of necking. The conventional way of data acquisition, involving strain gauges glued at the bars, have been presented, but the chapter included also more modern instrumentation protocols applying one or more high-speed cameras.

Acknowledgements This chapter reflects the expertise and technical know-how of tensile testing in split-Hopkinson bar systems of Dr Carlo Albertini, who was a researcher at the Joint Research Centre (JRS) of the European Commision in Ispra, Italy. He was instrumental in the design of the SHTB system that was installed at Structural Impact Laboratory (SIMLab) at the Norwegian University of Science and Technology (NTNU) in Trondheim, Norway, in 2001. The engagement of Professors Magnus Langseth, Odd Sture Hopperstad and Tore Børvik has also been crucial. The technical assistance of Mr Trond Auestad, Mr Olav Haldorsen, Mr Trygve Meltzer and Mr Tore Wisth during installation and subsequent improvements of the test rig is acknowledged. Last, but not least, important contributions to this work have been provided through the theses of Ph.D. Yan Chen, Ph.D. Vincent Vilamosa and MSc Henning Oppedal.

References

1. Osman TM, Rigney JD (2000) Introduction to the mechanical behaviour of metals. In: Kuhn H, Medlin D (eds) ASM Handbook, vol 8. Mechanical testing and evaluation. ASM International, Materials Park, OH, pp 1–12
2. Dieter GE (1988) Mechanical metallurgy. McGraw-Hill
3. Gromada M, Mishuris G, Öchsner A (2011) Correction formulae for the stress distribution in round tensile specimens at neck presence. Springer
4. Valiente A (2001) On Bridgman's stress solution for a tensile neck applied to axisymmetric blunt notched tension bars. J Appl Mech 68:412–419
5. Bridgman PW (1964) Studies in large plastic flow and fracture. Harvard University Press, Cambridge, MA
6. Vilamosa V, Clausen AH, Fagerholt E, Hopperstad OS, Børvik T (2014) Local measurement of stress-strain behaviour of ductile materials at elevated temperatures in a split-hopkinson tension bar system. Strain 50:223–235
7. Kolsky H (1963) Stress waves in solids. Dover Publications
8. Graff KF (1975) Wave motion in elastic solids. Dover Publications
9. Nemat-Nasser S (2000) Introduction to high strain rate tension and compression tests. In: Kuhn H, Medlin D (eds) ASM Handbook, vol 8. Mechanical testing and evaluation. ASM International, Materials Park, OH, pp 429–446
10. Gray GT (2000) Classic split-hopkinson pressure bar testing. In: Kuhn H, Medlin D (eds) ASM Handbook, vol 8. Mechanical testing and evaluation. ASM International, Materials Park, OH, pp 462–476
11. Chen W, Song B (2011) Split-Hopkinson (Kolsky) bar—design, Testing and Applications. Springer
12. Lindholm US, Yeakley LM (1968) High Strain-rate Testing: Tension and Com-pression. Exp Mech 8:1–9

13. Nicholas T (1981) Tensile Testing of Materials at High Rates of Strain. Exp Mech 21:177–185
14. Haugou G, Markiewicz E, Fabis J (2006) On the use of the non direct tensile loading on a classical split Hopkinson bar apparatus dedicated to sheet metal specimen characterisation. Int J Impact Eng 32:778–798
15. Mohr D, Gary G (2007) M-shaped specimen for the high-strain rate tensile testing using a split hopkinson pressure bar apparatus. Exp Mech 47:681–692
16. Dunand M, Gary G, Mohr D (2013) Load-inversion device for the high strain rate tensile testing of sheet materials with hopkinson pressure bars. Exp Mech 53:1177–1188
17. Roth CC, Gary G, Mohr D (2015) Compact SHPB system for intermediate and high strain rate plasticity and fracture testing of sheet metal. Exp Mech 55:1803–1811
18. Harding J, Wood EO, Campbell JD (1960) Tensile testing of materials at impact rates of strain. J Mech Eng Sci 2:88–96
19. Harding J, Welsh LM (1983) A tensile testing technique for fibre-reinforced composites at impact rates of strain. J Mater Sci 18:1810–1826
20. Thakur A, Nemat-Nasser S, Vecchio KS (1996) Dynamic bauchinger effect. Acta Mater 44:2797–2807
21. Ogawa K (1984) Impact-tension compression test by using a split-Hopkinson bar. Exp Mech 24:81–86
22. Guzman O, Frew DJ, Chen W (2011) A Kolsky tension bar technique using a hollow incident tube. Meas Sci Technol 22:045703
23. Van Slycken J, Verleysen P, Degrieck J, Bouquerel J, De Cooman BC (2007) Dynamic response of aluminium containing TRIP steel and its constituent phases. Mater Sci Eng, A 460–461:516–524
24. Gerlach R, Kettenbeil C, Petrinic N (2012) A new split Hopkinson tensile bar design. Int J Impact Eng 50:63–67
25. Albertini C, Montagnani M (1974) Testing techniques based on the split Hopkinson bar. Inst Phys Conf 21:22–32
26. Albertini C, Montagnani M (1976) Wave propagation effects in dynamic loading. Nucl Engrg Des 37:115–124
27. Staab GH, Gilat A (1991) A Direct-tension split Hopkinson bar for high strain-rate testing. Exp Mech 31:232–235
28. Cadoni E, Solomos G, Albertini C (2009) Mechanical characterisation of concrete in tension and compression at high strain rate using a modified Hopkinson bar. Mag Concrete Research 61:221–230
29. Chen Y, Clausen AH, Hopperstad OS, Langseth M (2011) Application of a slip-Hopkinson tension bar in a mutual assessment of experimental tests and numerical predictions. Int J Impact Eng 38:824–836
30. Tarigopula V, Hopperstad OS, Langseth M, Clausen AH, Hild F (2008) A study of localisation in dual-phase high-strength steels under dynamic loading using digital image correlation and FE analysis. Int J Solids Struct 45:601–619
31. Huh H, Kang WJ, Han SS (2002) A tension split Hopkinson bar for investigating the dynamic behavior of sheet metals. Exp Mech 42:8–17
32. Follansbee PS, Frantz C (1983) Wave propagation in the split Hopkinson pressure bar. J Eng Mat Tech 105:61–66
33. Albertini C, Montagnani D (1977) Dynamic material properties of several steels for fast breeder reactor safety analysis. Report EUR 5787 EN, Applied Mechanics Division, Joint Research Centre, Ispra, Italy
34. Verleysen P, Degrieck J (2004) Experimental investigation of the deformation of Hopkinson bar specimens. Int J Impact Eng 30:239–253
35. Li Y, Ramesh KT (2007) An optical technique for measurement of material properties in the tension Kolsky bar. Int J Impact Eng 34:784–798
36. Gilat A, Schmidt TE, Walker AL (2009) Full field strain measurement in compression and tensile split Hopkinson bar experiments. Exp Mech 49:291–302

37. Kajberg J, Wikman B (2007) Viscoelastic parameter estimation by high strain-rate experiments and inverse modelling—Speckle measurements and high-speed photography. Int J Solids Struct 44:145–164
38. Hopperstad OS, Børvik T, Langseth M, Labibes K, Albertini C (2003) On the influence of stress triaxiality and strain rate on the behaviour of a structural steel. Part I. Experiments. European J Mech A Solids 22:1–13
39. Lennon AM, Ramesh KT (1998) A technique for measuring the dynamic behaviour of materials at high temperatures. Int J Plasticity 14:1279–1292
40. Apostol M, Vuoristo T, Kuokkala VT (2003) High temperature high strain rate testing with a compressive SHPB. J de Physique IV 110:459–464
41. Lemanski SL, Petrinic N, Nurick GN (2013) Experimental Characterisation of Aluminium 6082 at Varying Temperature and Strain Rate. Strain 49:147–157
42. Clausen AH, Auestad T, Berstad T, Børvik T, Langseth M (2006) High-temperature tests on aluminium in a split-Hopkinson bar—experimental set-up and numerical predictions. J Phys IV France 134:603–608
43. Rosenberg Z, Dawicke D, Strader E, Bless SJ (1986) A new technique for heating specimens in split-Hopkinson-bar experiments using induction-coil heaters. Exp Mech 26:275–278
44. Børvik T, Lange H, Marken LA, Langseth M, Hopperstad OS, Aursand M, Rørvik G (2010) Pipe fittings in duplex stainless steel with deviation in quality caused by sigma phase precipitation. Mat Sci Eng A 527:6945–6955
45. Johnsen J, Grytten F, Hopperstad OS, Clausen AH (2016) Experimental set-up for determination of the large-strain tensile behavior of polymers at low temperatures. Polym. Test. 53:305–313

Chapter 3
Shear Testing Using the Kolsky-Hopkinson Bar Machine

Patricia Verleysen

3.1 Why Shear Tests?

The use of uniaxial tension and compression tests to characterise the mechanical behaviour of materials is widespread, also at dynamic strain rates. Indeed, multiple techniques which impose an axial load on a specimen have been developed. The strain rates which can be achieved range from extremely low values ($\dot{\varepsilon} = 10^{-8}/s$) for creep testing up to highly dynamic rates ($\dot{\varepsilon} > 10^5/s$) in explosion experiments [1, 2].

Assessment of the deformation and strength of structures or structural components is based on material parameters such as stiffness, strength, deformation capacity, energy absorption, etc. These parameters can be derived from tensile or compression test data in a straightforward fashion. In addition, most constitutive material models heavily rely on tensile tests. For example: one single tensile test is sufficient to define the yield surface and hardening curve required to describe Mises (or J_2-) plasticity. For a visco-elastic material, the creep or relaxation function can be derived from one tensile test as well. These examples demonstrate why uniaxial tensile tests are by far the most commonly used experiments for static material characterisation. Unsurprisingly, split Hopkinson bar *tensile and compression* tests are the most widely used experiments to characterise materials at strain rates between $200/s$ and $10^5/s$.

Although very widely used, tension and compression tests do exhibit some inherent shortcomings. These shortcomings are even more pronounced at high strain rates. Indeed, in high strain rate compression tests, the material characterisation is hampered by both radial inertia and barrelling of the sample. In high strain

P. Verleysen (✉)
Department EEMMECS, Materials Science and Technology – DyMaLab research group, Ghent University, Ghent, Belgium
e-mail: patricia.verleysen@UGent.be

R. Othman (ed.), *The Kolsky-Hopkinson Bar Machine*,
https://doi.org/10.1007/978-3-319-71919-1_3

rate tensile tests, the higher yield stress level which is often observed for metal alloys encourages earlier necking of the specimen. As soon as necking develops, both the heterogeneous strain distribution over the gauge section and the non-axial stresses in the neck complicate the interpretation of the experimental results. Indeed, the measured force and corresponding elongation can no longer be directly translated into the material's stress-strain response. This is in particular a problem for metals with a low strain hardening ability, such as some titanium and aluminium alloys. In addition, the high stress triaxiality and high localised strain in the necked section accelerate failure.

Shear tests, on the other hand, do not suffer from the disadvantages mentioned above. Indeed, no thinning or other changes of the gauge cross section occur in a well-designed shear specimen. As a consequence, diffuse and localised necking is avoided. Since there is no reduction in cross section, an unambiguous relation between true and engineering values of the stress can be derived. Moreover, shear tests provide valuable data on material fracture parameters. Unlike uniaxial tensile tests, where ductile metal failure results from highly localized deformations after necking, fracture during a shear test can be attributed to material failure instead of geometric instability. In a shear test, damage growth is retarded thanks to the lower stress triaxiality and generally larger strains—including stage IV hardening strains—are obtained in a shear test, which enables more accurate material modelling [3, 4]. However, materials susceptible to shear localization still fail at lower deformations due to strain localization rather than damage. For such materials, shear tests provide a useful means to investigate the mechanisms leading to shear localization, such as adiabatic shear banding.

Also from a more fundamental point of view, shear tests have their merits. Indeed, regardless of the deformation mode, a macroscopic strain imposed on a ductile material is accommodated by shear deformation in the microstructure. Shear tests not only introduce a fundamental deformation mode, but are representative for many applications in which metals are loaded in shear, e.g. machining, punching, extrusion, severe plastic deformation processing, ….

Several experimental techniques for shear testing have been developed. Shear tests can be performed by direct shear loading (e.g. torsion of thin-walled tubular samples), or appropriate sample design to convert a tensile or compressive load into shear (e.g. simple shear tests).

In the next sections, dynamic shear tests based on the split Hopkinson bar principle are discussed in detail. After a section dealing with shear deformation modes (Sect. 3.2), Sect. 3.3 elaborates on shear testing techniques for sheet or plate materials. Section 3.4 discusses dynamic shear testing of bulk materials by performing torsion tests on thin-walled tubes (Sects. 3.4.1 and 3.4.2), and compression tests on hat-shaped (Sect. 3.4.3) and purpose designed sample geometries (Sect. 3.4.4). Both the plate shear tests and the torsion test are material characterisation tests. Indeed, these tests introduce a well-controlled shear deformation in a sample and the shear stress-strain response of the material can be derived from measurements during the test.

3.2 Shear Deformation Modes

From a theoretical point of view, two shear deformation states are distinguished: simple shear and pure shear. These deformation states are illustrated in Fig. 3.1 for a square with initial side lengths of 1. In simple shear (Fig. 3.1a), two parallel directions are shifted against each other over a distance u_x whilst preserving the initial distance Δy. The area of an initial square does not change, but the angle α between the initially perpendicular sides changes by

$$\alpha = \tan^{-1}(\gamma) = \tan^{-1}(u_x/\Delta y) \tag{3.1}$$

and the directions of the minor and major strain axes rotate towards the fixed direction. The engineering strain is defined as $\gamma = u_x/\Delta y$. For small deformations, the engineering strain is equal to the angular change: $\gamma \approx \alpha$.

In pure shear, on the other hand, two originally perpendicular directions remain perpendicular, as shown in Fig. 3.1b. These directions correspond with the minor and major directions of the strain and remain constant. The highest shear values in the square are obtained at 45°. As is the case for simple shear, the area of an initial square does not change.

The deformation mode in a shear test is determined by the sample geometry, the boundary conditions and the imposed load or deformation. Generally, the deformation pattern in shear tests is closer to simple shear, however, also techniques have been developed which aim at pure shear.

Also in real shear deformation processes, most often a combination of simple and pure shear is found. This observation led geologists to the conclusion that 'pure shear is pure nonsense and simple shear is simple nonsense'.

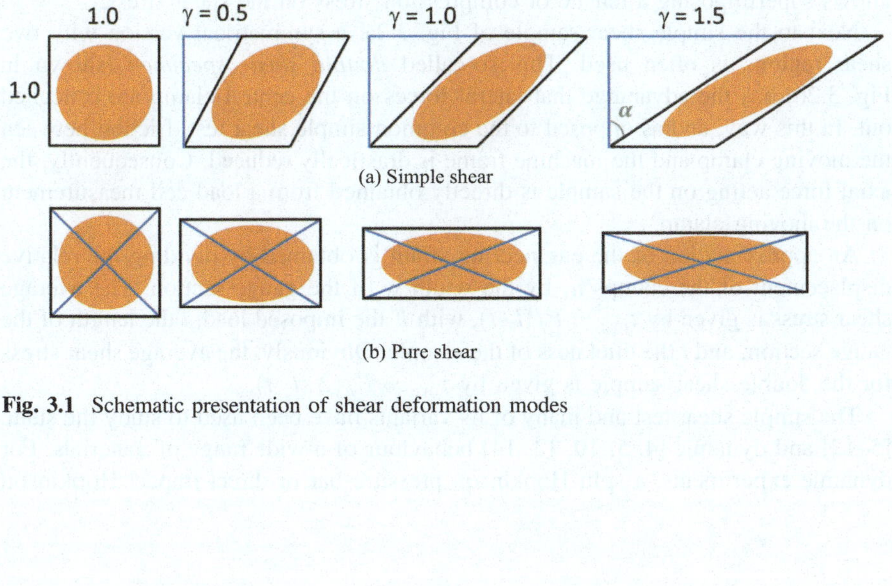

Fig. 3.1 Schematic presentation of shear deformation modes

3.3 Hopkinson Shear Testing of Sheet Materials

As is the case for static shear tests, Hopkinson shear tests on sheet materials are generally performed on a compression or tensile setup in a simple shear configuration. However, the simple shear test techniques developed for static testing cannot be used directly in a Hopkinson setup. Indeed, the specific nature of Hopkinson tests, and more specifically the use of mechanical waves to deform the sample, imposes limitations. Major challenges to be tackled are related to the selection of appropriate sample dimensions and the sample clamping mechanism. In addition to these experimental challenges, the simple shear test suffers from some inherent shortcomings. The most important drawback is that the full shear deformation potential of the material is not revealed by the test because of premature damage or fracture initiation at the free sample edges. Therefore, a new generation of test techniques has been developed which relies on the use of enhanced, yet also more complex, sample geometries. The simple shear tests are presented in Sect. 3.3.1. Techniques with optimised sample geometries are elaborated in Sect. 3.3.2. In Sect. 3.3.3 shear test results are presented for Ti6Al4V sheet.

3.3.1 Simple Shear Tests

The principle of the in-plane simple shear test is schematically presented in Fig. 3.2a. A rectangular sample is mechanically clamped at two opposite sides. Subsequently, the shear deformation is introduced by moving the clamps parallel to each other in opposite directions. Generally, the distance between the clamps, and thus the width of the sample gauge section, remains fixed. However, an additional load or deformation may be introduced in the lateral direction. This additional load allows superimposing a tensile or compression stress on the shear stress.

Next to the simple shear sample of Fig. 3.2a, a symmetrical version with two shear regions is often used. This so-called *double shear specimen* (shown in Fig. 3.2b) has the advantage that lateral forces on the central clamp are cancelled out. In this way, and as opposed to the common simple shear test, friction between the moving clamp and the machine frame is drastically reduced. Consequently, the axial force acting on the sample is directly obtained from a load cell measurement on the moving clamp.

An approximation of the engineering strain is obtained by dividing the relative displacement of the clamps u_x by the width w of the gauge section. The average shear stress is given by $\tau_{avg} = F/(l \cdot t)$, with F the imposed load, l the length of the gauge section, and t the thickness of the sample. Obviously, the average shear stress for the double shear sample is given by $\tau_{avg} = F/(2 \cdot l \cdot t)$.

The simple shear test and many of its variants have been used to study the static [5–12] and dynamic [4, 5, 10, 13, 14] behaviour of a wide range of materials. For dynamic experiments, a split Hopkinson pressure bar or direct impact Hopkinson

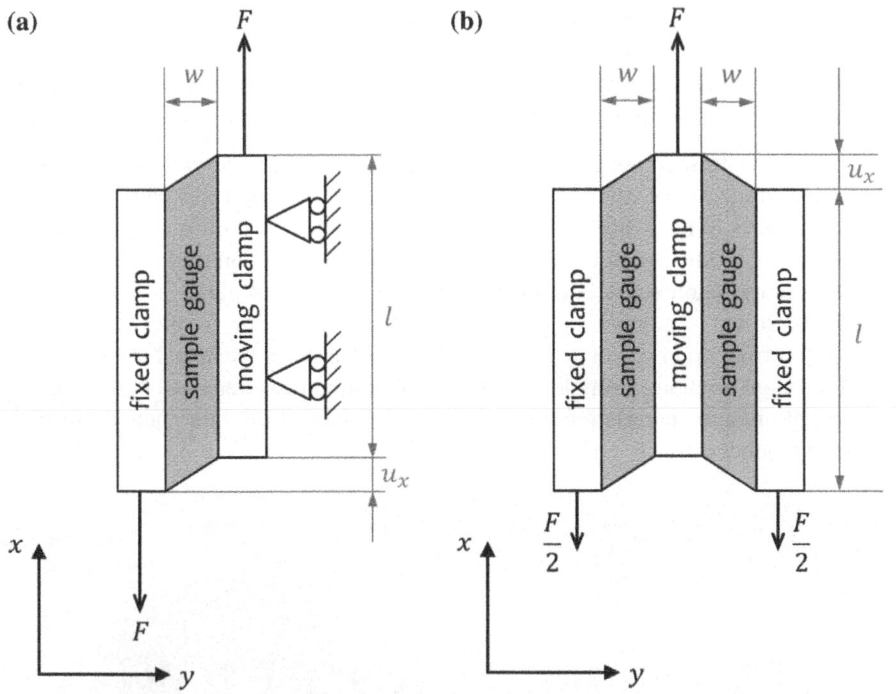

Fig. 3.2 Schematic representation of a simple shear test with a single (**a**) and double (**b**) gauge section. l and w are respectively the length and width of the sample gauge section, and u_x is the relative displacement of the clamps

bar setup is most commonly used with a double shear sample. Already in 1970, Campbell and Ferguson reported the use of a double-notched sample loaded in a Hopkinson setup with a solid input bar and hollow output bar [15]. The double shear sample is preferred to maintain axial loads in the setup.

Figure 3.3a shows the clamping device used by Merle and Zhao to test sheet metals in a split Hopkinson bar compression setup [13, 16]. The rectangular double shear specimen is connected to the input bar using a slitted cylindrical insert like shown in Fig. 3.3a. The sample is connected to the insert by four bolts, and with an adapter to the output bar by six bolts. The sample/clamp assembly is presented in Fig. 3.3a, right. To reduce the reflections at the interfaces between the clamps and the bars, Hopkinson bars with a diameter comparable to the clamp dimensions are used. However, waves propagating in the clamp parts introduce significant oscillations. Alternatively, Rusinek and Klepaczko [4] developed a direct impact Hopkinson bar test technique on a modified double shear specimen. This setup is presented in Fig. 3.3b. Since no input bar is used, the sample displacement is no longer obtained from strain reading on the input bar. Instead, the specimen displacement is measured using an advanced optical extensometer. Comprehensive studies on dynamic shear testing techniques can be found in the papers of Klepaczko and co-workers [4, 5, 10, 17, 18].

Despite the good results that have been obtained with these methods, the simple shear test with both single or double gauge sections exhibits some shortcomings:

1. The shear stress is not homogeneously distributed over the gauge section [19]. The average stress calculated by $\tau_{avg} = F/(l \cdot t)$ is only representative for the shear stress in the centre of the gauge section if the l/w ratio is sufficiently high and the w/t ratio is sufficiently small. Moreover, in addition to shear, also tensile and compression stress components exist in the shear region.
2. Uncontrolled specimen deformation is likely to occur in the clamps. The engineering strain calculated using u_x/w overestimates the actual shear strain in the gauge section. Complementary local strain measurements based on e.g. digital image correlation or other techniques [20] are hence strongly recommended.

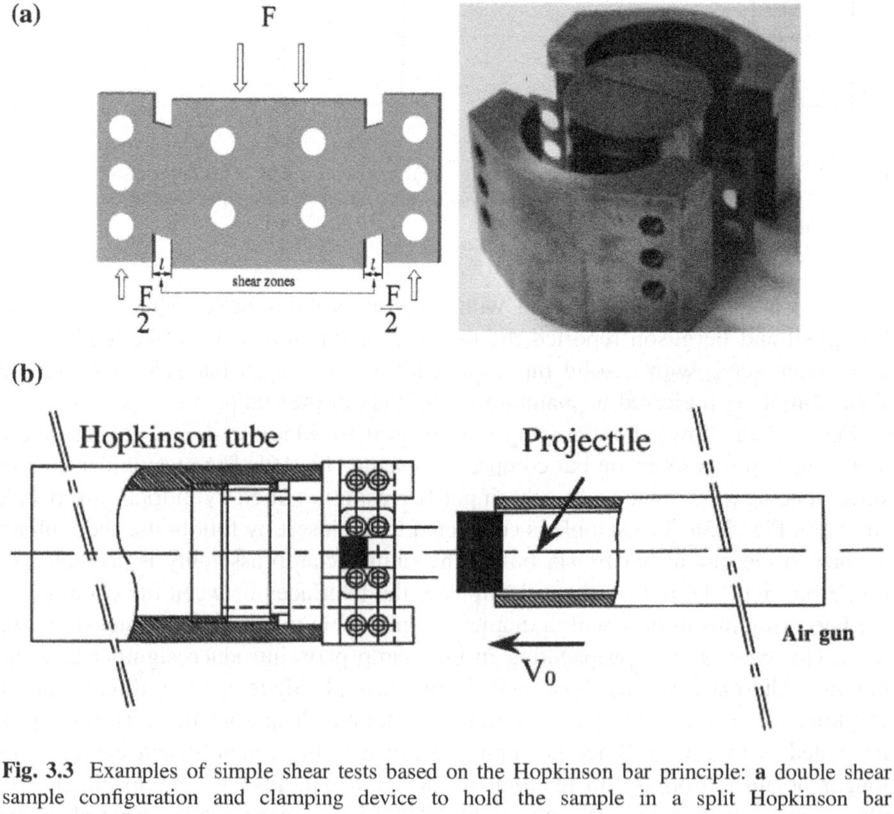

Fig. 3.3 Examples of simple shear tests based on the Hopkinson bar principle: **a** double shear sample configuration and clamping device to hold the sample in a split Hopkinson bar compression setup [16] and **b** direct impact Hopkinson setup with adapted double shear specimen [5] (reproduced with permission)

3. Frequently, buckling and cracking occurs at the borders of the shear region near the clamps. This leads to an early specimen failure initiated at the edges, instead of the targeted failure in the central part of the specimen [7, 19]. Decreasing the width over thickness ratio [19] or introducing slits at the edges [11] reduce the stress perturbations and hence reduce the risk of early specimen failure. However, these mitigation measures have a detrimental effect on the strain homogeneity [20].
4. Using the double shear specimen, an *average* of the material behaviour in the two shear regions is obtained.
5. The stress and strain distribution in the sample gauge section is strongly dependent on the dimensions of the sample. A satisfactory homogeneity until high levels of deformation is only obtained for sufficiently slender samples. For metallic materials, different authors [5, 19, 21] recommend an aspect ratio of $l/w \geq 10$. However, an essential prerequisite to obtain homogeneous stresses and strains in a Hopkinson experiment is that a state of quasi-static equilibrium is established in the dynamically loaded sample. The time required to reach quasi-static equilibrium is proportional with the length of the sample, which thus imposes strict limits on the total sample length.
6. On top of that, in split Hopkinson bar devices, where the specimen is deformed by mechanical waves, the devices used to fix the sample (clamps, screws, etc.) induce errors. Indeed, the elastic stress waves interact with the clamping devices, which affects both the sample deformation and applied force, as well as the accuracy of the load and deformation history calculated from strain gauge measurements on the bars. The influence of the clamping devices on the test results can be reduced by using larger diameter Hopkinson bars [13]. However, the use of larger diameter bars introduces other difficulties, such as excessive wave distortion and violation of the plane wave assumption on which the traditional calculation of the sample stress, strain and strain rate is based.

3.3.2 Tests with Optimised Sample Geometries

To alleviate the shortcomings associated with the traditional simple shear test, in recent years, an evolution towards more complex sample geometries is clear. Most of the geometries are designed to be loaded in tension. Due to the specific sample geometry, the tensile load is converted into a shear stress in the gauge section of the sample. A well-designed shear sample fulfils (most of) the following requirements:

1. Homogeneous stresses and strains in a considerably large region of the sample. The shear region should be large compared to the intrinsic material scale, such as the grain size for metals. Only in that case, the response of the sheared region will be representative for the macroscopic material response.

2. Crack initiation within the central part of the gauge section rather than on the free edges. Indeed, the plane stress condition existing at the free edges of the sheared gauge section results in a significantly higher stress triaxiality there. The higher stress triaxiality, combined with the imposed deformation, is likely to initiate premature fracture at the edges.
3. No out-of-plane displacements and shear buckling of the sample.
4. Straightforward relation between measured force and shear stress. This allows extracting the local stress in the shear zone from the measured force.
5. Easy and reproducible sample production.

Several optimised sample geometries have been developed for static shear testing. Examples can be found in Fig. 3.4. The geometry with butterfly gauge section (Fig. 3.4a), proposed by Bao and Wierzbicki [22], was used to study fracture of an aluminium alloy at low stress triaxiality. The thickness of the gauge section is reduced which requires careful machining in order to avoid residual stresses and micro-cracks at the surface. Thinning of the gauge section also disturbs the through-thickness plate structure. Therefore, Tarigopula et al. [3] proposed an alternative butterfly geometry (Fig. 3.4b) which does not require thinning of the gauge section. The sample geometry was used to assess the work hardening behaviour at large plastic strain and the ductile fracture of a dual phase steel. The butterfly design also inspired Reyes et al. [23] to derive the geometry presented in Fig. 3.4c to study the yield and fracture properties of an aluminium alloy. Interestingly, to study ductile fracture of a high strength dual phase steel, in [24] Gao et al. present a shear geometry with an asymmetric gauge section (Fig. 3.4d). The asymmetry was introduced to decrease the tensile stresses due to the unavoidable rotation of the shear zone during deformation. Consequently, a stress state close to pure shear is reached in the shear gauge section.

For most of the samples shown in Fig. 3.4, no precautions were taken to avoid fracture initiation at the edges. As a result, it is highly unlikely that the full deformation capacity of the material will be revealed by experiments using these geometries. Recently, Roth and Mohr [25] presented a geometry which was specifically optimized to maximize the likelihood that fracture initiates in the centre of the gauge section, rather than at the free boundaries. The sample geometry of Roth and Mohr is presented in Fig. 3.4e.

Unfortunately, all samples presented in Fig. 3.4 are developed for static testing and cannot be used directly for dynamic testing in a split Hopkinson tensile bar setup. Indeed, a major problem associated with the geometries of Fig. 3.4 is that their dimensions exceed the limits imposed by split Hopkinson bar tests. In particular, the samples are too long to reach a state of quasi-static equilibrium within a reasonable time interval. In addition, in the test setup design no attention is paid to issues which may affect the waves propagating in the Hopkinson bars. Summarizing, the use of a split Hopkinson bar facility for shear testing imposes the following additional requirements:

6. Establishment of a state of quasi-static equilibrium in the sample from the early stage of deformation. As a rule of thumb, the time needed to reach quasi-static

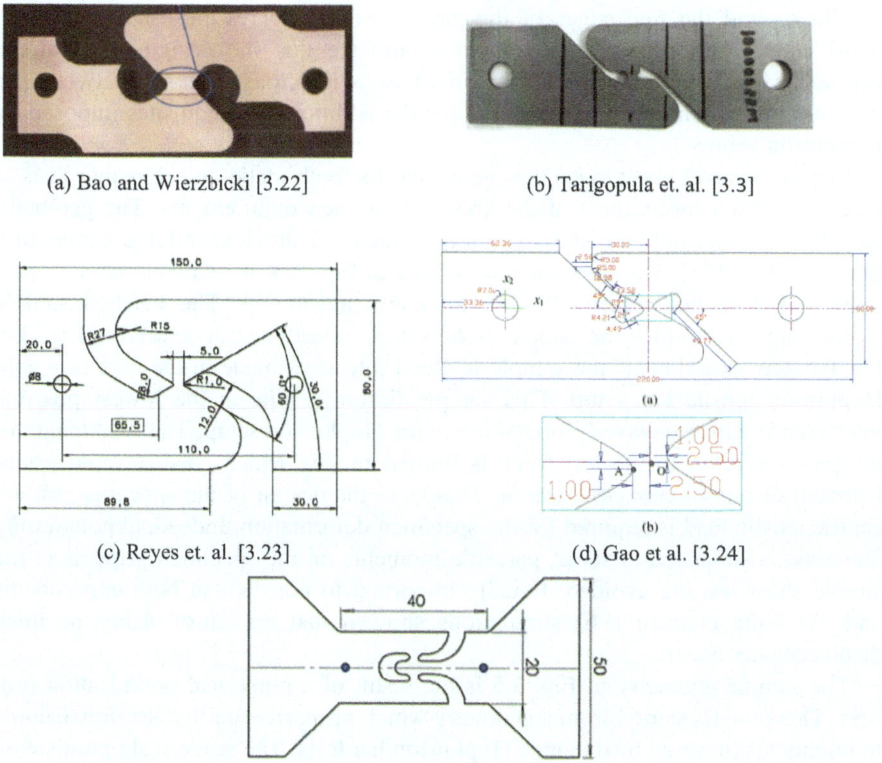

(a) Bao and Wierzbicki [3.22]

(b) Tarigopula et. al. [3.3]

(c) Reyes et. al. [3.23]

(d) Gao et al. [3.24]

(e) Roth and Mohr [3.25]

Fig. 3.4 Samples geometries used for static shear testing (reproduced with permission)

equilibrium can be estimated by the time needed for the wave to travel three times back and forth in the sample. This prerequisite imposes an upper bound on the length of the sample.

7. Limited or no use of clamps, screws or sample fixtures that could disturb the waves in the Hopkinson bars. Indeed, these mechanical devices not only introduce oscillations in the force and elongation history of the sample, but they also affect the accuracy of the force and the elongation histories obtained from the measured waves.
8. The load needed to deform the sample should be aligned with the bars. Parasitic non-axial forces or moments give rise to an uncontrollable, non-axial displacement of the sample/bar interfaces and bending waves in the Hopkinson bars.

Meeting the requirements 6–8 for dynamic testing can jeopardize fulfilment of the general requirements 1–5 and vice versa. For instance, the use of heavy rigid clamps to reach homogeneous shear strains clearly interferes with requirement 7. Increasing the length of the shear region to reduce the effect of the boundary conditions and the free edges of the gauge section, delays the establishment of quasi-static equilibrium. These examples indicate that the design of a sample geometry for dynamic shear testing involves a judicious trade-off between the general requirements for shear testing, and the additional prerequisites imposed by Hopkinson setups.

In [26] a novel sheet specimen geometry for both static and dynamic tests is proposed which meets most of the above mentioned requirements. The geometry was used to determine the stress-strain and fracture behaviour at large strains of a titanium alloy [27]. The geometry, presented in Fig. 3.5, is relatively easy to produce and does not require through thickness machining. The hatched sample regions are used to glue the sample in the setup. Indeed, as can be seen in Fig. 3.6, for dynamic experiments the sample is glued into slots made in the bars of a split Hopkinson tensile bar setup. This sample fixture results in the lowest possible interference with the waves propagating in the Hopkinson setup. The total length of the part subjected to the test force is limited in size which ensures rapid establishment of quasi-static equilibrium. Thanks to the design of the specimen, only a centric tensile load is required for the specimen deformation. Indeed, although only one zone is subjected to shear, parasitic moments on the specimen grips as in the simple shear test are avoided. Equally important to note is that both experiments and 3D finite element (FE) simulations showed that no out-of-plane specimen displacements occur.

The sample geometry of Fig. 3.5 is the result of a numerical optimisation process. This process started from a geometry which primarily fulfilled the dimensional requirements imposed by dynamic Hopkinson bar tests. The numerical optimisation by means of finite element calculations mainly aimed at obtaining homogeneous stresses and low stress triaxialities in the gauge section. In the design process, the following geometrical features of the sample have been optimised:

- length and width of the shear region,

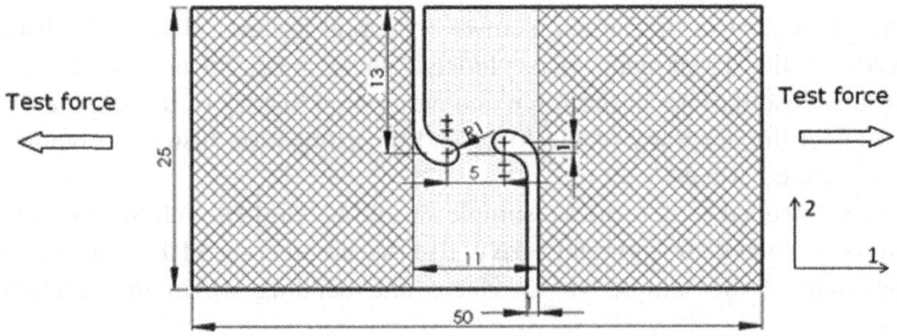

Fig. 3.5 Shear specimen geometry and dimensions as optimised for dynamic shear testing of Ti6Al4V in a split Hopkinson bar setup [26]. The key feature of the sample is the eccentricity of the round notches which delimit the shear zone. As a result of the eccentricity, the stress triaxiality in the shear region is significantly reduced. The hatched area is glued in the slits of the Hopkinson bars (reproduced with permission)

Fig. 3.6 Shear sample glued between the Hopkinson bars before testing. During a test, the deformation of a line grating attached to the sample is recorded by a high speed camera. This grating allows obtaining the local strain in the gauge section (reproduced with permission)

- size and geometry of the free edges ('notches') of the shear region,
- position of the notches relative to the axis of symmetry,
- overall dimensions of the specimen.

Smooth, rounded *notches* have a favourable effect on the homogeneity of the shear strains. They also give rise to relatively low stress concentrations. After analysis of several notch configurations, easy-to-manufacture round notches were selected. The size of the radius was chosen such that a sufficiently large region of the sample is subjected to shear. Indeed, the geometrical width of the shear zone is equal to twice the notch radius. Obviously, the actual width of the shear region differs from the geometrical width, and is dependent on both the material behaviour and the sample geometry. A larger width also simplifies the use of optical methods to measure the strain.

Application of a *length-over-width* ratio l/w of at least 10, as recommended for a simple shear sample [5, 19, 21], requires a length which is unacceptable for samples to be tested in a split Hopkinson bar setup. Fortunately, finite element simulations have shown that, if no clamps are used, a limited length of the shear region is beneficial for the homogeneity of the stress and strain in the shear region [26]. Since the force required to deform the sample is proportional to the length of the shear region, lower forces are needed for shorter samples. This also limits the deformation of the material adjacent to the shear zone and the risk of shear buckling.

The key feature of the sample is the eccentric *position of the notches,* which leads to an almost pure shear stress state up to large strains. Indeed, the position of the notches was optimised to lower the stress triaxiality and thus to reach a state of pure shear. Figure 3.7 represents simulated stress triaxialities η for a Ti6Al4V sample with three different notch configurations at an imposed displacement of $u_x \approx 0.75$ mm. Comparing the sample with eccentric positioned notches (Fig. 3.7c, with $e = 1$ mm) with the sample with symmetrical notches (Fig. 3.7, with $e = 0$

mm), shows a reduction of the average stress triaxiality in the shear region with a factor higher than 10. The simulated average stress triaxiality during plastic deformation up to a logarithmic shear strain of $e_{12} = 0.3$ never exceeds $\eta = 0.02$. The use of an undercut in the notch (see Fig. 3.7b, and also Fig. 3.4b), reduces the triaxiality with only 20%. The reduced triaxiality stems from compressive stress in the shear region induced by the eccentric notch position. This compressive stress compensates the tensile stress caused by elastic deformation of the material adjacent to the shear region. For symmetrically placed notches, the shear zone is subjected to both shear and tensile stresses. The magnitude of the eccentricity required to obtain a pure shear stress state depends on the overall geometry of the sample and the mechanical properties of the material.

Finite element simulations also reveal that the round shape of the notches, combined with their eccentric position, results in a strain distribution with higher plastic strains around the shear gauge centre than at the notches. Moreover, the zone where the highest strains appear at the tensile side of the notch (left side for the upper notches of Fig. 3.7, right side for the lower notches) is shifted away from the shear region. Both factors increase the likelihood that fracture will initiate around the centre of the shear region. An estimate of this likelihood can be made following a procedure described in [25].

The local stress in the shear zone of the sample shown in Fig. 3.5 can be approximated by dividing the force F by the area of the shear region, i.e. by $\tau_{avg} = F/(l \cdot t)$, with t the thickness of the specimen and l the length of the shear region.

The optimal combination of geometrical parameters depends on the mechanical behaviour of the material under investigation. For Ti6Al4V, the optimised sample has a length of the shear region of $l = 3$ mm, a notch radius of $r = 1$ mm and an

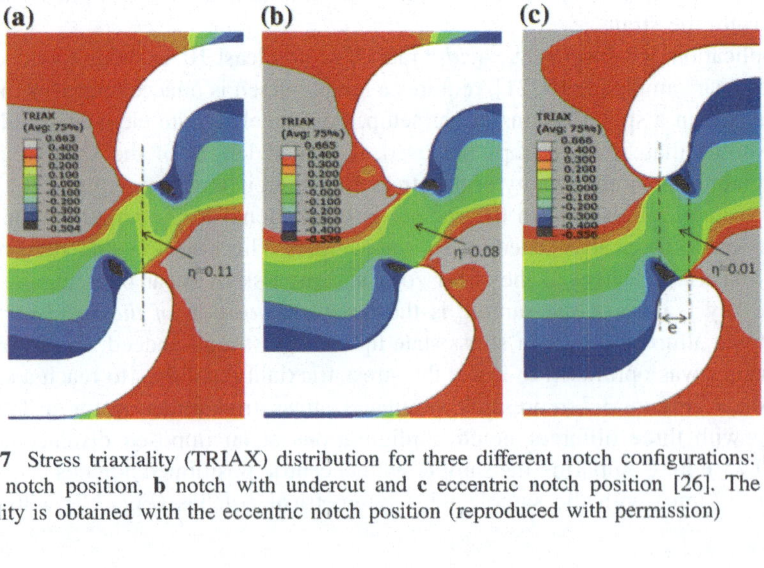

Fig. 3.7 Stress triaxiality (TRIAX) distribution for three different notch configurations: **a** symmetric notch position, **b** notch with undercut and **c** eccentric notch position [26]. The lowest triaxiality is obtained with the eccentric notch position (reproduced with permission)

eccentricity of $e = 1$ mm. With these dimensions, good results were also obtained for a magnesium alloy and a dual phase steel [28]. A similar sample geometry has also been adopted to study the shear deformation of a TWIP steel with a pronounced strain hardening. During plastic deformation, the yield strength of the TWIP steel increased from 375 to 2000 MPa. For such materials, a sample with a slightly lower width of the shear radius ($r = 0.75$ mm instead of $r = 1$ mm) and a lower eccentricity ($e = 0.25$ instead of $e = 1$ mm) is recommended.

3.3.3 Example: Shear Testing of Ti6Al4V Sheet

This section presents the results of static and dynamic shear tests on Ti6Al4V using the geometry presented in Fig. 3.5. In particular, results which contribute to a better understanding of shear tests are presented. More details about the material, the experimental program and the numerical modelling can be found in [26, 27].

The Ti6Al4V sheet material has an initial thickness of 0.6 mm. The samples are tested in a traditional tensile bench for the static experiments, and in a tensile split Hopkinson bar setup for the dynamic tests.

Figure 3.8 shows the sample in the static setup. In order to have the same boundary conditions as for the high strain rate experiments, the specimen is glued in slotted bars. These bars have the same diameter (25 mm) and are made of the same aluminium alloy as the Hopkinson bars. The flattened ends of these bars are clamped in the test bench. During the static experiment, the clamps separate with a velocity of 0.0033 mm/s. The applied load is measured by a load cell. The relative displacement of the specimen/bar interfaces is calculated as the average of the displacements measured by three LVDTs placed around the specimen (see Fig. 3.8). In addition, the deformation of a speckle pattern or line grating attached to the sample is recorded by means of digital cameras, including a high speed camera. From the recorded images, the local strain is calculated using digital image correlation (DIC) or Moiré techniques respectively. Figure 3.9 shows images of a deforming sample with a line grating during a static test. The angle between the grid lines and the vertical direction is a measure for the local shear angle. Just before fracture, a local maximum engineering shear strain of ≈ 0.85 is reached. Assuming simple shear (cfr. Sect. 3.5), this corresponds to a logarithmic strain of $e_{12} \approx 0.38$. From Fig. 3.9, it is clear that the plastic deformation concentrates in the region between the notches. The width of the plastically deforming region is slightly smaller than the notch diameter and remains relatively constant during the test. However, the strain gradient along the width of the shear region increases. Close to fracture initiation, sharp corners arise in the notches, fortunately in the regions where compression stresses exist. The exact location of fracture initiation is not captured by the high speed images. However, it is most likely that the additional compressive stresses at the free edges due to the eccentric notches prevent the onset of fracture at these locations. The final fracture is likely to start in the centre of the specimen, aligned with the tensile direction.

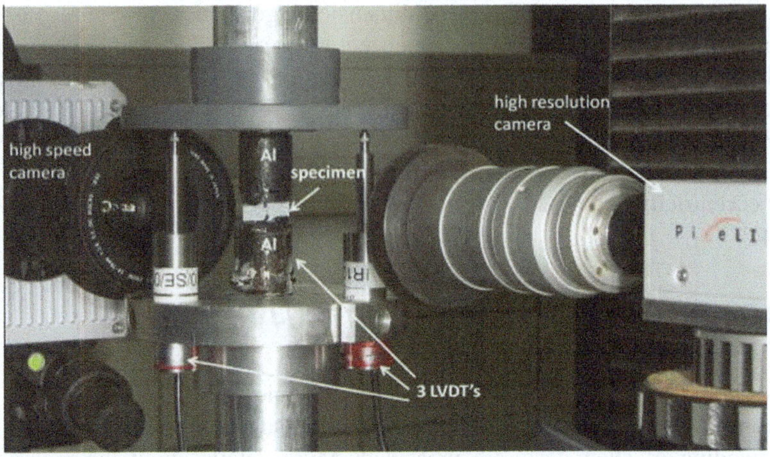

Fig. 3.8 Picture of a Ti4Al4V shear sample in the static test setup. The sample displacements are measured by 3 LVDTs. The deformation of a speckle pattern attached to the sample is recorded by means of digital cameras, including a high speed camera [26] (reproduced with permission)

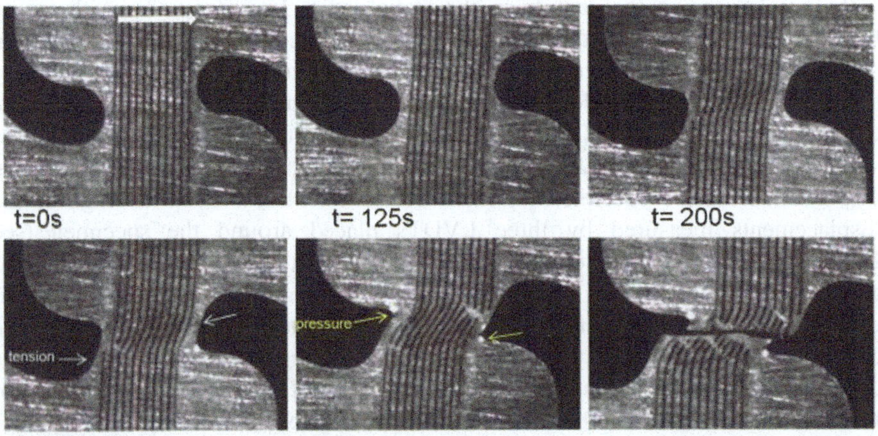

Fig. 3.9 Sequence of images of the deforming shear region with line grid during static test [26] (reproduced with permission)

In Fig. 3.6, a sample glued between the Hopkinson bars is presented. During the experiment, the deformation of a line grating or speckle pattern attached to the sample is recorded by a high speed camera. These measurements allow obtaining the local strain in the gauge section during the test. Successive high speed camera images of the shear region during a dynamic test are shown in Fig. 3.10 (left column). For this test, the specimen/bar interfaces separate with a speed of approximately 2.7 m/s. The resolution of the images is 256 by 176 pixels. The figure clearly shows that the line connecting the notch tips is rotating clockwise. From these pictures, the local shear strain is calculated using digital image correlation. The results are shown in the middle column of Fig. 3.10. Next to the DIC strain, the strain field from 2D FE simulations is also represented in the right column. A good qualitative and quantitative agreement between the measured and the simulated strain can be seen.

The force-displacement curves of two typical dynamic experiments and one static experiment are shown in Fig. 3.11. The corresponding local logarithmic shear strains in the centre of the shear region, calculated with DIC and FE simulations, are also provided. Two parts in the experimental load-displacement curves can easily be distinguished. At the onset of loading, the specimen mainly deforms elastically.

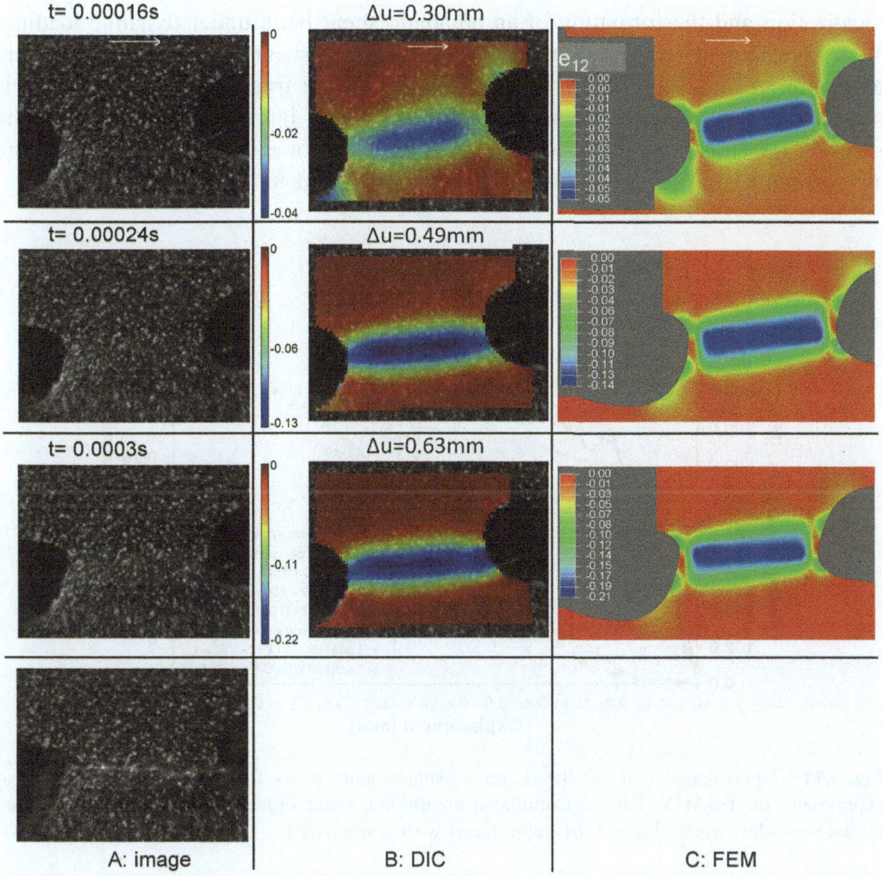

Fig. 3.10 **a** Sequence of images of the deforming shear region with speckle pattern during a dynamic test, **b** corresponding logarithmic shear strain calculated with DIC and **c** FE simulated logarithmic shear strain [26] (reproduced with permission)

The strain in the shear region is limited and plastic deformation only occurs at the notches. At a displacement of $u_x \approx 0.2$ mm, the material in the shear region starts to yield and further plastic deformation concentrates in the shear region.

All experiments finish by an abrupt drop in force. No necking occurs before fracture. Figure 3.11 reveals important differences between the dynamic and static tests. The stress level in the static test is much lower than in the dynamic test. Furthermore, the local logarithmic shear strain at fracture obtained by DIC is around 25% for dynamic tests and above 35% for static tests. The shear stress-strain data of Fig. 3.11 can be used straightforwardly to model the constitutive behaviour at large strains [27]. In a static tensile test on the same material, reliable stress-strain data are only obtained till necking, which sets in at a strain value around 9%. At dynamic strain rates, the maximum uniform strain is further reduced to 2–3%.

The fracture strain in the static experiment is 10% higher than in the dynamic experiment. The strain rate dependent fracture behaviour can be explained by strain localization and the formation of an adiabatic shear band under dynamic loading. Very high strains occur within the narrow adiabatic shear band. These strains are not captured by DIC. The higher displacement before fracture in the static tests is not only due to a higher macroscopic fracture strain but also to the fact that the strain is less localised in the static test. Simulated force-displacement curves are also presented in Fig. 3.11, more details can be found in [26].

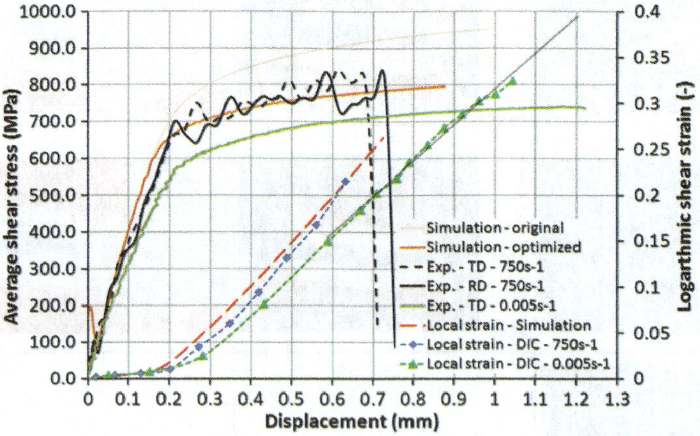

Fig. 3.11 Experimental and simulated force-displacement curves for dynamic and static shear experiments on Ti6Al4V. DIC and simulated logarithmic shear strains in the centre are presented on the secondary vertical axis [26] (reproduced with permission)

3.4 Hopkinson Shear Testing of Bulk Materials

As elucidated in the previous sections, dynamic shear testing of sheet materials relies on the conversion of an axial load, generated in a split Hopkinson bar *tensile* or *compression* setup, into a shear load by means of an appropriate sample geometry. For bulk materials, a more straightforward way of shear testing consists of imposing a *torsional* deformation onto a thin-walled tubular sample. Dynamic loading conditions are generally achieved in a split Hopkinson bar torsion setup. A well-designed sample ensures homogeneous shear stresses and strains—prior to strain localisation—in the gauge section of the sample with a small (or even non-existent) hydrostatic stress. As is the case for compression or tensile Hopkinson test, the strain, strain rate and stress history in the sample can straightforwardly be derived from strain wave measurements on the bars. Furthermore, the shear zone is accessible for high speed camera observation. This allows using full field deformation measurement techniques based on e.g. digital image correlation or Moiré interferometry. As such, torsion tests on thin-walled tubular samples provide an unambiguous means for the experimental characterisation of the shear behaviour of materials. In Sect. 3.4.1, split Hopkinson bar torsion tests on thin-walled tubular samples are extensively discussed, including the equations used to calculate the stress, strain and strain rate in the sample, and the design of the setup.

Notwithstanding the fact that torsion of thin-walled tubes is generally accepted as an accurate and reliable test technique, torsion tests are not that widely used. Obviously, testing of thin-walled tubes requires a torsion setup, which is not that common. Several shear tests have therefore been developed for compression setups. The most straightforward technique uses the compression of cylindrical samples, which gives rise to shear stresses with maximum value in planes inclined at an angle of 45° to the sample axis. Since the shear stress is combined with a hydrostatic pressure, high strains can be reached in metallic samples. Therefore, dynamic compression of cylindrical samples in a split Hopkinson bar compression setup has been extensively used to study dynamic shear failure phenomena such as adiabatic shear banding. However, the quantitative assessment of the sample's stress-strain response is hampered by different factors, such as barrelling of the sample due to friction in the bar/sample interfaces, radial inertia and the lack of visual observation.

Aiming at a predominant shear deformation mode, punch tests for Hopkinson compression setups have been developed [29–31]. In these tests, a flat circular punch at the end of the input Hopkinson bar pushes a flat plate sample into a cylindrical die (most often a hollow output bar). A small clearance exists between the punch and the die. The major disadvantage of this technique is that the results are highly dependent on the geometry of the punch and the die. Especially the clearance between punch and die, and the sharpness of the edges of both the punch and the die have a significant influence on the measured results [64].

An alternative technique which makes use of hat-shaped samples has seen wide application. Localised shear strains are obtained in the specimen by pushing the shaft into the brim of the material sample. Generally, the outer diameter of the shaft is slightly higher than the inner diameter of the brim. As a result, compression stresses arise in the sheared region. This compressive stress state delays failure. Due to the oversized shaft, both the orientation and the amount of material involved in the shearing process continuously change during the test. Even before localisation, the extent of the shear zone is determined by both the sample's material response and geometry. Unlike torsion tests, hat-shaped sample tests do not allow to control the amount of material involved in the shear deformation. Moreover, the stresses needed for the expansion of the brim are the result of the complex interaction between material and sample response. Consequently, the exact values of the shear stress and shear strain cannot be derived from the measured force and displacement. Tests on hat-shaped samples are no substitute for material characterisation tests. However, compression of hat-shaped samples has proven to be valuable to study microstructural shear phenomena, such as adiabatic shear banding. The use of hat-shaped samples is discussed in Sect. 3.4.3.

Advanced shear tests have been developed using purpose-designed geometries to obtain an enhanced control over the stress and deformation state imposed on the sample material (cfr. Sect. 3.4.4). These tests are mainly used to study shear localisation phenomena such as adiabatic shear banding. Still, when combined with more advanced measurement techniques and numerical simulations, these tests also allow obtaining quantitative information on the shear stress-strain response of the material under investigation.

3.4.1 Hopkinson Torsion Tests

In line with the principle of split Hopkinson bar setups, a torsion setup basically consists of two long bars—the input and output bar—between which a sample is sandwiched. In a split Hopkinson bar torsion setup—unlike tensile or compression setups—the incident loading wave is a torsion wave. This torsion wave is achieved by the sudden release of a pre-torqued part of the input bar. Indeed, a torsional wave, with a well-defined and controllable length and amplitude, cannot be generated by the impact of a projectile on the free end of the input bar.

A design drawing of the split Hopkinson bar torsion device at Ghent University is presented in Fig. 3.12. To pre-torque a certain length of the input bar, that part is clamped, and subsequently twisted. The incident torsion wave twists the thin-walled tubular specimen, and is partly reflected back into the input (or incident) bar and partly transmitted into the output (or transmitter) bar. Shear strain gauges attached to the input bar measure the incident and reflected wave. On the output bar, the transmitted wave is measured.

Similar to the traditional tensile and compression Hopkinson tests, the time histories of the torsional load, deformation and deformation rate imposed on the sample can be calculated from the measured incident, reflected and transmitted torsion waves. From these histories, the corresponding average shear stress, strain

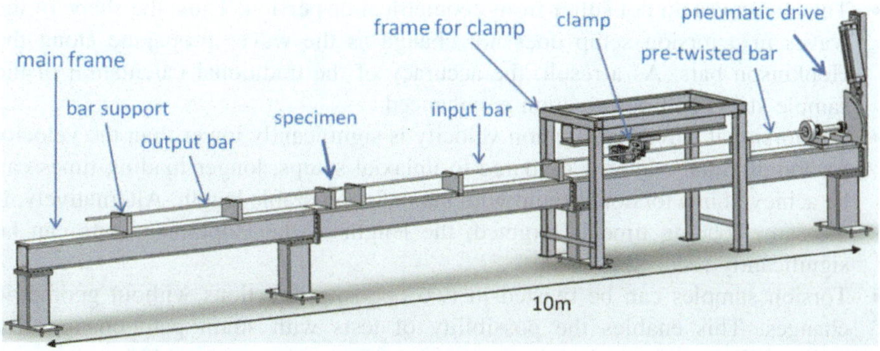

Fig. 3.12 Design drawing of a split Hopkinson bar torsion setup

and strain rate histories in the sample can be derived (cfr. Section *Equations for stress, strain rate and strain* below).

Baker and Yew [32] were the first to apply the principle of split Hopkinson bar tests to torsion. Their main motivation for the development of a torsional setup was the elimination of errors introduced by radial inertia and friction in the specimen/bar interfaces in dynamically compressed cylindrical samples [33]. Since the introduction in 1966, Hopkinson torsion tests have improved significantly. An enhanced understanding of torsion testing, and the introduction of more advanced measurement and processing techniques have extended their use from traditional materials towards less conventional, demanding materials such as human tissue [34]. Furthermore, torsion test setups have been designed which allow the simultaneous application of a torsion and an axial—tensile or compression—load. To allow the microstructural investigation of the sample material at different stages of the deformation process, techniques to interrupt the test have been developed [35].

Torsion split Hopkinson bar tests have been used to characterise the response to high shear rates of many materials. The low stress triaxiality in the sample combined with the absence of features which may give rise to structural instabilities, render torsion tests particularly suited for brittle materials or metals with low strain hardening which cannot straightforwardly be characterised by tensile tests. Indeed, for these materials, the dogbone-shaped tensile sample gives rise to an almost immediate sample failure at the shoulders or early necking. Significantly higher uniform strains are obtained in torsion tests compared to tensile tests. Torsion tests are also highly useful to study the mechanisms that govern shear failure phenomena. In the pioneering work of Marchand and Duffy [36], the phenomenon of adiabatic shear bands in structural steels is studied extensively by Hopkinson torsion tests. Later, Hopkinson torsion tests were also used by amongst others Gilat [37], Klepaczko [38] and Ramesh [39–41].

In addition to all the benefits inherent to shear testing (see Sect. 3.1), Hopkinson torsion tests have some additional advantages over Hopkinson tensile and compression tests:

- Torsion waves do not suffer from geometrical dispersion. Thus, the shape of the waves in a torsion setup does not change as the waves propagate along the Hopkinson bars. As a result, the accuracy of the traditional calculation of the sample stress and deformation is enhanced.
- The torsional wave propagation velocity is significantly lower than the velocity for longitudinal waves. Compared to uniaxial setups, longer loading times can be achieved in a torsional setup with bars of comparable length. Alternatively, if the same loading time is targeted, the length of the Hopkinson bars can be significantly reduced.
- Torsion samples can be twisted in two opposite directions without geometric changes. This enables the possibility of tests with strain path changes for advanced material characterisation, such as Bauschinger tests [42].

Compared to the shear sheet test discussed in Sect. 3.3, torsion tests also have distinct merits:

- A low triaxiality stress state close to pure shear is obtained in the sample, regardless of the material under investigation. As opposed to the sheet shear samples discussed in Sects. 3.3.1 and 3.3.2, the geometry of a torsion sample does not need to be optimised for each material individually to obtain the desired stress state.
- Torsion samples are axisymmetric with—prior to damage or fracture induced strain localisation—homogeneous stresses and strains in the entire gauge section. Unlike the sheet shear specimen, the torsion specimen does not have notches or free edges which disturb the stress and strain fields. Torsion samples are hence not prone to undesirable, premature fracture of the sample.
- Hopkinson torsion setups allow superimposing a well-controlled normal stress to the shear stress. The thus obtained material test data provides valuable information, complementary to data obtained by compression and tensile tests, to study for example stress-state dependent plasticity [43] and failure under multiaxial stress states [44–46]. Adding a compressive stress, material failure is delayed and phenomena typically occurring at high strains, such as grain fragmentation and shear band formation, can be studied. Dynamic torsion combined with a static axial load is relatively easy to realise. Combined dynamic torsion/compression is much more demanding. A technique to introduce combined torsion and compression in dynamic experiments can be found in [41].

The disadvantages of torsion testing are mainly related to the use of thin-walled tube samples. Obviously, tubular samples can only be made from bulk materials and are more difficult to manufacture than compression or tensile samples. Proper design of the sample geometry should guarantee that the differences between the stresses (and strains) at the inner and outer diameter are sufficiently low. Additionally, the way that the sample is attached between the Hopkinson bars requires special attention. Here again, interaction between the sample grips and the waves should be minimal.

Equations for stress, strain rate and strain

The basic principle of a torsional split Hopkinson bar setup is identical to the traditional tensile or compression Kolsky device. The chief difference is that *torsional* waves are introduced instead of tensile or compression waves. The wave equation in torsion reads

$$C_b^2 \frac{\partial^2 \theta(x,t)}{\partial x^2} = \frac{\partial^2 \theta(x,t)}{\partial t^2} \tag{3.2}$$

where

$$C_b = \sqrt{\frac{G_b}{\rho_b}} \tag{3.3}$$

is the propagation speed of the torsional waves, with G_b the shear stiffness and ρ_b the density of the Hopkinson bar material. The rotation angle at time t of a section at location x along the bar is denoted as $\theta(x,t)$. The propagation speed of torsion waves is significantly lower than the propagation speed of the longitudinal waves encountered in tensile or compression Hopkinson setups: for aluminium and steel $C_b \approx 3\,100$ m/s, while the longitudinal wave speed $C_l \approx 5000$ m/s.

Hopkinson bars are generally solid, circular bars. Elastic strains due to torsion in circular bars are not constant through the cross section. Indeed, the strain is zero in the centre and varies linearly to reach its highest value at the outer radius r_b of the bar. In case of an elastically deforming bar, a shear strain γ measured at the surface of the bar is related to the torque T via the shear stiffness $G_b J_b$ of the bar:

$$T = \gamma \frac{G_b J_b}{r_b} \tag{3.4}$$

with J_b the polar moment of inertia of the bar which can be calculated as

$$J_b = \frac{\pi r_b^4}{2} \tag{3.5}$$

For specific applications (see Section *Design of a split Hopkinson bar torsion setup* below), hollow Hopkinson bars provide a better alternative. In a hollow bar, the elastic strain also varies linearly over the wall thickness. The relation (3.4) still holds with

$$J_b = \frac{\pi \left(r_{b,o}^4 - r_{b,i}^4 \right)}{2} \tag{3.6}$$

the polar moment of inertia of a hollow bar with outer radius $r_{b,o}$ and inner radius $r_{b,i}$.

In a torsion Hopkinson setup, a thin-walled tubular sample is placed between the input bar and the output bar. The tube gauge section has a length L_s and is defined by the inner radius r_i and outer radius r_o or, alternatively, by the average radius $r_s = \frac{r_i + r_o}{2}$ and wall thickness $h_s = r_o - r_i$.

A torque wave with amplitude T gives rise to a rotational speed $\dot{\theta}$ of a bar section given by

$$\dot{\theta} = \frac{T}{Z_b} \tag{3.7}$$

with

$$Z_b = \rho_b C_b J_b \tag{3.8}$$

the torsional impedance. A wave with equal amplitude travelling along the bar in the opposite direction gives rise to an equal rotational speed though with opposite sign.

An elastic torque wave with amplitude T_i is generated in the input bar. This incident wave travels along the input bar towards the sample and interacts with the sample. If the sample is short enough, multiple reflections of the waves in the sample lead to a state of *quasi-static equilibrium* within the sample. This ensures a uniform stress state along the length of the gauge section. Following the interaction with the sample, a part of the incident wave is transmitted into the output bar (T_t), and another part is reflected back into the input bar (T_r).

Using Eq. (3.4) combined with Eq. (3.5) or Eq. (3.6) for respectively solid or hollow Hopkinson bars, the torque histories T_i, T_r and T_t of respectively the incident, reflected and transmitted waves, can be calculated from the corresponding shear strain waves γ_i, γ_t and γ_r measured at the surface of the Hopkinson bars.

To calculate the stress and deformation history of the sample, the incident, reflected and transmitted waves are shifted from the measurement location towards the sample/bar interfaces. From the shifted incident and reflected waves, the rotational speed at the interface of the sample with the input bar is obtained using Eq. (3.7) as

$$\dot{\theta}_{in} = \frac{T_i - T_r}{Z_{b,in}} \tag{3.9}$$

with $Z_{b,in}$ the torsional impedance of the input bar. Similarly, the rotational speed at the interface of the sample with the output bar is calculated from the shifted transmitted wave as

$$\dot{\theta}_{out} = \frac{T_t}{Z_{b,out}} \tag{3.10}$$

with $Z_{b,out}$ the torsional impedance of the output bar. The relative rotational speed of the sample/bar interfaces is then obtained as

$$\dot{\theta}_{in} - \dot{\theta}_{out} = \frac{T_i - T_r}{Z_{b,in}} - \frac{T_t}{Z_{b,out}} \tag{3.11}$$

Using Eq. (3.11) and assuming that the entire deformation is concentrated in the sample gauge section with length L_s, the average engineering strain rate in a thin-walled tubular sample with radius r_s becomes

$$\bar{\dot{\gamma}}_s = \frac{\dot{\theta}_{in} - \dot{\theta}_{out}}{L_s} \cdot r_s = \left(\frac{T_i - T_r}{Z_{b,in}} - \frac{T_t}{Z_{b,out}} \right) \cdot \frac{r_s}{L_s} \tag{3.12}$$

The average shear strain $\bar{\gamma}_s$ is then obtained by integration of $\bar{\dot{\gamma}}_s$.

The shear stress resulting from a torque T imposed on a thin-walled tube with radius r_s and wall-thickness h_s is given by

$$\tau(t) = \frac{T}{2\pi r_s^2 h_s} \tag{3.13}$$

Using Eq. (3.13), the shear stress in the tubular sample at the side of the input bar can be calculated from the torque $T_i + T_r$ acting on the input bar/sample interface:

$$\tau_{s,in}(t) = \frac{T_i + T_r}{2\pi r_s^2 h_s} \tag{3.14}$$

Similarly, the sample shear stress at the output bar/sample interface becomes

$$\tau_{s,out}(t) = \frac{T_t}{2\pi r_s^2 h_s} \tag{3.15}$$

Using Eqs. (3.14) and (3.15), the average shear stress in the sample gauge section can be calculated as

$$\bar{\tau}_s(t) = \frac{\tau_{s,in}(t) + \tau_{s,out}(t)}{2} = \frac{T_i + T_r + T_t}{4\pi r_s^2 h_s} \tag{3.16}$$

If the sample is sufficiently short, quasi-static equilibrium conditions can be assumed in the sample, i.e.

$$T_t = T_i + T_r \tag{3.17}$$

In that case, prior to the eventual strain localisation or fracture, the shear stress $\tau_s(t)$, strain rate $\dot{\gamma}_s(t)$ and strain $\gamma_s(t)$ are homogeneous in the gauge section of the sample and equal to their averaged values. Combining Eqs. (3.16) and (3.17), the shear stress in the sample reduces to

$$\tau_s(t) = \frac{T_t(t)}{2\pi r_s^2 h_s} \tag{3.18}$$

As is the case for tensile and compression Hopkinson tests, Eq. (3.18) shows that also for torsion tests the sample stress is proportional to the transmitted wave.

Usually, the diameter and the material used for the input and output bars are the same. In that case, or more generally if the input and output bars have the same torsional impedance Z_b, the shear strain rate in the sample can be calculated by combining Eqs. (3.12) and (3.17):

$$\dot{\gamma}_s(t) = \frac{-2 \cdot T_r(t)}{Z_b} \cdot \frac{r_s}{L_s} \tag{3.19}$$

The shear strain in the sample can subsequently be obtained by integration

$$\gamma_s(t) = -\frac{2r_s}{L_s}\frac{1}{Z_b}\int_0^t T_r(\tau)d\tau \tag{3.20}$$

Here again, as for tensile and compression Hopkinson tests, the sample strain rate is proportional to the reflected wave.

For pure torsion, assuming von Mises plasticity, the effective Mises stress σ_s and equivalent strain ε_s are related to the shear stress τ_s and shear strain γ_s as follows [47–49]

$$\begin{cases} \sigma_s = \sqrt{3}\tau_s \\ \varepsilon_s = \frac{1}{\sqrt{3}}\gamma_s \end{cases} \tag{3.21}$$

Design of a split Hopkinson bar torsion setup

The following paragraphs provide guidelines for the selection of the material and dimensions (length and cross section) of the Hopkinson bars and the sample geometry. Additionally, the clamping of the input bar and the way to support the bars are discussed.

Length of the Hopkinson bars

The minimum length of the Hopkinson bars is mainly dictated by the desired duration of the sample deformation. This duration is equal to the duration of the incident wave, i.e. the length of the incident wave divided by the torsional wave propagation velocity.

The generation of an incident wave with a well-determined length and amplitude, its propagation in the input bar and the subsequent interaction with the sample can be understood by means of a Lagrange time-space diagram, as presented in Fig. 3.3. Before a test, the left part of the input bar is clamped at a distance L_l away from the bar end, where a torque mechanism is mounted. The torque mechanism (for example, the pneumatic drive in Fig. 3.12) applies a pre-torque of magnitude T_L to the clamped bar. At the moment the clamp is opened—at time $t = 0s$ in Fig. 3.13—a torsional release wave with amplitude $T_i = T_L/2$ starts to propagate towards the sample. Meanwhile, another release wave of the same amplitude but opposite sign propagates in the other direction along the pre-stressed section. This wave reflects at the rigid torque mechanism at $t = L_l/C_b$ while its sign remains the same, hence bringing the torque in the pre-stressed section to zero whilst propagating towards the sample. As a result, the length of the incident wave is $2L_L$, i.e. twice the length of the pre-stressed section. In the implementation represented in Fig. 3.12, the clamp is attached to a separate frame. Its position along the input bar can be change, thus allowing the adjustment of the incident wave length and, hence, the duration of the sample loading.

The incident wave deforms the specimen dynamically and is partly reflected. From the Lagrange diagram in Fig. 3.13 it is clear that, in order to be able to measure the incident (γ_i) and reflected (γ_r) wave separately, the length of the input bar should be at least $2L_l$. The length of the input bar in Fig. 3.13 is slightly longer. As a result, a strain gauge attached to the input bar at the location indicated by the vertical dotted line in the Lagrange diagram measures both waves separately.

In the output bar, the transmitted wave (γ_t) can be measured at a short distance a ($\approx$ 5 to 10 times the bar diameter) away from the sample. Consequently, in order to measure the transmitted wave without interference with its reflection at the free end of the output bar, a minimal length of $L_l + a$ is required for the output bar.

The setup presented in Fig. 3.12, for example, is designed to achieve loading times of 1.5 ms. Steel Hopkinson bars are used with a torsional wave propagation speed of approximately $C_b \approx 3100$ m/s. Consequently, the length of the incident wave should be equal to $1.5\,\text{ms} \cdot 3100\,\text{m/s} = 4.65\,\text{m}$. Accordingly, the length of the pre-torqued section should be $4.65/2 = 2.325\,\text{m}$. As the length of the input bar should be slightly longer than the incident wave length, $L_i = 4.8$ m is recommended. The length of the output bar should be at least $L_o = 2.5$ m.

Material and diameter of the Hopkinson bars

The versatility of a torsion setup—the range of materials that can be tested and the maximum strain rates which can be achieved—is to a large extent dependent on the material and diameter of the Hopkinson bars.

For most materials, the highest strain rate in the sample is obtained for the highest possible amplitude of the incident wave. Indeed, when the strain rate sensitivity of the material under investigation is not too pronounced, the shear stress τ_s given by Eq. (3.18) does not vary significantly over the course of the experiment. Consequently, the amplitude of the transmitted wave does not vary too much. If quasi-static equilibrium is achieved, a higher amplitude of the incident loading

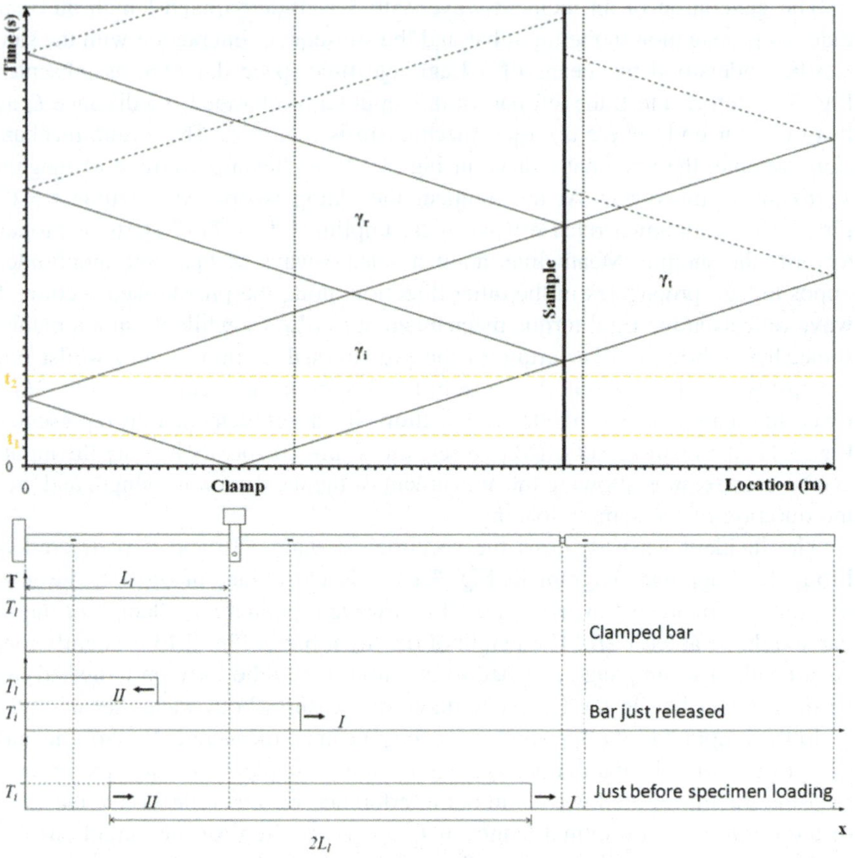

Fig. 3.13 Principle and Lagrange time-space diagram used for the generation of a torsional wave. The left part of the input bar with length L_l is pre-torqued. When the clamp is opened, the stored energy of the pre-torqued section is released and generates the incident wave. By interaction with the sample a part of the incident wave is reflected, another part is transmitted

wave directly results in a higher amplitude of the reflected wave, as obvious from Eq. (3.17). According to Eq. (3.19), this leads to a higher strain rate in the sample. Obviously, the Hopkinson bars have to remain elastic during the experiment. This requirement imposes an upper bound on the allowable amplitude of the incident wave. In the input bar, the maximum shear stress is reached in the pre-torqued part just before the clamp opening. Consequently, using $\tau = G\gamma$, Eqs. (3.4) and (3.21), the shear stress τ_L applied during pre-torque of the input bar has to be limited to

$$\tau_L = \frac{T_L r_b}{J_b} < \tau_{y,b} = \frac{\sigma_{y,b}}{\sqrt{3}} \tag{3.22}$$

where $\sigma_{y,b}$ is the yield stress of the Hopkinson input bar.

Since $T_i = T_L/2$, the maximum amplitude of the incident torque wave that can be allowed is

$$T_i^{\max} < \frac{\sigma_{y,b}}{2\sqrt{3}} \cdot \frac{J_b}{r_b} \tag{3.23}$$

In case solid Hopkinson bars with the same diameter and material for the input and output bar are used, the highest strain rate which can be imposed on the sample can be calculated starting from Eq. (3.19), using Eqs. (3.5), (3.17), (3.18) and (3.23):

$$\bar{\dot{\gamma}}_s(t)_{\max} = \frac{1}{L_s\sqrt{\rho_b G_b}} \left[\frac{\sigma_{y,b}}{\sqrt{3}} \frac{r_s}{r_b} - 8\left(\frac{r_s}{r_b}\right)^3 \frac{h_s}{r_b} \tau_s(t) \right] \tag{3.24}$$

This equation clearly shows that geometrical (r_b, r_s, h_s, L_s) and material properties (ρ_b, G_b, $\sigma_{y,b}$, τ_s) of both the Hopkinson bars and the sample affect the maximum strain rate. If high strain rates are targeted, *bar materials* which combine a low density ρ_b and stiffness G_b with a high yield strength $\sigma_{y,b}$, such as high strength titanium and aluminium alloys, are preferred. However, because the pre-torque needed to reach a certain strain rate in the sample is higher for titanium Hopkinson bars than for aluminium bars, titanium bars require a heavier clamp and torque mechanism. As such, it is not surprising that many torsion Hopkinson facilities can be found with aluminium bars [37, 39, 50–54]. To obtain a higher wear resistance, certainly for the surfaces in contact with the clamp and the specimen, Hopkinson bars made out of steel can also be used. Because of the high yield stress, hardness and ductility, maraging steel bars are often applied [55, 56].

In order to deform a sample dynamically, the amplitude of the incident torque wave should be sufficiently higher than the torque $T_t(t)$ needed to deform the sample. This requirement determines the diameter of the input bar section. For a solid input bar, the maximum amplitude of the incident wave given by Eq. (3.23) is proportional with r_b^3. Consequently, to retain the possibility of testing high strength materials, i.e. materials with high $\tau_s(t)$ in Eqs. (3.18) and (3.24), the Hopkinson input bar diameter should be sufficiently large:

$$T_i^{\max} \gg T_t \Rightarrow \left[r_b^3 \gg \frac{8\sqrt{3}}{\sigma_{y,b}} r_s^2 \tau_s \right] \tag{3.25}$$

Again, the higher pre-torque required for the larger bar diameters complicates the design of the setup. As a compromise between sufficiently high test torques and practical feasibility, bar diameters for solid round Hopkinson bars are typically 20–30 mm [37, 39, 50–53, 55, 56].

The lower bound for the *diameter of the output bar* is determined by the requirement that the transmitted wave should not exceed the yield strength of the output bar. If $\tau_s^{\max}$ is the highest stress level that will be reached in a sample with certain dimensions, then the output bar diameter can be calculated from Eqs. (3.18), (3.4) and (3.21):

$$T_t^{\max} = 2\pi r_s^2 h_s \tau_s^{\max} < \frac{\sigma_{y,b}}{\sqrt{3}} \cdot \frac{J_b}{r_b} \tag{3.26}$$

At the same time, the output bar diameter should be limited in order to allow accurate measurement of the transmitted wave.

It is interesting to note that, for setups with the same input and output bar, the maximum strain rate in the sample is not affected by the bar diameter as long as the sample diameter and wall-thickness are scaled up or down with the bar diameter [see Eq. (3.24)]. However, increasing the bar diameter while maintaining the sample dimensions will lower the maximum strain rate.

For specific cases, asymmetrical Hopkinson bar setups, i.e. with different input and output bars, might be appropriate. For instance, when testing soft or low strength materials, the amplitude of the transmitted wave might be too low to be captured with a sufficiently high accuracy. In that case, the use of a hollow or smaller diameter output bar and/or an output bar material with a lower stiffness can be appropriate [54, 57]. Since clamping and torqueing of a hollow bar is not straightforward, a solid input bar is preferred.

Sample geometry

Traditionally, samples for Hopkinson torsion tests consist of a thin-walled tubular gauge section bounded by flanges. The most important design requirement for a torsion sample is that homogeneous stresses and strains are obtained in the gauge section. However, when a thin-walled tube is twisted, the shear strain unavoidably varies along the radius. Consequently, in order to obtain homogeneous stresses and strains, the *wall thickness* should be small compared to the diameter. In [58], a radius to wall thickness ratio of $r_s/h_s \geq 25$ is recommended and a wall thickness $h_s \leq 0.5$ mm. However, the wall thickness cannot be arbitrarily small. Indeed, to obtain results representative for the sample material, the thickness should be a multiple of the materials intrinsic scale, such as the grain size in metals. Small wall thicknesses are also more sensitive to errors introduced by production tolerances and may give rise to wrinkling during testing.

The wall thickness also appears in Eq. (3.20) for the maximum shear strain rate which can be obtained in the sample. Indeed, the higher torque needed to deform a sample with higher wall thickness reduces the reflected wave and thus the strain rate.

More important than the wall thickness, is the *radius of the sample gauge section*. As obvious from Eq. (3.19) and Eq. (3.20), both the strain and strain rate are linearly dependent on the sample radius. However, also the reflected wave T_r in these equations is dependent on the radius. Indeed, the transmitted wave is proportional to r_s^3, and thus a higher sample radius will give rise to a higher transmitted wave, and hence a lower reflected wave [Eq. (3.17)]. For each material and bar configuration, the sample radius can be selected to maximize the strain and strain rate. For instance, to test a sample with wall thickness $h_s = 0.4$ mm and a constant shear stress of $\tau = 900$ MPa in a setup with steel Hopkinson bars with a diameter of $D_b = 20$ mm and a shear yield limit of $\tau_{y,b} = 290$ MPa, the optimal sample diameter can be calculated as $D_s = 2r_s = 13$ mm.

For a given bar configuration and sample section $\{r_s, h_s\}$, the maximum strain rate in the sample given by Eq. (3.24) can be increased by decreasing the length of the gauge section. Using short samples, very high strain rates can be reached. However, here again, in order to obtain test results representative for the sample material, the gauge section should still be sufficiently large. In [59] results of torsion tests using samples with a gauge length as short as $h_s = 0.15$ mm are reported. Using such short samples, shear strain rates well above $10^4 s^{-1}$ are reached. However, as shown in [60] the test results for short samples are affected by the multi-axial stress state and deformation of the sample material adjacent to the gauge section. In the gauge section, stress components different from pure shear arise. Experimental and numerical results have shown that the stress-strain curves from the tests with the short specimens are about 10% below the ones from tests using longer specimens. Additionally, the plastic zone extends beyond the gauge length of the specimen. The strain and strain rate calculated from the Hopkinson bar waves, [Eqs. (3.19) and (3.20)], overestimate the actual values.

Table 3.1 gives an overview of sample geometries found in literature. Wall thicknesses varying between $h_s = 0.2$ mm and 0.64 mm and typically around 0.4 mm are used. The gauge diameters vary between 9 mm and 25 mm. A commonly used gauge length is.

Mounting of the sample

To provide a mechanical coupling between the sample and the Hopkinson bars, most often flanges—at both sides of the gauge section—are used. A sufficiently large fillet radius at the transition between the flange and the gauge section should be implemented to prevent premature sample fracture. The flanges, and the way they are connected to the bars, should guarantee smooth and efficient transfer of the torque load and deformation from the bars to the sample. Interference between the sample attachments and the waves travelling in the setup should be avoided. Consequently, the difference in torsional impedance between the Hopkinson bars and the attachments should be kept minimal. Indeed, mechanical clamps, bolts, screws, … give rise to oscillations in the shear stress and strain imposed on the sample and introduce inaccuracies in the obtained results.

Minimal interference is obtained with the glued connection between the sample and the bars used by Chichili et al. [41]. For this purpose, round flanges were manufactured, as shown in Fig. 3.14. However, even when large diameter flanges are used, the limited strength of the glue allows testing merely low strength materials or gauge sections with limited dimensions. Additional disadvantages of a glued connection are that the chemical bonding takes time, and that sample removal after the test is cumbersome.

Table 3.1 Data on dimensions (r_s = average radius of gauge section, h_s = wall thickness gauge section, L_s = length gauge section and R_s = transition zone radius) of thin-walled tubular torsion samples and flanges shapes found in literature (dimensions in mm)

Sample material	r_s	h_s	L_s	$R_s n$	Flange shape	References
Ti6Al4V	4.94	0.38	2.5	–	Hexagonal	Shih-Chieh et al.
	–	0.3	2.5	0.15	Hexagonal	Macdougall et al.
	5.02	0.45	3.2	–	Round	DaSilva et al.
	4.70	0.4	2.0	–	Hexagonal	Ranc et al.
	6.59	0.48	3.18	>1	Round	Khan et al.
1100-O Al	8.26	0.33	0.15	–	Round	Gilat et al.
	7.85	0.5	2.5	–	Round	Leung et al.
Low carbon steel	8.15	0.4	1.3	~0	Round	Stevenson et al.
HY-100 steel	4.95	0.4	2.5	–	Hexagonal	Duffy et al.
EN2A steel	7.69	0.38	5.1	–	Round	Lewis et al.
α-Ti	5.13	0.46	3.15	3.18	–	Chichili et al.
Remco Fe	–	0.4	0.5	0.25	Hexagonal	Fellows et al.
Inconel 718	12.25	0.51	2	–	Internal Hexagonal	Kobayashi et al.
2024 Al	6.67	0.64	1.27–12.7	<0.08	Round	Nicholas et al.
Tungsten	–	0.30	2.5	–	Hexagonal	Kim et al.

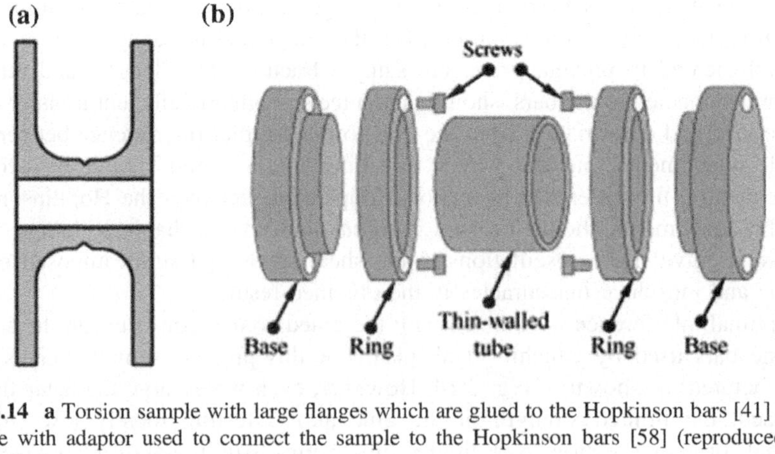

Fig. 3.14 **a** Torsion sample with large flanges which are glued to the Hopkinson bars [41] **b** tube sample with adaptor used to connect the sample to the Hopkinson bars [58] (reproduced with permission)

The drawbacks of a glued connection are alleviated by the mounting technique used by, amongst others, Marchand and Duffy [36], Macdougall and Harding [61] and Ranc et al. [53]. Instead of relying on a glued connection, these authors use samples with hexagonal flanges that fit into sockets machined in the Hopkinson bars. While such samples require more efforts to machine, the sample mounting and removal is faster and easier. Generally, because of the technical difficulties related to the accurate machining of internal shapes in long bars, the sample is not directly connected to the Hopkinson bars, but adapter pieces are used.

Table 3.1 provides an overview on the shape of the sample flanges found in literature. In Fig. 3.15, the sample geometry with hexagonal flanges used at Ghent University is presented, together with a Ti6Al4V sample.

Both the round and hexagonal flanges require a considerable volume of sample material, which might not be available. Additionally, for materials with a low machinability, such as brittle materials or composite materials, production of the samples may be time consuming and thus costly. Ensuring a uniform wall-thickness in these samples, a prerequisite for uniform stresses and strains, is cumbersome. In [58] Li et al. present a novel sample geometry which addresses these issues. The thin-walled tubular sample has no flanges and is connected to the bars with

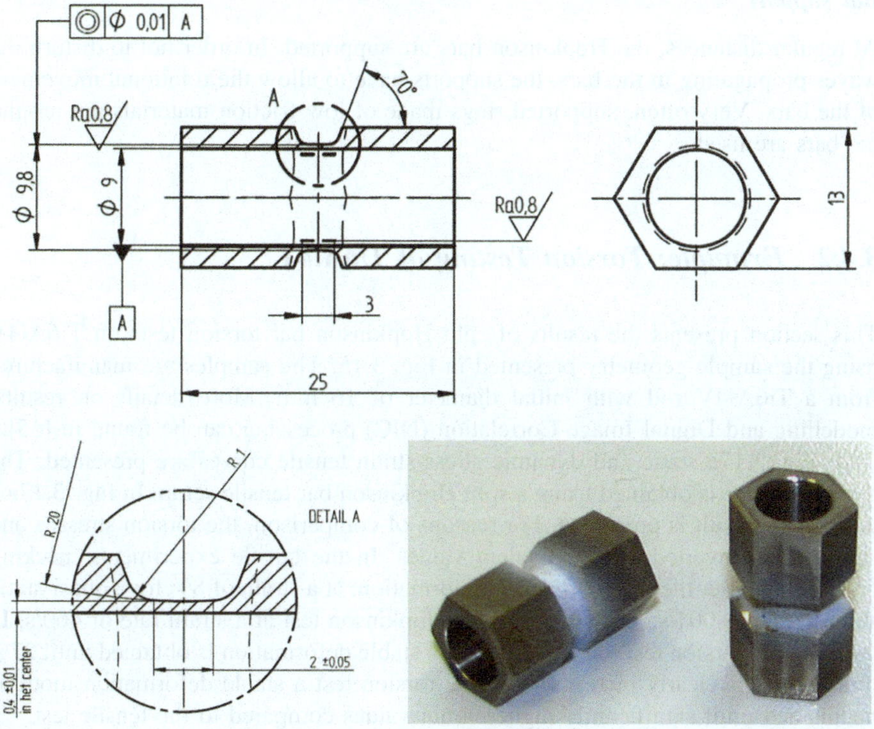

Fig. 3.15 Sample geometry with hexagonal flanges used at Ghent University, including a picture of Ti6Al4V samples before testing. White paint and black speckles are applied to the gauge section to obtain the local strains from high speed camera images of the deforming sample using digital image correlation

adapters, like shown in Fig. 3.14b. Sample, bases and rings are glued together. The screws are used to align the rings with respect to the bases prior to glueing. The base diameter is identical to the Hopkinson bar diameter and is glued to the bars.

Exceptionally, sample geometries are used with inner hexagons [62] or crossed grooves [63] to connect with the Hopkinson bars.

Clamp

The clamp is one of the most critical components of a torsional Hopkinson setup. Firstly, the clamp has to hold the desired torque without slipping. Secondly, the clamp has to open fast and at the desired moment. Indeed, a fast release of the torque is of paramount importance to introduce a torsional wave with a short rise time. Typically, release times between 20 to 30 microseconds are required. Thirdly, the clamp should be designed such that the torque is released symmetrically, i.e. without the introduction of bending waves. Most clamping mechanisms hold the bar by friction and open by fracture of a notched bolt. Figure 3.16 shows the clamp design used at Ghent University. Details on alternative clamping devices can be found in [64].

Bar support

At regular distances, the Hopkinson bars are supported. In order not to disturb the waves propagating in the bars, the supports have to allow the rotational movement of the bars. Very often, supported rings made of low friction materials put around the bars are used.

3.4.2 Example: Torsion Testing of Ti6Al4V

This section presents the results of split Hopkinson bar torsion tests on Ti6Al4V using the sample geometry presented in Fig. 3.15. The samples are manufactured from a Ti6Al4V rod with initial diameter of 16 mm. More details on results, modelling and Digital Image Correlation (DIC) processing can be found in [65].

In Fig. 3.17a static and dynamic stress-strain tensile curves are presented. The dynamic curve is obtained using a split Hopkinson bar tensile setup. In Fig. 3.17a a torsion test result is presented. For reasons of comparison, the torsion stresses and strains are converted into equivalent values. In the tensile experiments, necking occurs soon after the onset of plastic deformation: at a strain of 8% for a quasi-static strain rate of 0.001/s, and at 4% for the Hopkinson test at a strain rate of 460/s. In the dynamic torsion test, on the contrary, a stable deformation is obtained until 27% of strain. This clearly proves that in the torsion test a stable deformation mode is maintained until significantly higher strain values compared to the tensile test.

Figure 3.18 presents torsion stress-strain curves obtained at different strain rates. Here again, in all experiments, strain (and strain rate) hardening is observed until fracture. Fracture initiation occurs in the centre of the gauge section, immediately after shear strain localization. Scatter in the fracture strain is attributed to

(a)

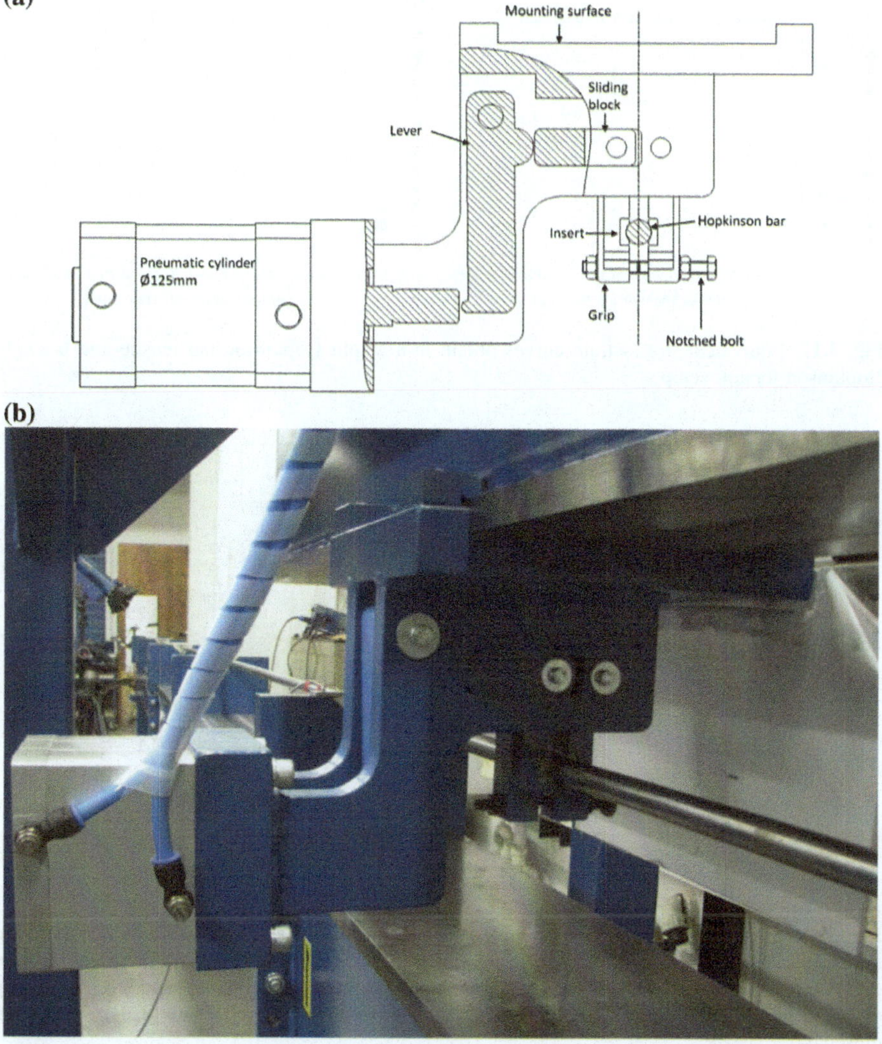

(b)

Fig. 3.16 **a** Design drawing and **b** picture of the clamp used to hold the input bar of the split Hopkinson bar torsion device at Ghent University

geometrical imperfections of the sample. Indeed, it is known that strain localization is sensitive to geometrical variations [51].

The result of the local strain measurement calculated from high speed camera images of a deforming speckle pattern applied to the sample processed with the MatchID™ digital image correlation software is also shown in Fig. 3.18. The DIC-strain values represent the average strain in a zone with a width of 0.5 mm around the centre of the gauge section of the sample deformed at 830 s^{-1}. The

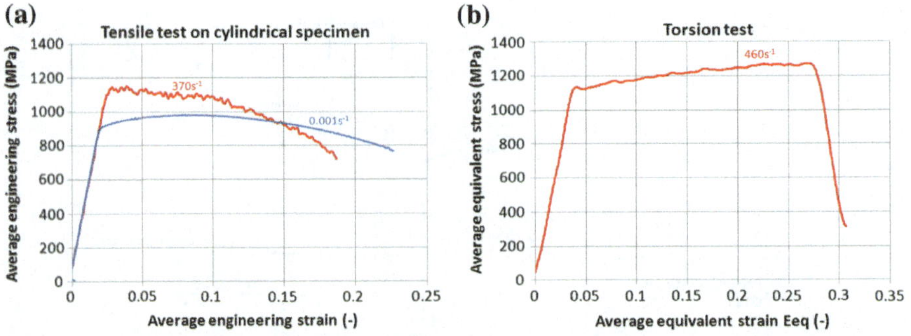

Fig. 3.17 Equivalent stress-strain curves obtain in **a** a split Hopkinson bar tensile and **b** a split Hopkinson torsion setup

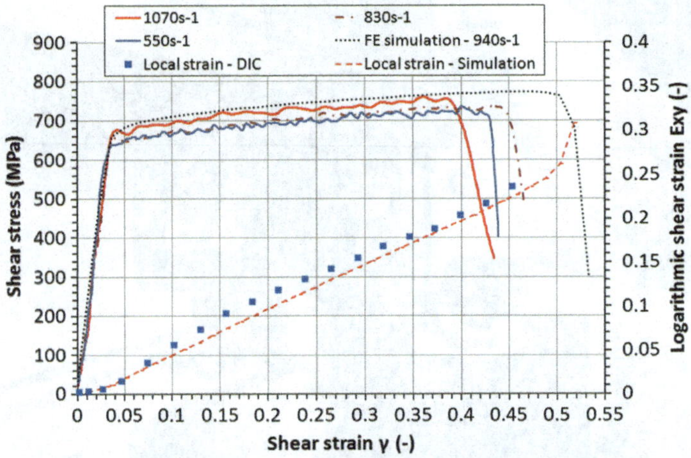

Fig. 3.18 Shear stress-strain curves obtained from dynamic torsion tests. The strain rate indicated in the legend is the average shear strain rate $\dot{\gamma}_s$. On the secondary vertical axis, the local strain in the central section of the sample determined by both FE simulation of the test and DIC processing of high speed camera recording of the deforming speckle pattern attached to the sample are presented

maximum shear strain measured with DIC just before fracture is 23.6%, which corresponds with an equivalent strain of 27%.

3.4.3 *Compression of Hat-Shaped Samples*

Compression of hat-shaped samples is frequently used to study shear phenomena in bulk materials. Dynamic loading conditions are commonly achieved in a split

Hopkinson pressure bar setup, see Fig. 3.19. The axisymmetric sample consists of a shaft and a brim. When the shaft is pushed into the brim, the material in the narrow zone between shaft and brim—denoted as shear zone in Fig. 3.19b—mainly deforms in shear. As a result, even materials that do not spontaneously show strain localization, can be forced to fail in shear.

Hat-shaped samples were originally proposed by Meyer and Manwaring in 1986 [67] and were more recently used by Bronkhorst to study shear localization in tantalum and 316L stainless steel [68]. Tantalum was also studied by Nemat-Nasser [69]. However, although both researchers performed similar experiments, they used other sample dimensions, which resulted in differences in observed behaviour. Results presented by Kad et al. [70], Meyers et al. [71], Beatty et al. [72], Couque et al. [73], Longère et al. [74] confirm that the outcome of tests on hat-shaped samples is not only affected by the tested material, but also by the sample geometry.

Figure 3.20 schematically presents the hat-shaped sample geometry, while Table 3.2 provides the corresponding sample dimensions found in literature. In order to delay failure, generally compressive stresses are aimed at in the shear zone. A compressive stress state can be obtained by choosing a sample geometry with a slightly oversized shaft ($r_2 > r_1$). However, the resulting hydrostatic pressures are not homogeneously distributed in the shear zone and strongly evolve during the deformation process. Especially after the formation of a shear band, high values for the hydrostatic pressure are reached [75]. Moreover, the major shear direction and the material involved in the deformation process are continuously changing during the test.

As the nucleation and propagation of adiabatic shear bands (ASBs) depends on the stress condition, the relation between sample geometry and stress state can be exploited for an in-depth study of ASBs [73, 76]. Indeed, it appears that both normal, compressive stress and hydrostatic pressure delay shear fracture. Yet, the

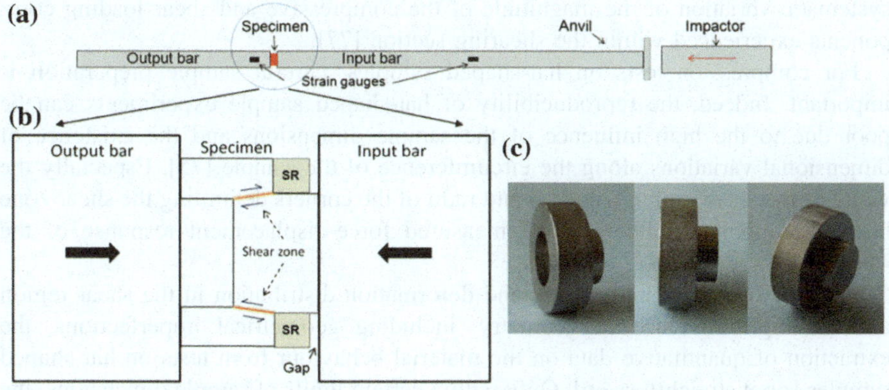

Fig. 3.19 **a** Schematic presentation of a hat-shaped specimen loaded in a split Hopkinson pressure bar setup, **b** detail of a sample with stopper ring (SR) between the Hopkinson bars and **c** sample images [66]. Arrows represent the loading direction (reproduced with permission)

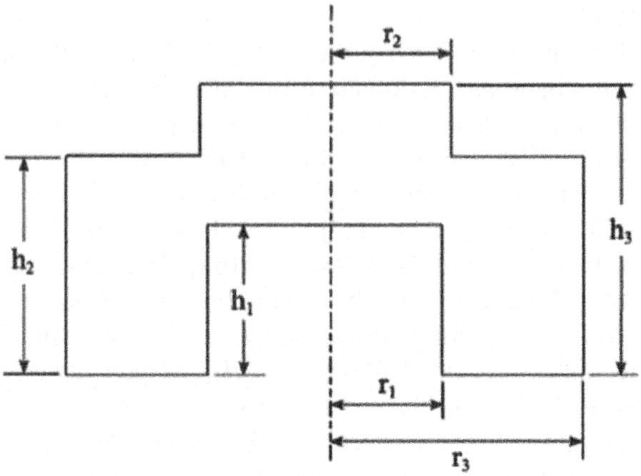

Fig. 3.20 Schematic representation of a hat-shaped sample geometry [75]

Table 3.2 Dimensions of hat-shaped samples used in literature (all in mm)

Source	r_1	r_2	r_1/r_2	r_3	h_1	h_2	$h_2–h_1$	h_3
[68]	2.09	2.28	0.92	4.30	2.60	3.47	0.87	5.11
[69]	4.31	4.44	0.97	6.35	3.30	4.55	1.25	7.85
[70]	4.99	5.08	0.98	9.55	4.80	–	–	10.00
[72]	4.76	4.76	1	9.53	4.75	5.97	1.22	11.81
[73]	6.2	7.8	0.8	10	7	9.5	2.5	15.5
[78]	4.85	5.08	0.95	9.53	6.97	8.00	1.03	15.00
[79]	4	4.05	0.99	–	3.5	5.5	2	8.5

question how these stresses affect the susceptibility of materials to adiabatic shear banding is not fully resolved [64]. Modification of the sample dimensions—changing the radius of the shaft and the inner radius of the brim—allows the systematic variation of the magnitude of the compressive and shear loading components experienced within the shearing section [77].

For compression tests on hat-shaped samples, careful sample preparation is important. Indeed, the reproducibility of hat-shaped sample experiments can be poor due to the high influence of the sample dimensions and the existence of dimensional variations along the circumference of the sample [75]. Especially the width of the shear region ($r_2 - r_1$) and radii of the corners delimiting the shear zone have a pronounced effect on the measured force-displacement response of the sample.

Because of the complex stress and deformation distribution in the shear region and the high influence of geometry, including geometrical imperfections, the extraction of quantitative data on the material behaviour from tests on hat-shaped samples is not straightforward. Only within certain limits of sample dimensions, the shear stress τ in the centre of the shear region can be approximated by dividing the measured force F by the surface $A_{sz} = \pi(r_1 + r_2)(h_2 - h_1)$ of the geometric shear zone: $\tau = F/A_{sz}$. Using this equation, good results are obtained for Ti6Al4V samples with $r_1/r_2 = 0.975$. However, as soon as the r_1/r_2 ratio deviates from this value, significant errors are introduced [75].

Unfortunately, no unambiguous relation between the measured sample displacement and the shear strain can be established. Since the shear zone and possible shear bands develop subsurface, it not possible to track the shear deformation, nor to identify possible shear bands or to measure the temperatures involved. However, split Hopkinson bar compression tests can easily be interrupted by the use of a stopper ring, see SR in Fig. 3.19b, which allows observing the microstructure at different stages of the deformation process. Consequently, compression tests on hat-shaped samples have been extensively used to study shear localisation phenomena, including the formation of adiabatic shear bands, the initiation and propagation of microcracks. Examples can be found in [71, 77, 80].

In Fig. 3.21, the results of two interrupted tests on hat-shaped Ti6Al4V samples are presented. The test denoted with Exp1 is interrupted just before the drop of the force. In the second test, denoted with Exp2, the sample is deformed beyond the peak stress in the stress–displacement curve, see Fig. 3.21a. Figure 3.21b, c shows light optical images of the shear region in the sample of Exp1, though at different locations along the sample circumference. In Fig. 3.21b, a ±30 μm wide band with elongated grains can be seen, while in Fig. 3.21c, a very narrow and faint band of ±2 μm (indicated with arrows on the figure) is observed. In the shear zone of the more deformed sample of Exp2, a band with a width of ± 6 μm is found, see Fig. 3.21d. More information, including extensive TEM analysis of the shear zones, can be found in [66].

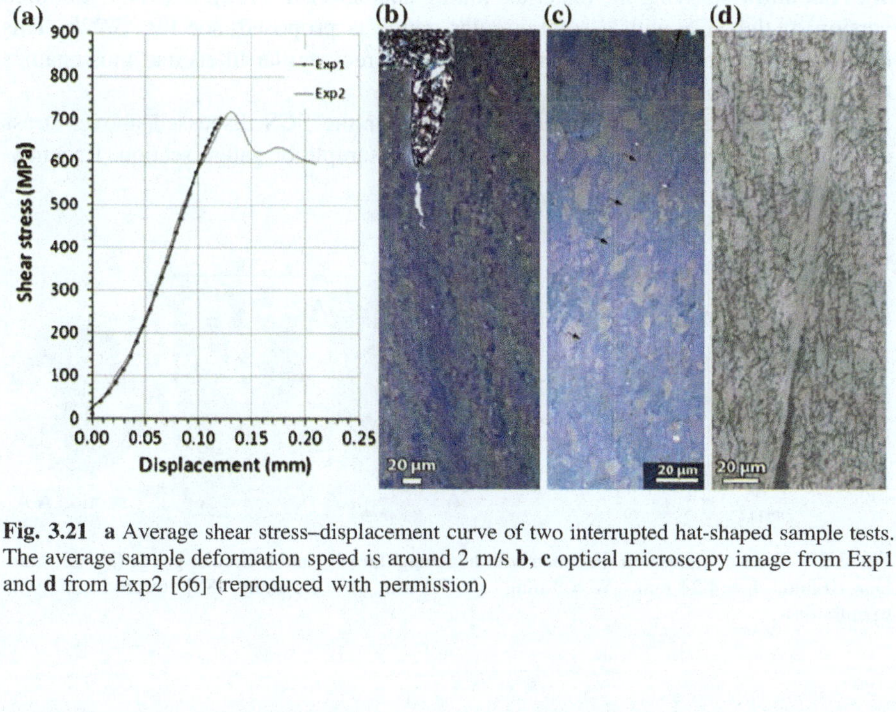

Fig. 3.21 **a** Average shear stress–displacement curve of two interrupted hat-shaped sample tests. The average sample deformation speed is around 2 m/s **b**, **c** optical microscopy image from Exp1 and **d** from Exp2 [66] (reproduced with permission)

3.4.4 Tests with Purpose Designed Sample Geometries

In recent years, also for bulk materials samples geometries have been developed which, when loaded in a compression setup, introduce shear in the material. Some of these geometries, because of their compact dimensions, can be dynamically loaded in a split Hopkinson bar compression device.

In an attempt to reach a higher contribution of the shear deformation in a compression test, Meyer et al. used cylindrical samples with an axis inclined against the loading direction [80, 81]. Compared to the compression of fully aligned samples, a higher shear to normal stress ratio is obtained. This ratio can be increased using a higher inclination [80, 82]. However, compression remains the dominant stress component. Moreover, the shear to compression stress ratio is not homogeneously distributed over the geometry and changes during the deformation.

A dominant shear component is obtained in the shear-compression specimen (SCS) developed by Rittel [83]. This specimen consists of a cylinder with an inclined gauge section created by two diametrically opposed rectangular slots which are machined at 45° with respect to the longitudinal axis, see Fig. 3.22a. Provided the friction in the sample/bar interfaces—which may introduce bending in the gauge—is minimised, the specimen can be used to determine the dynamic stress-strain characteristics of metals. Indeed, equivalent stresses and strains are reasonably uniform within the gauge section and can be calculated from the measured forces and displacements using a procedure described in [84, 85]. Specimen manufacturing is critical. Certainly, the radii of the notches are important. The main limitation of the SCS is that failure occurs in the fillet regions before the full deformation capacity of the gauge material has been reached. Thus, the SCS does not allow studying the materials failure characteristics [86]. In [87], a modified version of the SCS with a semi-circular gauge is proposed, see Fig. 3.22b. This modified specimen reduces the influence of the rectangular fillets and thus enables to reach higher plastic strains.

Using the tilted axis cylindrical geometry or the SCS discussed above, stress components different from shear coexist in the sample or gauge section. Recently,

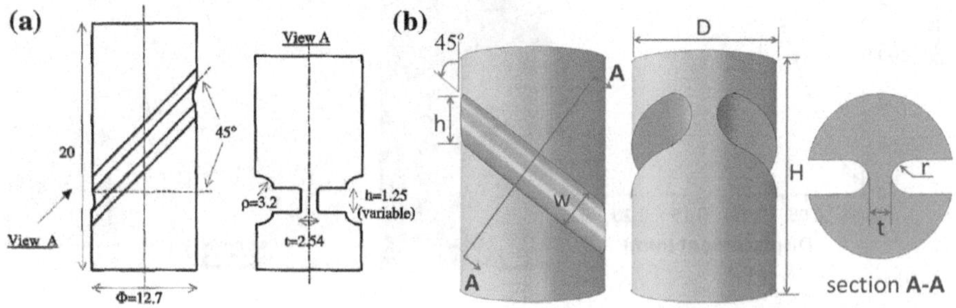

Fig. 3.22 Shear compression specimen with **a** rectangular [84] and **b** circular gauge (H = 20 mm, D = 10 mm, h = 4.24 mm, W = 3 mm, r = 1.5 mm, t = 1.6 mm) [88] (reproduced with permission)

Gray III et al. [88] introduced the compact forced-simple-shear specimen (CFSS) presented in Fig. 3.23. The CFSS imposes a simple shear deformation to the gauge material. The major advantage of the CFSS geometry is that the shearing zone can be aligned with any desired direction, e.g. crystallographic texture direction, and that this direction is maintained during testing. The sample can be loaded in compression or tension. For the latter, the sample is extended with threaded ends. By inserting dimension wedges in the slots adjacent to the gauge section or the use of a cylinder around the sample, a compression test can be interrupted at a

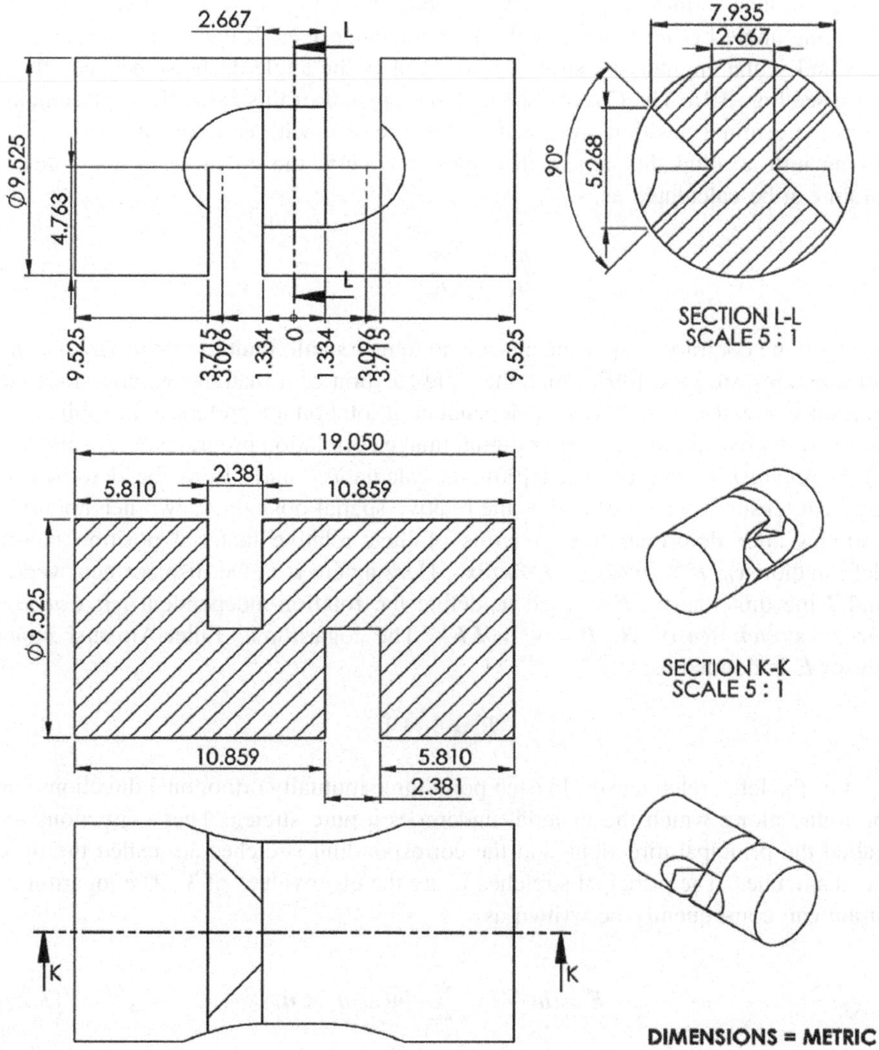

Fig. 3.23 Compact forced-simple-shear specimen [88] (reproduced with permission)

pre-defined level of deformation. Obviously, this is also possible for other sample geometries.

3.5 Shear Strain Formulations

Different formulations exist for shear strains in materials, especially when large deformations are involved. When strains as well as rotations are small, shear strain equations from the *small or infinitesimal strain theory* are commonly adopted. In this case shear strains have a clear physical meaning: they are linked with the changing angle between two initially perpendicular axes. If these axes are denoted as x and y, the engineering strain γ is defined as the angle decrease between these axes, see Fig. 3.1a. The Cauchy shear strain ***e*** is half of this value. If displacements from the original position are given by the vector ***u*** with components (u_x, u_y, u_z), taking into account that for small angles $\alpha \approx \tan\alpha$, the engineering and Cauchy strain can be calculated as:

$$\gamma = \frac{\partial u_y}{\partial x} + \frac{\partial u_x}{\partial y}; \quad e = \gamma/2 \tag{3.27}$$

If, on the contrary, displacements are no longer small, multiple *finite shear strain* formulations are used [89]. Since the rigid rotation of a material volume does not introduce stresses, formulations independent of rotation are preferred. In application software such as finite element or digital image correlation programs, very often the *Euler-Almansi strains* are used [90]. Its calculation starts from the deformation gradient tensor ***F*** which describes the relative spatial position of two neighbouring particles after deformation ***x*** in terms of their relative material position before deformation $\boldsymbol{x_0}$: $\boldsymbol{F} = \partial \boldsymbol{x}/\partial \boldsymbol{x_0} = \boldsymbol{I} + \partial \boldsymbol{u}/\partial \boldsymbol{x_0}$. Here again, ***u*** is the displacement vector and ***I*** the unit matrix. ***F*** is used to define the rotation-independent *left Cauchy-Green stretch tensor* ***B***: $\boldsymbol{B} = \boldsymbol{V}^2 = \boldsymbol{F}\boldsymbol{F}^T$. The logarithmic Euler-Almansi strain tensor ***E*** is defined as:

$$\boldsymbol{E} = ln(\boldsymbol{V}) \tag{3.28}$$

V is the left stretch tensor. In each point three mutually orthogonal directions can be found along which the material undergoes a pure stretch. These directions are called the principal directions and the corresponding stretches are called the principal stretches. The principal stretches λ_i are the eigenvalues of **V**. The logarithmic strain can consequently be written as:

$$\boldsymbol{E} = ln(\boldsymbol{V}) = \sum_{i=1}^{3} ln(\lambda_i)\boldsymbol{n}_i \otimes \boldsymbol{n}_i \tag{3.29}$$

with $\boldsymbol{n}_i$ the principal stretch directions in the deformed configuration.

In case of simple shear in the x–y plane (Fig. 3.1a) the deformation gradient tensor takes the form:

$$F = \begin{bmatrix} 1 & \gamma \\ 0 & 1 \end{bmatrix} \quad (3.30)$$

The shear component of the logarithmic strain **E** can then be calculated as [91]:

$$(e_{12})_{Log} = \frac{\ln\left(2+\gamma^2+\gamma\sqrt{4+\gamma^2}\right) - \ln\left(2+\gamma^2-\gamma\sqrt{4+\gamma^2}\right)}{2\sqrt{4+\gamma^2}} \quad (3.31)$$

For small strains $(e_{12})_{Log} \approx \gamma/2$, for larger strains $(e_{12})_{Log} < \gamma/2$. For simple shear $(e_{12})_{Log}$ has a maximum value of 0.66 for $\gamma \approx 3$.

Besides the logarithmic strain, several other finite strain tensors are used. Moreover, the strains can be calculated in the undeformed or deformed reference frame, i.e. the right $C = F^TF = U^2$ or left *stretch tensor* $B = FF^T = V^2$. A generalised formulation was introduced by Doyle and Ericksen [92]. The so-called Doyle-Ericksen generalised strain tensor describes a whole family of strain measures through the parameter m:

$$\begin{array}{lll} \frac{1}{m}(\mathbf{U}^m - \mathbf{I}), & \frac{1}{m}(\mathbf{V}^m - \mathbf{I}) & \text{for } m \neq 0 \\ \ln(\mathbf{U}), & \ln(\mathbf{V}) & \text{for } m = 0 \end{array} \quad (3.32)$$

The Green-Lagrange strain tensor is obtained for m = 2, the Biot nominal strain tensor for m = 1 and the logarithmic strains for m = 0. In the case of simple shear [Eq. (3.30)], the following relations between γ and the strain definitions of Eq. (3.32) exist:

$$\begin{aligned} (e_{12})_{Green-Lagrange} &= \frac{\gamma}{2} \\ (e_{12})_{Biot} &= \frac{\sqrt{4+2\gamma^2+2\gamma\sqrt{4+\gamma^2}} - \sqrt{4+2\gamma^2-2\gamma\sqrt{4+\gamma^2}}}{2\sqrt{4+\gamma^2}} \end{aligned} \quad (3.33)$$

For metals, also the work-conjugated finite strain to the Cauchy stress is often used as strain measure. Assuming von Mises plasticity, this so-called equivalent or Mises strain is generally derived from γ using [see Eq. (3.21)] $\varepsilon_{eq} = \frac{1}{\sqrt{3}}\gamma$. In finite element codes, however, the equivalent strain is obtained by integration of the symmetric part of the velocity gradient. If the antisymmetric part of the velocity gradient, i.e. the vorticity, is different from zero, the integration should be made in a reference frame where the vorticity is zero. The thus obtained cumulated Eulerian strain in a simple shear sample becomes [16]:

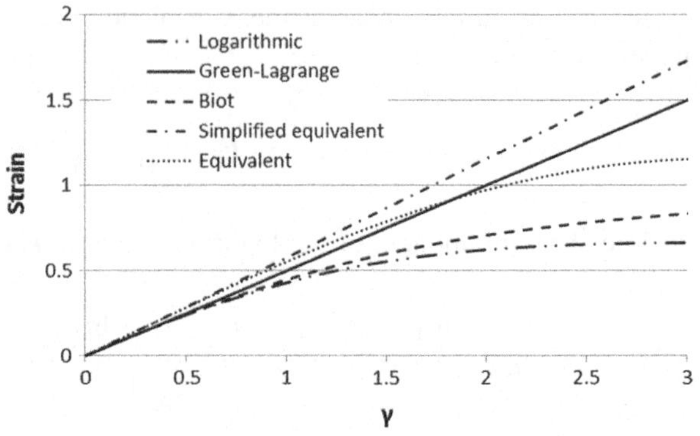

Fig. 3.24 Comparison of shear strain formulations as a function of engineering strain γ for in-plane simple shear

$$\varepsilon_{eq} = \sqrt{\frac{2}{3}(1 - \cos\gamma)} \tag{3.34}$$

If the strain formulation of Eq. (3.34) is used, good agreement is found between simulated values and values calculated from gamma.

In Fig. 3.24 the Green-Lagrange [Eq. (3.33)], Biot [Eq. (3.33)] and logarithmic Euler-Almansi [Eq. (3.31)] finite simple shear strains, together with the equivalent [Eq. (3.34)] and simplified equivalent [Eq. (3.21)] strains, are represented as a function of the engineering strain γ. Note that for low strains, the Green-Lagrange, Biot and logarithmic strain formulations quasi-coincide, as do the equivalent and simplified equivalent strain. With increasing deformation the strain curves diverge.

References

1. Meyers MA (1994) Dynamic behaviour of materials. Wiley, New York, pp 296–322
2. Field JE, Walley SM, Proud WG, Goldrein HT, Siviour CR (2004) Review of experimental techniques for high rate deformation and shock studies. Int J Impact Eng 30:725–775
3. Tarigopula V, Hopperstad OS, Langseth M, Clausen AH, Hild F, Lademo OG, Eriksson M (2008) A study of large plastic deformations in dual phase steel using digital image correlation and FE analysis. Exp Mech 48:181–196
4. Rusinek A, Klepaczko JR (2001) Shear testing of a steel sheet at wide range of strain rates and a constitutive relation with strain rate and temperature dependence of the flow stress. Int J Plast 17(1):87–115
5. Klepaczko JR, Nguyen HV, Nowacki WK (1999) Quasi-static and dynamic shearing of sheet metals. Eur J Mech A Solids 18(2):271–289
6. Rauch EF (2009) Plastic behaviour of metals at large strains: experimental studies involving simple shear. J Eng Mater Technol 3.1(1), 8
7. Bouvier S, Gardey B, Haddadi H, Teodosiu C (2006) Characterization of the strain induced plastic anisotropy of rolled sheets by using sequences of simple shear and uniaxial tensile tests. J Mater Process Technol 174(1–3):115–126
8. Inal K, Wu PD, Neale KW (2002) Large strain behaviour of aluminium sheets subjected to in-plane simple shear, modelling and simulation. Mater Sci Eng 10(2):237–252

9. Bacroix B, Genevois P, Teodosiu C (1994) Plastic anisotropy in low carbon steels subjected to simple shear with strain path changes. Eur J Mech A Solids 13(5):661–675
10. Klepaczko JR (1994) An experimental technique for shear testing at high and very high strain rates—the case of a mild steel. Int J Impact Eng 15(1):25–39
11. Miyauchi K (1984) A proposal of a planar simple shear test in sheet metals. Sci Pap Inst Phys Chem Res 78(3):27–40
12. Gasperini M, Pinna C, Swiatnicki W (1996) Microstructure evolution and strain localization during shear deformation of an aluminium alloy. Acta Mater 44(10):4195–4208
13. Merle R, Zhao H (2004) An experimental study of sheet metals under dynamic double shear at large strains. Adv Eng Plast Appl Parts 1–2 274–276:787–792
14. Harding J, Huddart J (1980) The use of the double notch shear test in determining the mechanical properties of uranium at very high rates of strain. Proc Conf Mech Prop Mater High Rates Strain 47:49–61
15. Campbell JD, Ferguson WG (1970) The temperature and strain-rate dependence of the shear strength of mild steel. Philos Mag 21:63–82
16. Shi FF, Merle R, Hou B, Liu JG, Li YL, Zhao H (2014) A critical analysis of plane shear tests under quasi-static and impact loading. Int J Impact Eng 74:107–119
17. Bonnet-Lebouvier AS, Klepaczko JR (2002) Numerical study of shear deformation in Ti-6Al-4V at medium and high strain rates, critical impact velocity in shear. Int J Impact Eng 27(7):755–769
18. Chalik P, Klepaczko JR, Rusinek A (2003) Impact shear—numerical analyses of ASB evolution and failure for Ti-6Al-4V alloy. Proceedings of the 7th International Conference on Mechanical and Physical Behaviour of Materials under Dynamic Loading, Oporto, Portugal
19. Bouvier S, Haddadi H, Levée P, Teodosiu C (2006) Simple shear tests: experimental techniques and characterization of the plastic anisotropy of rolled sheets at large strains. J Mater Process Technol 172(1):96–103
20. An YG, Vegter H, Heijne J (2009) Development of simple shear test for the measurement of work hardening. J Mater Process Technol 209:4248–4254
21. Ziolkowski A (2006) Simple shear test in identification of constitutive behaviour of materials submitted to large deformations—hyperelastic materials case. Eng Trans 54(4):251–269
22. Bao YB, Wierzbicki T (2004) On fracture locus in the equivalent strain and stress triaxiality space. Int J Mech Sci 46(1):81–98
23. Reyes A, Eriksson M, Lademo OG, Hopperstad OS, Langseth M (2009) Assessment of yield and fracture criteria using shear and bending tests. Mater Des 30:596–608
24. Gao F, Gui L, Fan Z (2011) Experimental and numerical analysis of an in-plane shear specimen designed for ductile fracture studies. Exp Mech 51:891–901
25. Roth CC, Mohr D (2015) Ductile fracture experiments with locally proportional loading histories. Int J Plast, In Press, Corrected Proof, pp 1–27
26. Peirs J, Verleysen P, Degrieck J (2012) Novel technique for static and dynamic shear testing of Ti6Al4V sheet. Exp Mech 52:729–741
27. Peirs J, Verleysen P, Van Paepegem W, Degrieck J (2011) Determining the stress-strain behaviour at large strains from high strain rate tensile and shear experiments. Int J Impact Eng 38(5):406–415
28. Abedini A, Butcher C, Anderson D, Worswick M, Skszek T (2015) Fracture characterization of automotive alloys in shear loading. SAE Int J Mater Manf 8(3):2015
29. Dowling AR, Harding J, Campbell JD (1970) The dynamic punching of metals. J Inst Met 98:215–224
30. Harding J, Huddart J (1979) Mechanical properties of materials at high rates of strain. Inst Phys Conf Ser 47:49–61
31. Hartmann KH, Kunze HD, Meyer LW (1981) In: Meyers MA, Murr LE (ed) Shock waves and high strain phenomena in metals, Plenum Press, New York, pp 325–337
32. Baker W, Yew CH (1966) Strain rate effects in the propagation of torsional plastic waves. J Appl Mech 33:917–923
33. Mechanical Testing and Evaluation ASM Handbook Vol 8 (2000) ASM International

34. Nie X, Prabhu R, Chen W et al (2011) A Kolsky bar technique for characterization of dynamic shear response of soft materials. Exp Mech 51:1527–1534
35. Bai Y, Xue Q, Xu Y, Shen L (1994) Characteristics and microstructure in the evolution of shear localization in Ti-6Al-4V. Mech Mater 17:155–164
36. Marchand A, Duffy J (1988) An experimental study of the formation process of adiabatic shear bands in a structural steel. J Mech Phys Solids 36(3):251
37. Gilat A, Wu X (1994) Elevated-temperature testing with the torsional split Hopkinson bar. Exp Mech 34(2):166–170
38. Klepaczko JR (2002) Visco-plasticity of polycrystalline tantalum—rational phenomenology in constitutive modelling. Arch Mech 54(5–6):459–478
39. Chichili DR, Ramesh KT (1999) Recovery experiments for adiabatic shear localization: a novel experimental technique. J Appl Mech-Trans ASME 66(1):10–20
40. daSilva MG, Ramesh KT (1997) The rate-dependent deformation and localization of fully dense and porous Ti-6Al-4V. Mater Sci Eng A Struct Mater Prop Microstruct Process 232(1–2):11–22
41. Chichili DR, Ramesh KT, Hemker KJ (2004) Adiabatic shear localization in alpha titanium: experiments, modeling and microstructural evolution. J Mech Phys Solids 52(8):1889–1909
42. Peirs J, Verleysen P, Verbeken K, Coghe F, Degrieck J (2012) High strain rate torsion and Bauschinger tests on Ti6AI4V. Mater Sci Forum 706–709:774–779
43. Gao X, Zhang T, Zhou J, Graham S, Hayden M, Roe C (2011) On stress-state dependent plasticity modelling: significance of the hydrostatic stress, the third invariant of stress deviator and the non-associated flow rule. Int J Plast 27:217–231
44. Graham S, Zhang T, Gao X, Hayden M (2012) Development of a combined tension-torsion experiment for calibration of ductile fracture models under conditions of low triaxiality. Int J Mech Sci 54:172–181
45. Papasiderao J, Doquet V, Mohr D (2014) Determination of the effect of stress state on the onset of ductile fracture through tension-torsion experiments. Exp Mech 54:137–151
46. Barsoum I, Faleskog J (2007) Rupture mechanisms in combined tension and shear—micromechanics. Int J Solids Struct 44(17):5481–5498
47. Shrivastava SC, Jonas JJ, Canova G (1982) Equivalent strain in large deformation torsion testing: theoretical and practical considerations. J Mech Phys Solids 30(1–2):75–90
48. Graber A, Pholandt K (1990) State of the art of the torsion test for determining flow curves. Steel Res 61(5):6
49. Kobayashi T, Simons JW, Brown CS, Shockey DA (2008) Plastic flow behaviour of Inconel 718 under dynamic shear loads. Int J Impact Eng 35(5):389–396
50. Lewis JL, Campbell JD (1972) Development and use of a torsional Hopkinson-bar apparatus. Exp Mech 12(11):520
51. Shih-Chieh L, Duffy J (1998) Adiabatic shear bands in a Ti-6Al-4V titanium alloy. J Mech Phys Solids 46(11):2201–2231
52. Xue Q, Shen LT, Bai YL (1995) A modified split Hopkinson torsional bar in studying shear localization. Meas Sci Technol 6(11):1557–1565
53. Ranc N, Taravella L, Pina V, Herve P (2008) Temperature field measurement in titanium alloy during high strain rate loading—adiabatic shear bands phenomenon. Mech Mater 40:255–270
54. Nicholas T, Lawson JE (1972) Determination of mechanical properties of materials at high shear-strain rates. J Mech Phys Solids 20(2):57
55. Khan AS, Suh YS, Kazmi R (2004) Quasi-static and dynamic loading responses and constitutive modeling of titanium alloys. Int J Plast 20(12):2233–2248
56. Shockey DA, Simons JW, Brown CS et al (2007) Shear failure of Inconel 718 under dynamic loads. Exp Mech 47(6):723–732
57. Lee WS, Lin CF, Xiea GL (1998) Dynamic shear deformation and failure behaviour of pure polycrystalline tungsten. Mater Sci Eng A Struct Mater Prop Microstruct Process 247(1–2):102–112

58. Li YL, Ramesh KT, Chin ESC (2004) The mechanical response of an A359/SiCp MMC and the A359 aluminium matrix to dynamic shearing deformations. Mater Sci Eng A Struct Mater Prop Microstruct Process 382(1–2):162–170
59. Gilat A, Cheng CS (2000) Torsional split Hopkinson bar tests at strain rates above 10^4 s^{-1}. Exp Mech 40:54–59
60. Gilat A, Cheng C-S (2002) Modeling of torsional split Hopkinson bar tests at strain rates above 10,000 s^{-1}. Int J Plast 18:787–799
61. Macdougall DAS, Harding J (1998) The measurement of specimen surface temperature in high-speed tension and torsion tests. Int J Impact Eng 21(6):473–488
62. Giovanola J (1988) Adiabatic shear banding under pure shear loading part I: direct observation of strain localization and energy dissipation measurement. Mech Mater 7:59–71
63. Lee WS, Lin CF, Lin YC (2006) Effect of temperature and strain rate on the shear properties of Ti-6Al-4V. J Mech Eng Sci Part C 22(127–3):6
64. Bai Y, Dodd B (1992) In: Dymat (ed) Adiabatic shear localization, 2 edn
65. Peirs J (2012) Experimental characterisation and modelling of the dynamic behaviour of the titanium alloy Ti6Al4V. PhD dissertation, Ghent University. ISBN: 978-90-8578-493-7
66. Peirs J, Tirry W, Amin-Ahmadi B, Coghe F, Verleysen P, Rabet L, Schryvers D, Degrieck J (2013) Microstructure of adiabatic shear bands in Ti6Al4V. Mater Charact 75:79–92
67. Meyer LW, Manwaring S (1986) Critical adiabatic shear strength of low alloyed steel under compressive loading. In: Staudhammer KP, Meyers MA, Murr LE (eds) Metallurgical applications of shock-wave and high-strain-rate phenomena. Marcel Dekker Inc., New York, pp 657–674
68. Bronkhorst CA, Cerreta EK, Xue Q, Maudlin PJ, Mason TA, Gray GT (2006) An experimental and numerical study of the localization behavior of tantalum and stainless steel. Int J Plast 22(7):3.04–35
69. Nemat-Nasser S, Isaacs JB, Liu MQ (1998) Microstructure of high-strain, high-strain rate deformed tantalum. Acta Mater 46(4):3.07–25
70. Kad BK, Gebert JM, Perez-Prado MT, Kassner ME, Meyers MK (2006) Ultrafine-grain-sized zirconium by dynamic deformation. Acta Mater 54(16):4111–4127
71. Meyers MA, Xu YB, Xue Q, Perez-Prado MT, McNelley TR (2003) Microstructural evolution in adiabatic shear localization in stainless steel. Acta Mater 51(5):3.07–25
72. Beatty JH, Meyer LW, Meyers MA, Nemat-Nasser S (1990) Formation of controlled adiabatic shear bands in AISI 4340 high strength steel. Technical report MTL, pp 90–54
73. Couque H (2003) A hydrodynamic hat specimen to investigate pressure and strain rate dependence on adiabatic shear band formation. J Phys IV 110:423–428
74. Longère P, Dragon A, Trumel H, Deprince X (2005) Adiabatic shear banding-induced degradation in a thermo-elastic/viscoplastic material under dynamic loading. Int J Impact Eng 32(1–4):285–320
75. Peirs J, Verleysen P, Degrieck J, Coghe F (2010) The use of hat-shaped specimens to study the high strain rate shear behaviour of Ti–6Al–4V. Int J Impact Eng 37:703–714
76. Li JR, Yu JL, Wei ZG (2003) Influence of specimen geometry on adiabatic shear instability of tungsten heavy alloys. Int J Impact Eng 28(3):303–314
77. Cerreta EK, Bingert JF, Gray GT III, Trujillo CP, Lopez MF, Bronkhorst CA, Hansen BL (2013) Microstructural examination of quasi-static and dynamic shear in high-purity iron. Int J Plast 40:23–38
78. Lins JFC, Sandim HRZ, Kestenbach H, Raabe D, Vecchio KS (2007) A microstructural investigation of adiabatic shear bands in an interstitial free steel. Mater Sci Eng A Struct Mater Prop Microstruct Process 457(1–2):205–218
79. Lee WS, Liu CY, Chen TH (2008) Adiabatic shearing behavior of different steels under extreme high shear loading. J Nucl Mater 374(1–2):33–39
80. Meyer LW, Staskewitsch E (1992) Adiabatic shear failure of the titanium alloy Ti-6Al-4V under biaxial dynamic compression/shear loading. Titanium 1992, science and technology, vol 1993. The Minerals, Metals & Materials Society, USA, pp 1939–1946

81. Meyer LW, Halle T (2004) Comparison of different methods to measure shear and combined shear under high speed loading. Fifth International Symposium of Impact Engineering, Cambridge
82. Wei Z, Yu J, Si Hu, Li Y (2000) Influence of microstructure on adiabatic shear localization of pre-twisted tungsten heavy alloy. Int J Impact Eng 24:747–758
83. Rittel D, Lee S, Ravichandran G (2002) A shear compression specimen for large strain testing. Exp Mech 42:58–64
84. Dorogoy A, Rittel D (2005) Numerical validation of the shear compression specimen (SCS) part II: dynamic large strain testing. Exp Mech 45:178–185
85. Dorogoy A, Rittel D (2006) A numerical study of the applicability of the shear compression specimen to parabolic hardening materials. Exp Mech 46:355–366
86. Osovski S, Rittel D, Landau P, Venkert A (2012) Microstructural effects on adiabatic shear band formation. Scripta Mater 66:9–12
87. Dorogoy A, Rittel D, Godinger A (2015) Modification of the shear-compression specimen for large strain testing. Exp Mech 55:1627–1639
88. Gray GT III, Vecchio KS, Livescu V (2016) Compact forced simple-shear sample for studying shear localisation in materials. Acta Mater 103:12–22
89. Hiermaier S (2008) Structures under crash and impact: continuum mechanics, discretization and experimental characterization. Springer, Berlin
90. Lava P, Cooreman S, Coppieters S, Debruyne D (2013) Sources of systematic errors in the determination of heterogeneous strain fields obtained via DIC. In: Application of imaging techniques to mechanics of materials and structures, vol 4. Springer, Berlin, pp 271–281
91. Xiao H, Bruhns OT, Meyers A (1997) Logarithmic strain, logarithmic spin and logarithmic rate. Acta Mech 124:89–105
92. Doyle TC, Eriksen JL (1956) Non-linear elasticity. Adv Appl Mech 4:53–115

Chapter 4
Dynamic Brazilian Test Using the Kolsky-Hopkinson Bar Machine

Pengwan Chen, Baoqiao Guo and Jingjing Chen

4.1 Introduction

Brittle materials, such as rock, concrete and ceramic, break without significant deformation when subjected to static or dynamic load. The tensile strength, which is much smaller than the compressive strength, is a key mechanical parameter for the brittle materials. The tensile fracture is an important failure mode. Due to its brittleness, a direct uniaxial tension test using a conventional dumbbell sample is difficult to be conducted for a brittle material, and this may become even more difficult in dynamic conditions.

Alternatively, the tensile strength of brittle materials can be measured using indirect tensile tests. The Brazilian test, or diametric compression test, is a typical indirect tensile test, which was initially used to determine the tensile strength of brittle materials. With the recommendation of the International Society for Rock Mechanics [1], the Brazilian test became a standard method to determine the tensile strength of rocks. The Brazilian test also became a standard to determine the tensile strength of concrete [2] and rock [3]. To reduce the stress concentration near the loading point, Mellor and Hawkes [4] proposed the arc Brazilian Test, in which the contact surface of the indenter is in the form of arc. Furthermore, Wang et al. [5] proposed the so-called flattened Brazilian disc (FBD) test, in which two small pieces were cut off from the disc to form two parallel planes for the loading surfaces, which reduces the local contact stress concentration. The FBD test can be used to measure both the tensile strength and fracture toughness [5, 6]. Using the Brazilian test, Ye et al. studied the tensile elastic modulus [7] and the bi-modulus of rock material [8].

P. Chen (✉) · B. Guo · J. Chen
State Key Laboratory of Explosion Science and Technology,
Beijing Institute of Technology, Beijing 100081, China
e-mail: pwchen@bit.edu.cn

R. Othman (ed.), *The Kolsky-Hopkinson Bar Machine*,
https://doi.org/10.1007/978-3-319-71919-1_4

Due to the simplicity of the sample preparation and experimental test, the Brazilian test has also been used to measure the dynamic tensile strength of brittle materials such as rock, concrete and ceramics [9–11]. Through numerical analysis, Antonn et al. [12] found that the stress distribution in the dynamic Brazilian test was similar to the static test; therefore the dynamic Brazilian test was also considered as an effective method to measure the dynamic tensile properties. The dynamic tensile strength of a glass ceramic was measured using the dynamic Brazilian test to study the rate-dependent effect [13]. The tensile strength of rocks was also studied using the dynamic Brazilian test [14], and the experimental results of the two tested rocks showed that the tensile strength is rate dependent. The dynamic Brazilian disc test has been selected by ISRM as the suggested method for determining the dynamic strength and toughness of rock materials [15]. The above dynamic tests were performed by aide of an apparatus called Kolsky bar, which is also widely known as a split Hopkinson pressure bar (SHPB). This technique is the most used tool for the dynamic mechanical studies on materials [16–18] at high strain rates (10^2–10^4 s^{-1}). The tensile strength of brittle materials is sensitive to strain rate. Using a modified Kolsky bar and direct tensile test, the loading related dynamic tensile strengths were determined [19]. Nevertheless, the Brazilian disc test, as a common and favorite indirect tensile test, is mostly used to study the tensile behaviors of brittle materials. The loading rate dependence on the tensile strengths were found for both dry and saturated sandstone Brazilian disc samples [20]. Zhang and Zhao presented the major developments in split Hopkinson bar testing on dynamic mechanical behavior of rock materials [21].

Strain gauges were conventionally used to measure the deformation in a Brazilian test. As a contact measurement method, strain gauges have advantages such as high sensitivity and low cost, however they can only measure one or several points at the same time, and cannot conduct a full field deformation measurement. In addition, for non-uniform deformation, the effect of the gauge size is not negligible. As an alternative method, some optical non-contact techniques are available and widely used in various domains, which can provide full-field deformation information for evidence analysis. Field et al. provided a summary of the optical full-filed measurement techniques for high rate deformation and shock studies [22]. Among the optical techniques, the high-sensitivity interferometry techniques have good accuracy and can measure very tiny deformation. But the application of these techniques is often limited due to the experimental conditions. Besides the interferometry techniques, the digital image correlation (DIC) method is a powerful optical non- interferometry full-field measurement technique. Due to its various advantages such as simple low-cost experimental setup, low requirements on measurement environment, DIC has been applied in various fields in the past decades [23–25]. Combined with DIC, Liu determined the elastic constants by using the Brazilian disc test [26]. By measuring the strain with both the laser gap gage technique and high-speed DIC method, Chen et al. studied the dynamic tensile strength of brittle materials [27]. The dynamic Brazilian tests were also performed on polymer bonded explosives by using the SHPB and high speed photography [28, 29]. Bhattacharya and Goulbourne [30] studied the high strain rate deformation mechanisms of a ceramic using a dynamic full-field deformation measurement by DIC. Zhang and Zhao [31] investigated the dynamic mechanical properties of rock using the DIC measurement. Pierron et al. [32] studied the dynamic response and stress distribution of aluminum using the three-point bending test and ultra high speed DIC.

The recent rapid accent of the (ultra) high-speed photography techniques and optical techniques provide a new approach to investigate the dynamic mechanical behavior of materials. Using the Kolsky-Hopkinson bar apparatus and high speed DIC, the dynamic deformation and failure of Brazilian disc samples of brittle materials, as well as the dynamic tensile strength were investigated.

4.2 Principle of the Brazilian Disc Test

The Brazilian disc test is a circular disc sample with a pair of compressive load applied diametrically to the disc diameter. As shown in Fig. 4.1, a pair of vertical load P is applied to a disc with radius R and thickness t.

The sample is very compact and the testing configuration is simple. It's favorite for testing the mechanical behaviors of brittle materials. Furthermore, an analytical solution can be obtained for both the stress and deformation distributions. For planar deformation hypothesis (either plane strain or plane stress), according to the elasticity theory [26, 33], the stress components are expressed as

$$\begin{cases} \sigma_{xx} = -\dfrac{2P}{\pi t}\left[\dfrac{x^2(R-y)}{\left(x^2+(R-y)^2\right)^2} + \dfrac{x^2(R+y)}{\left(x^2+(R+y)^2\right)^2} - \dfrac{1}{2R}\right] \\ \sigma_{yy} = -\dfrac{2P}{\pi t}\left[\dfrac{(R+y)^3}{\left(x^2+(R+y)^2\right)^2} + \dfrac{(R-y)^3}{\left(x^2+(R-y)^2\right)^2} - \dfrac{1}{2R}\right] \\ \tau_{xy} = +\dfrac{2P}{\pi t}\left[\dfrac{x(R-y)^2}{\left(x^2+(R-y)^2\right)^2} - \dfrac{x(R+y)^2}{\left(x^2+(R+y)^2\right)^2}\right] \end{cases} \tag{4.1}$$

where the origin of the Cartesian coordinates (x, y) is located in the disc center. An example of a Brazilian disc dimensions and the load corresponding to Fig. 4.1 is given in Table 4.1. The distribution of the stress components therefore can be plotted.

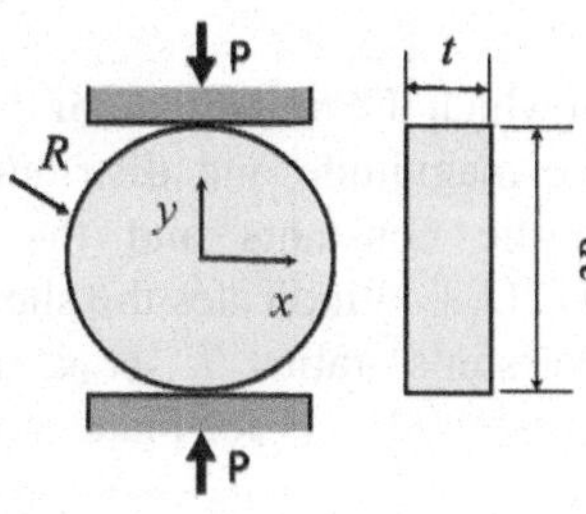

Fig. 4.1 Schematic of Brazilian disc under compression diametral load

Table 4.1 Brazilian disc sample dimensions and load

Radius	R = 25 mm
Thickness	t = 10 mm
Load	P = 2000 N

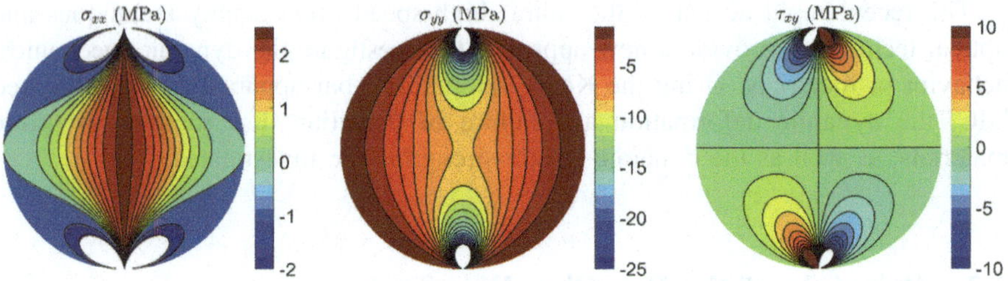

Fig. 4.2 Stress fields of Brazilian disc under compression diametral load

Figure 4.2 shows the contour plots of the three stress components. There is no material parameter in Eq. (4.1), so the distribution of the three stress components is not dependent neither to the materials elastic constants, nor to the hypothesis of plane strain or plane stress. The stress magnitude is just related to load P and disc thickness t. The color bars in Fig. 4.2 display the stress scale of the two normal and one shear stress components. The positive values for normal stress illustrate the tensile stress, and the negative values illustrate the compressive stress. As the diametral load is applied on the top and bottom point disc sample is charged through the vertical diameter, the stress concentration is found in the region near the two loaded points. The normal stress in Y direction σ_{yy} is mostly compressive stress on the whole circular surface, while the normal stress in X direction σ_{xx} is mostly tensile stress. The maxi tensile stress value locates on the vertical diameter. Due to the much weakness of the tensile strength compared to the compressive one, the Brazilian disc sample of brittle materials is split into two halves. The tensile strength in static or dynamic tests is therefore determined by this kind of indirect test.

From the elasticity theory, the displacement components are also expressed as

$$\begin{cases} u = -\frac{P}{4\pi t\mu}\left[(\kappa-1)\tan^{-1}\left(\frac{2Rx}{R^2-x^2-y^2}\right) - \frac{(\kappa-1)x}{R} - \frac{2x(R-y)}{x^2+(R-y)^2} - \frac{2x(R+y)}{x^2+(R+y)^2}\right] \\ v = -\frac{P}{4\pi t\mu}\left[\left(\frac{\kappa+1}{2}\right)\ln\left(\frac{x^2+(R+y)^2}{x^2+(R-y)^2}\right) - \frac{(\kappa-1)y}{R} - \frac{x^2-(R-y)^2}{x^2+(R-y)^2} + \frac{x^2-(R+y)^2}{x^2+(R+y)^2}\right] \end{cases} \quad (4.2)$$

in which the magnitude of displacement is related to the load per thickness P/t And the magnitude and distribution of displacement are also related to the material elastic constants and the planar deformation hypothesis. In Eq. (4.2), $\mu = E/2(1+v)$ indicates the shear modulus, where E is the elastic modulus and v is the Poisson's ratio. κ is a coefficient related to the planar hypothesis: $\kappa = (3-v)/(1+v)$ for plane stress and $\kappa = (3-4v)$ for plane strain. In this study the plane stress hypothesis is used. For the given parameters in Table 4.1, the contour maps of the displacement components shown in Fig. 4.3 are related to a typical rock with the elastic constants: E = 70 GPa, ν = 0.30.

The strain components, are the gradients of displacement, can be systematically deduced from Eq. (4.2). The equation and the map of the strain components which are presented in the literatures, are not shown here. The distribution and the contour lines of the strain components are quite similar to those of the stress components in Fig. 4.2.

Using a suitable optical technique, the stress or the deformation fields can be measured. The distribution of stress can be measured by photoelastic technique, and the displacement and strain fields can be measured by the interferometric techniques (moiré or speckle interferometry) or non-interferometry techniques (DIC or grid method) [34, 35]. Thanks to these full-field measurement techniques, the deformation and failure process during the loading steps can be observed and analyzed [21, 36], and the dynamic mechanical behaviors of rocks were investigated.

As expressed in Eq. (4.1) and shown in Fig. 4.1, stress σ_{xx} along the vertical diameter ($x = 0$) is a constant. Nevertheless, from the previous studies reported in the literatures, the crack initiates from the circular disc center due to its weak tensile strength. Actually in the experimental tests, the pair of compression load is commonly applied on the disc with aide of anvils (illustrated in Fig. 4.1). The disc is deformed under the load, and the contact condition is no more single-point contact. Thus the stress and displacement (strain) distribution changes. With the increase of contact length, from the normal stress profiles along the vertical diameter, the extreme compressive stress value near the contact point decreases, but there is little influence on the normal stress values in disc center [26]. For the Brazilian disc test on brittle materials, the disc sample is split into two halves along the vertical diameter where a pair of compression load is applied (Fig. 4.1). The maximum tensile stress is located at the disc center, where values of the stress components are

$$\begin{cases} \sigma_{xx}(0, \quad 0) = \sigma_t = +\frac{P}{\pi Rt} \\ \sigma_{yy}(0, \quad 0) = \sigma_c = -\frac{3P}{\pi Rt} \\ \tau_{yy}(0, \quad 0) = 0 \end{cases}, \quad \sigma_c = -3\sigma_t \tag{4.3}$$

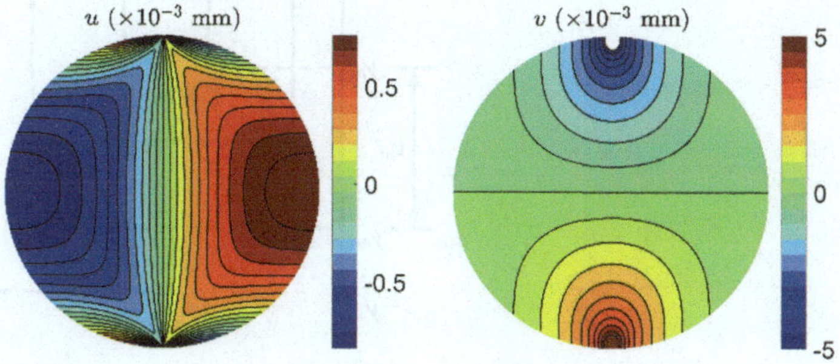

Fig. 4.3 Displacement fields of Brazilian disc

In the disc center, the compressive stress σ_c is 3 times of the tensile stress σ_t. In general, failure occurs along the horizontal loading axis in the Brazilian test when the tensile stress σ_t reaches the tensile strength of the material. The crack firstly initiates at the disk center, and then, extends toward the loading points along the horizontal axis.

4.3 Digital Image Correlation Method and High-Speed Photography

Digital image correlation (DIC) method is a widely used and useful optical method to measure deformation and strain field in experimental mechanics because of the advantages of non-contact, high accuracy, real time and full deformation field measurement. In the system, a digital camera and a computer were used to capture the movement of speckles in situ before and after a deformation. By comparing the speckle pattern in the objective configuration with the speckle pattern in the initial configuration, the displacement and strain full-field can be determined. The matching theory of this method is given in Fig. 4.4. Quadrangle S (solid line) is a reference (or undeformed) sub image and quadrangle S_1 (dash line) is the corresponding deformed sub image. In order to obtain the displacements u_m and v_m of point M, sub image S_1 is matched with sub image S using a correlation operation.

In DIC analysis, the correlation coefficient, which is a criterion for comparing two images, is given such as

$$C =' 1 - \frac{\sum f(x,y) \cdot \sum g(x^*,y^*)}{\sqrt{\sum f^2(x,y) \cdot \sum g^2(x^*,y^*)}} \tag{4.4}$$

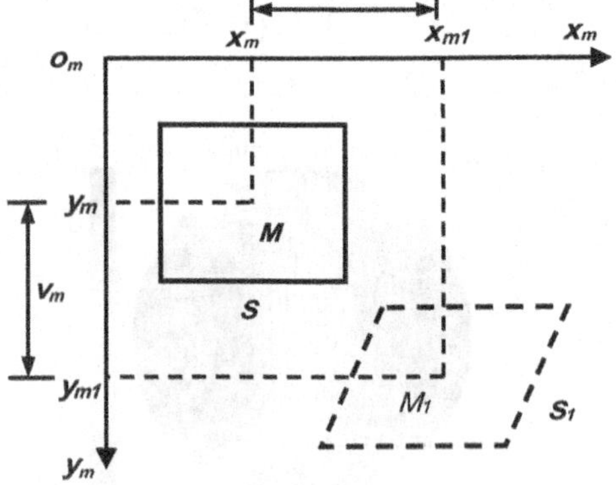

Fig. 4.4 Schematic diagram of deformed process for planar object

where (x, y) and (x^*, y^*) are Cartesian coordinates of a material point in the image of the undeformed and deformed patterns, respectively. $f(x, y)$ and $g(x^*, y^*)$ are light intensities at that point in the corresponding images. The correlation coefficient C varies from 0 to 1, with 0 signifying a perfect match between the two images [37]. When the damage or crack occurs in the local region during deformation, the local morphological character of the speckle pattern is changed, where The mismatching between the undeformed and deformation makes the correlation coefficient map becomes a good indicator of crack and fracture [38].

For the full-field measurement in quasi-static tests, the digital CCD/CMOS cameras are inexpensive and of good quality. Nevertheless, the duration from the deformation to the final failure of brittle materials under dynamic loading is very short. To obtain a detailed dynamic deformation and fracture process of the brittle materials in the dynamic tests using optical technique such as DIC, a high speed camera of very frame rate is indispensable. Various kinds of high speed cameras and its application for full-field measurement has been investigated in the previous studies [39, 40]. Classed by the frame rate, the ultra-high speed (UHS) cameras are considered with maxi frame rate more than one million *fps* (frame per second), such as rotating mirror cameras and intensified gated cameras. Commonly the UHS cameras have a satisfactory spatial resolution of 1000 × 1000 pixels or better, but the main disadvantage is the high cost and the limited number (from several to dozens) of images captured during one test. Recently the so-called in situ image sensor (ISIS) sensor UHS cameras which have good compromise among the frame rate, image resolution and image number, have a good potential for dynamic full-field measurement tests [41]. Nevertheless, the frame rate used in most dynamic tests is less than 1 M fps. The single CMOS/CCD high speed (HS) cameras with maxi frame rate from 1000 to 1 M fps have an increasing application in dynamic tests. Recently the performance of this technique has been rapidly increased. Due to the build-in memory, the main advantage is that a large number of frames can be captured by one test. Nevertheless, this technology is limited by the stockage speed from the CCD/CMOD sensor to memory. The frame rate at full image resolution (usually about 1 M pixels) is about 10^4 fps, which is usually not enough in dynamic tests. There is a compromise between the frame rate and image resolution.

In this work presented in this chapter, a HS camera (type Photron Fastcam SA5) is used with frame rate 7500 fps in full image resolution (1024 × 1000 pixels) or maxi frame rate 1 M fps at a very low resolution (64 × 16 pixels). The specification of Photron Fastcam SA5 high-speed camera is presented in Fig. 4.5. The storage speed keeps relatively unchanged at different frame rates [42]. A higher image resolution results in a lower frame rate. In the following experimental tests, various frame rates were used according to the tested material properties and sample size of observed area. The dynamic displacement and strain fields are processed using VIC-2D software [37].

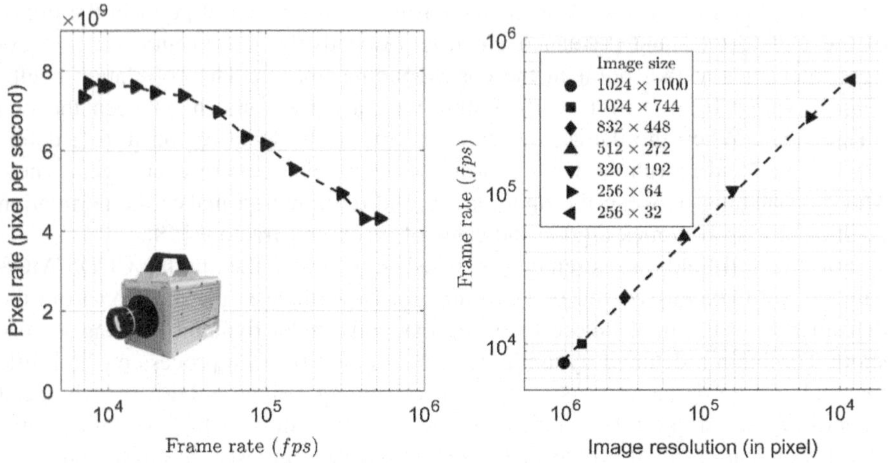

(a) Storage speed as a function of frame rate

(b) Relation of frame rate and image resolution

Fig. 4.5 Performance of Photron Fastcam SA5 high-speed camera

4.4 Experimentation

4.4.1 Materials and Sample Preparation

The experiments were performed on three brittle materials: Al_2O_3 ceramic, granite and PMMA. The principal parameters of the three tested materials and the dimensions of samples are listed in Table 4.2. Al_2O_3 ceramic samples were prepared by powder sintering, with porosity of 5.7%. The granite rock is produced from the mountain in Fangshan District in Beijing, and the crystal size is about 1 mm. The disc samples of granite were processed from a big rock plate of the same thickness. In the same way the disc samples of PMMA were prepared.

The original Brazilian disc samples made the three materials are shown in Fig. 4.6. In order to apply DIC method for full field measurement, the speckle pattern was fabricated by spraying the paint on the surface of disc samples. The surface was at first covered by an uniform white paint layer, and then the black random spots were prepared.

Table 4.2 Dimensions of samples, and density, elastic parameters of the three tested materials

Material	Sample dimensions	Density (g/cm3)	Young's Modulus (GPa)	Poisson's ratio
PMMA	Φ32 × 13 mm	1.20	3.6	0.40
Granite	Φ50 × 16 mm	2.80	60	0.34
Al_2O_3 ceramic	Φ40 × 18 mm	3.74	290	0.22

Fig. 4.6 Brazilian disc samples (from left to right: PMMA, Al_2O_3 ceramic, granite)

4.4.2 Experimental Setup

The experimental setup for the dynamic Brazilian test is shown in Fig. 4.7. A SHPB system is used for the dynamic loading device. The disc sample is placed between the incident bar and transmission bar. The signals of the incident and reflected waves are recorded by a strain gauge mounted on the incident bar. The transmitted wave signal is recorded by a strain gauge mounted on the transmitted bar. A pulse shaper made of a thin copper patch is attached to a section of the incident bar to obtain the desired loading pulse. A high speed camera is also used to record the dynamic deformation and failure process of the disc samples. Two SHPB apparatus with different diameters were used: SHPB with bars diameter 14.5 mm for PMMA

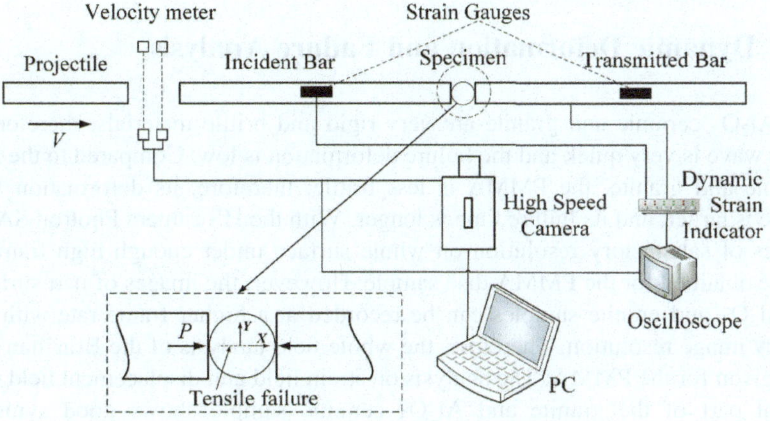

Fig. 4.7 Schematic of the experimental setup of the dynamic Brazilian test

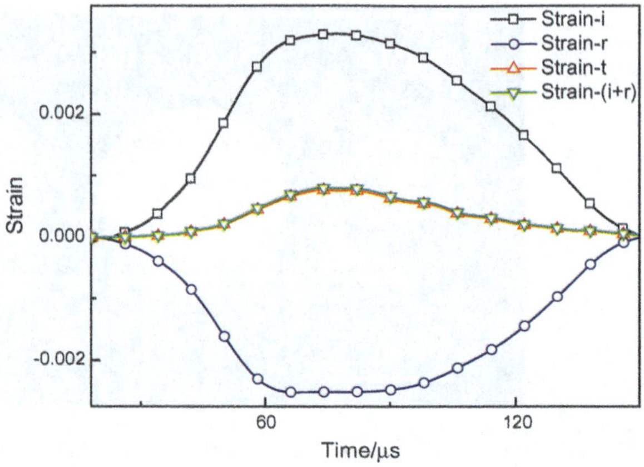

Fig. 4.8 Typical incident, reflected, and transmitted strain pulses of the Al_2O_3 ceramic

samples, SHPB with bars diameter 37 mm for the granite and Al_2O_3 ceramic samples.

When the projectile is launched and crosses the velocity meter, the shifting signal of the velocity meter is generated. The HS camera and dynamic strain indicator are simultaneously triggered. The digital images and stress wave signals are then recorded simultaneously by the high speed camera and dynamic strain indicator.

The typical incident (ε_i), reflected (ε_r), and transmitted (ε_t) pulses recorded in the Brazilian test of the Al_2O_3 ceramic are shown in Fig. 4.8. According to elastic wave theory, it is evident that stress equilibrium is achieved on the disc sample in the dynamic Brazilian test.

4.5 Dynamic Deformation and Failure Analysis

The Al_2O_3 ceramic and granite are very rigid and brittle materials; therefore, the stress wave is very quick and the failure deformation is low. Compared to the Al_2O_3 ceramic and granite, the PMMA is less brittle; therefore, its deformation before failure is larger, and its failure time is longer. With the HS camera Photron SA5, the images of satisfactory resolution on whole surface under enough high frame rate can be obtained for the PMMA disc sample. However, the images of part surface of the Al_2O_3 and granite samples can be recorded at a higher frame rate with satisfactory image resolution. Therefore, the whole field analysis of the Brazilian test is only given for the PMMA. The analysis on strain field and displacement field on the central part of the granite and Al_2O_3 ceramic sample shows good symmetry, demonstrating that stress equilibrium can also be obtained.

4.5.1 Dynamic Deformation of PMMA

A typical dynamic loading curve on the PMMA disc sample is shown in Fig. 4.9. The images were recorded at the frame rate of 50,000 fps with image resolution 512 × 272 pixels. the indicated five points with delay 20 μs between two neighbor points correspond to the moments when the images were captured.

Figure 4.10 shows the evolution of the displacement fields of the PMMA in the dynamic Brazilian test. The disc sample is impacted on the left side by the incident bar of the SHPB. The displacement fields are computed with a subset size of 21 pixels and step size of 5 pixels. The subfigures (a1–a5) and (b1–b5) respectively represent the experimental displacements in the X direction (u field) and Y direction (v field) at 20, 40, 60, 80 and 100 μs after the disc sample was impacted by the incident bar. From these fields, the stress wave propagation from left to right is obviously observed. The displacement fields in the X and Y directions appear symmetrical along the X and Y directions, respectively, at 80 μs after impact, which demonstrates that the stress equilibrium state is reached at this moment. A numerical simulation of the same condition is performed by using LS-DYNA for comparative analysis. A linear elastic model of the PMMA is used, and the elastic parameters are presented in Table 4.2. The subfigures (c1–c5) and (d1–d5) respectively represent the simulated displacements in the X direction (u field) and Y direction (v field) at 20, 40, 60, 77 and 100 μs. The subfigures (c4) and (d4) represent the simulated displacements when the stress equilibrium is observed. The simulated stress equilibrium time is approximately 77 μs, which agrees well with the experimental results that have a time resolution of 20 μs. It can also be seen

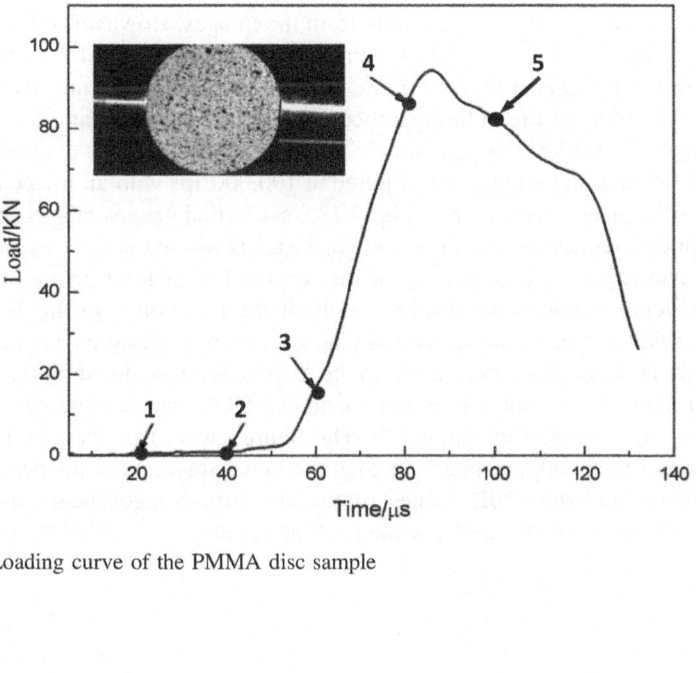

Fig. 4.9 Loading curve of the PMMA disc sample

from Fig. 4.10 that both the distribution and amplitude of the displacement fields obtained by DIC measurement are in good agreement with the results obtained by numerical simulation, implying that the assumption of linear elasticity for PMMA under dynamic Brazilian test is reasonable.

4.5.2 *Dynamic Deformation of Al_2O_3 Ceramic*

Although the stress wave speed in the Al_2O_3 ceramic and granite are higher than in the PMMA, the dynamic deformation and failure of the Al_2O_3 ceramic and granite disc samples can be analyzed using a higher image frame rate and lower image resolution. In this case, only part of the disc sample is observed. The experiments on the Al_2O_3 ceramic demonstrate the dynamic deformation and failure process. The frame rate is 175,000 fps with an image resolution of 320 × 96 pixels. Figure 4.11 shows the tensile strain fields ε_y in the local area along the loading line for the Al_2O_3 sample at several moments before failure during the dynamic loading process. The displacement fields are computed for a subset size of 17 pixels and step size of 2 pixels. At 51.4 μs, the strain concentration band appears along the horizontal diameter, and the band becomes symmetrical at 57.1 μs. The maximum strain band can be found in the center of the disc sample, which is in agreement with elastic theory. After this time, the crack initiates and propagates in the sample.

Figure 4.12 shows the displacement vector fields of the Al_2O_3 sample at different moments. At 57.1 μs, a global rightward movement can be observed along the X direction, which is due to the impact of the incident bar from the left side. At 62.9 μs, a horizontal crack appears in the disc center along the loading line. Then, the disc sample splits into two halves as the crack propagates. In Fig. 4.12c, d, the crack through the loading line is visible from the images. However, due to the high crack propagation speed of the Al_2O_3 ceramic and the limitation of the frame rate of the camera used in this study, the crack propagation velocity is not determined.

To further analyze the splitting process of the Al_2O_3 disc samples, the crack development is tracked by the crack opening displacements (CODs). In this experimental case, the images are acquired at 100,000 fps with an image resolution of 320 × 192 pixels. As shown in Fig. 4.13a, six virtual gauges labeled **E1–E6** are selected at different positions along the crack. E5 is located at the disc center, and gages E6 and E2 are symmetrical to the disc center. For each virtual gage, the COD is the difference between the displacements of the two points in the Y direction. Thus, the COD curves of the six virtual gages at different locations can be achieved from the dynamic displacement fields in the Y direction calculated by the DIC. The measured load versus time curve and calculated COD versus time curves for the Al_2O_3 disc are provided in Fig. 4.13b. The figure shows that the COD values of the six gages remain approximately zero at the initial stage. After the peak load, the crack initiates and the COD values of the six virtual gages begin to increase. The COD values increase rapidly while the load decreases. The COD curve of gage

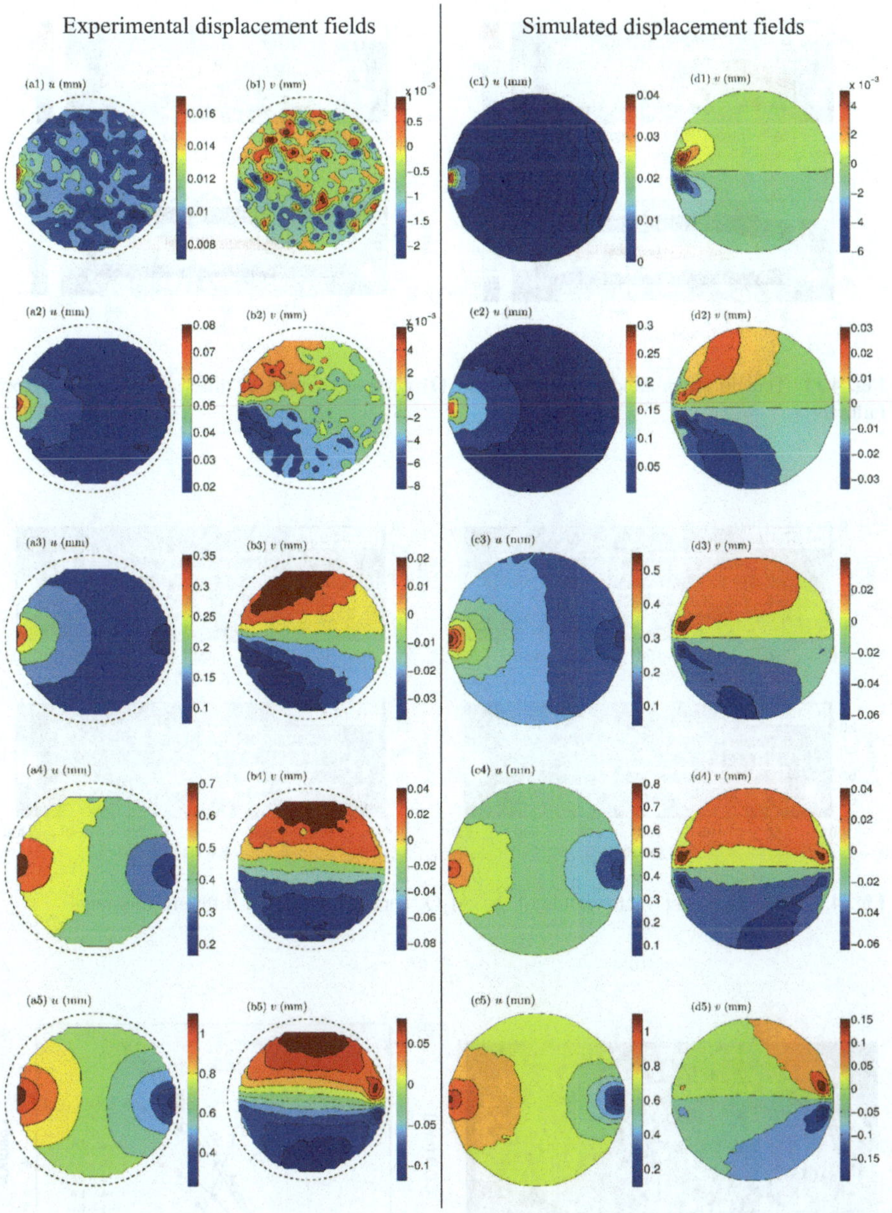

Fig. 4.10 Experimental and simulated displacement fields of PMMA under dynamic Brazilian test. The first column (a1–a5) and second column (b1–b5) represent the experimental displacements in X direction (u field) and in Y direction (v field) at 20, 40, 60, 80 and 100 μs after impact respectively. The dashed circle presents the edge of disc sample. The third column (c1–c5) and forth column (d1–d5) represent the simulated displacements in X direction and in Y direction at 20, 40, 60, 77 and 100 μs

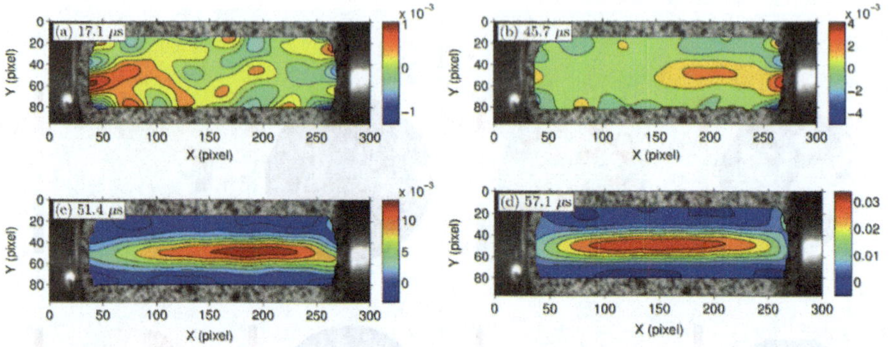

Fig. 4.11 Tensile strain fields (ε_y) of the Al_2O_3 ceramic sample at different moments before failure

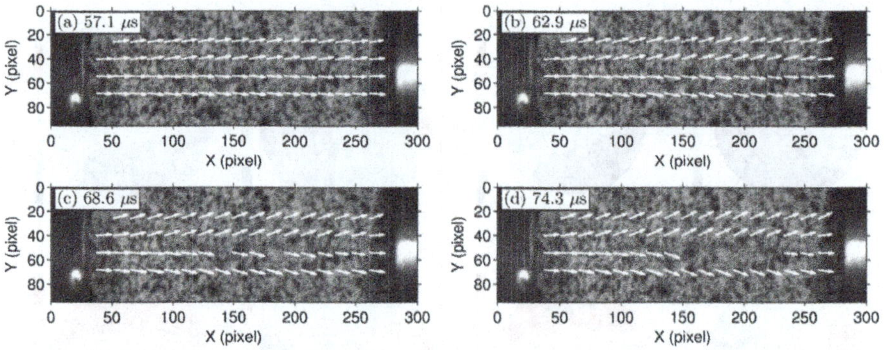

Fig. 4.12 Displacement vector fields of the Al_2O_3 ceramic sample at different moments

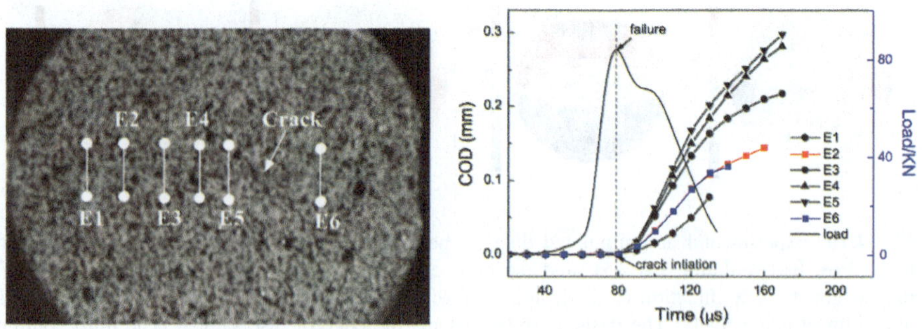

(a) COD gauges along the crack route (b) COD curves and loading curve

Fig. 4.13 Virtual gages along the crack route and their COD curves calculated by the DIC

E5 located in the center is the highest among all six gages. At each moment after failure, the closer the gauge is to the center of the sample, the higher the COD curve. It is obvious that the crack initiates at the disc center and propagates to two sides along the loading direction. The COD curves of gage E2 and E6 overlap because they are the same distance from the center. This signifies that the stress and deformation is symmetric in the dynamic Brazilian test.

The same tests are also performed on the granite disc samples with a frame rate of 100,000 fps. As typical for a brittle material, the dynamic deformation and failure process of the granite is similar to the Al_2O_3 ceramic. Based on the high speed photography technique and full field measurement by the DIC method of the dynamic Brazilian tests, the dynamic kinematic fields of the three tested brittle materials under the impact loading are calculated. The experiments show the stress wave propagation process in the disc samples. The COD curves also show that the crack initiates at the disc center. Based on elasticity theory, the maximum tensile load is located at the circle center, which is proven by these experimental tests.

4.5.3 Dynamic Deformation of Granite

The dynamic Brazilian disc test of granite samples is illustrated in Fig. 4.14. The region of interest (ROI) represents the observed area by HS camera. The frame rate is 100,000 fps with an image resolution of 256 × 128 pixels.

Figure 4.15 shows the tensile strain maps (normal strain in Y direction ε_y) at different moments after impact. The time interval is 10 μs and the time after impact is indicated in each subfigure. The strain concentration of granite disc sample illustrated the crack initiated and propagated from the disc center to side.

With the propagation of the crack, the failure of the sample and the crack is visible. The images in Fig. 4.16 shows the crack propagation of granite disc sample before its final failure. Furthermore, the evolution of crack is also analyzed by using the COD curves. As shown in Fig. 4.17a, 5 pairs of COD virtual gauges are taken:

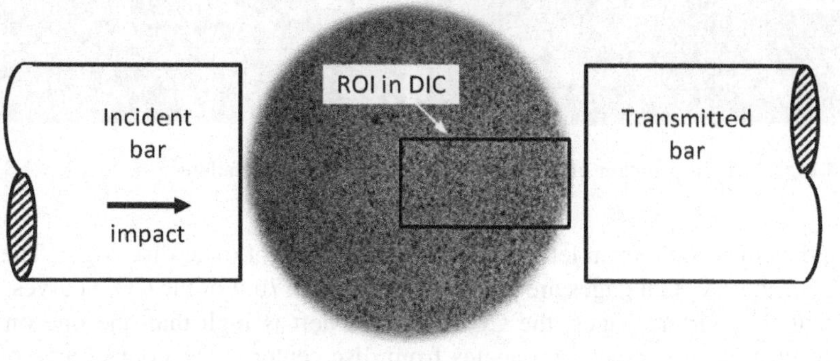

Fig. 4.14 ROI of granite disc sample during dynamic test

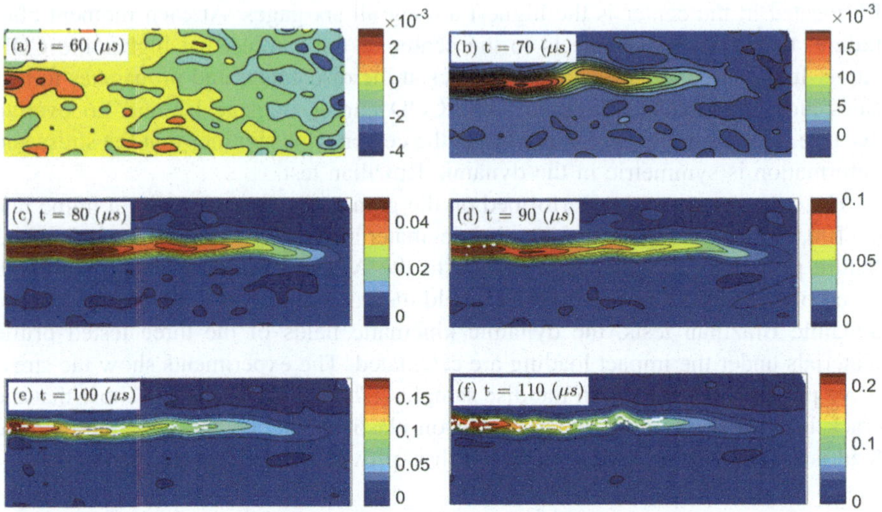

Fig. 4.15 Tensile strain maps at different moments after impact

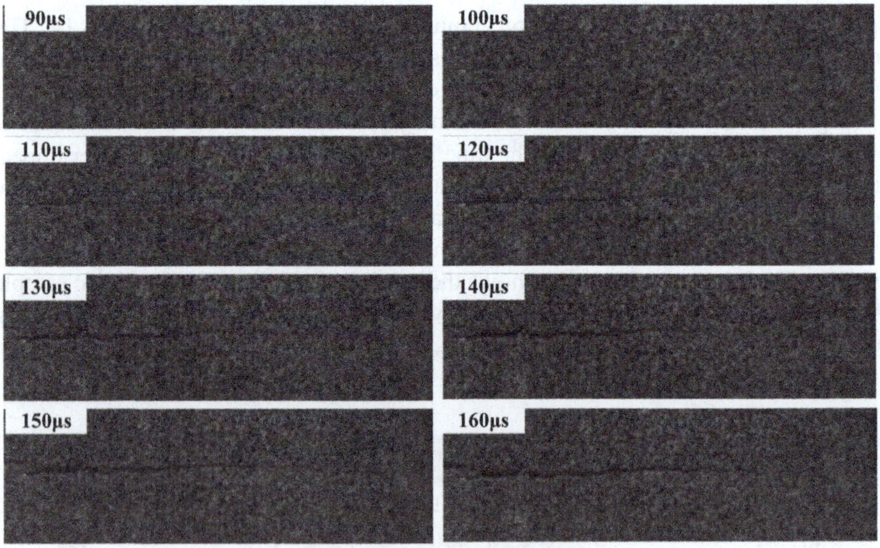

Fig. 4.16 Crack propagation of granite disc sample before final failure

along the crack route from left (near disc center) to right (near disc edge). The COD curves of the 5 virtual gages are presented in Fig. 4.17b. For the COD curves of any two neighbor virtual gages, the COD curve at left is high than the one on right, which illustrates the crack propagates from disc center to the contact area of load.

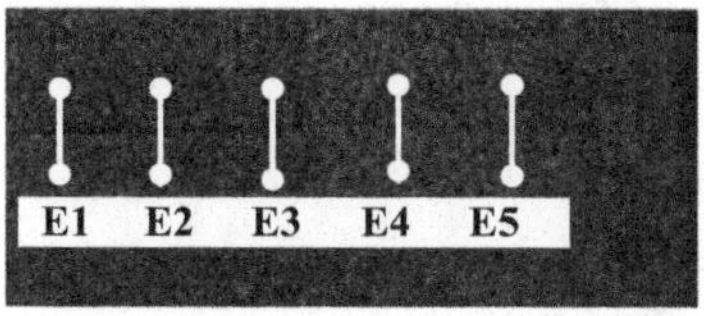

(a) COD virtual gauges along the crack

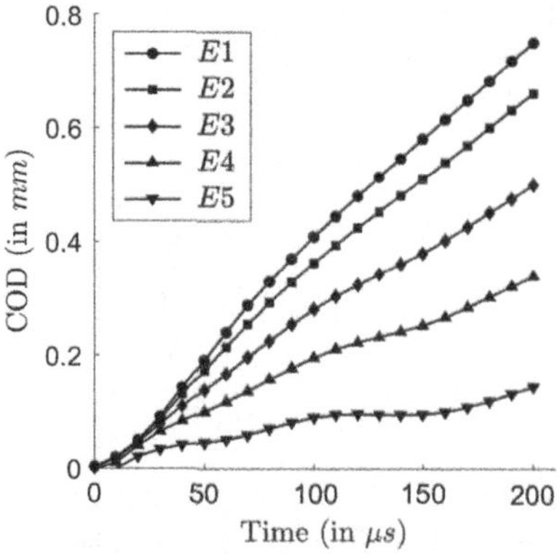

(b) COD curves

Fig. 4.17 Virtual gages along the crack route and their COD curves calculated by the DIC

4.6 Determination of the Dynamic Tensile Strength

The tensile strength of brittle materials is often much lower than the compressive strength. Due to the brittleness, a conventional tensile sample is difficult to prepare for brittle materials. The Brazilian test is an alternative method to determining the tensile strength of brittle materials, such as rocks and ceramics. In a conventional dynamic uniaxial compression or tension test, the stress and deformation are uniform; thus, the rate-related mechanical properties of the materials are often considered to be strain-rate dependent. However, in the dynamic Brazilian test, the stress and strain distributions are not uniform. Figure 4.18 shows a typical stress-time curve at the disc center in the dynamic Brazilian test, where the stress is the tensile one which is calculated by Eq. (4.3). A linear increment stage can be observed before the stress reaches a peak value, and the slope of the stage $k = \mathrm{d}\sigma/\mathrm{d}t$ is defined as the loading rate. Therefore, the rate-related strengths are studied by the loading-rate. In Fig. 4.18, a second stress peak can also be observed. This is because the load is supported by the two separated halves of the disc sample after splitting failure.

The tensile strain-stress curves of Al_2O_3 ceramic in two different stress rates are shown in Fig. 4.19. The curve represents the tensile stress and tensile strain in the sample center, where the strain values are calculated by DIC. The maxi stress value of the curve represents the dynamic tensile strength. By comparison of the two curves, it can be found that tensile strength increases with the increasing of stress rate.

Figure 4.20 shows the dynamic tensile strengths of the Al_2O_3 ceramic, granite and PMMA measured at various loading rates. The static tensile strengths of the three tested materials are given for comparison. The tensile strengths of the Al_2O_3 ceramic and granite measured by the static Brazilian test are 31.3 and 8.23 MPa, respectively. The PMMA shows a viscoelastic behavior in the static Brazilian test,

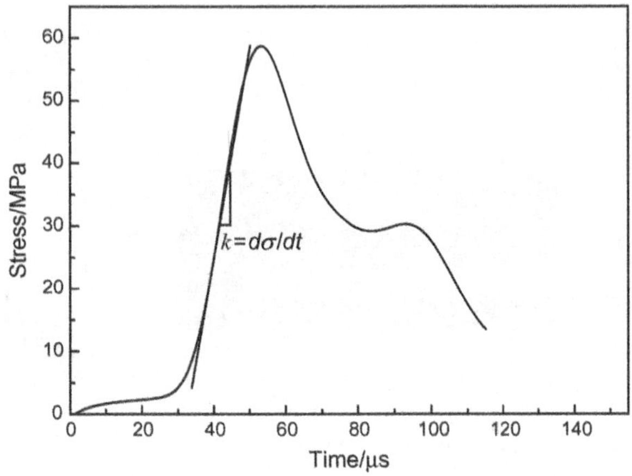

Fig. 4.18 Typical stress-time curve of the dynamic Brazilian test

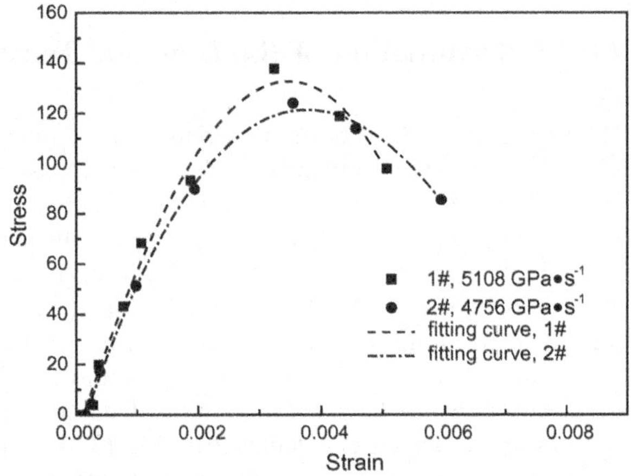

Fig. 4.19 Tensile strain-stress curves of Al_2O_3 ceramic

and splitting failure at the disc center cannot be observed. Therefore, the static tensile strength of the PMMA is not experimentally determined in this study. However a static tensile strength of 50 MPa for PMMA determined by the standard tensile test using a dumbbell sample [43] is given for comparison. Despite the difference among the experimental methods, the Brazilian test is an effective method to determine the dynamic tensile strength of brittle materials. It is shown that the dynamic tensile strengths are obviously higher than the static ones for the three tested materials. Strong loading rate dependence of the tensile strength can be found for the three materials. The dynamic tensile strength increases with the increase of loading rate. The quasi-linear relation can be fitted between the dynamic tensile strength versus the logarithm of loading rate for each material, as seen by the fitted dashed lines in Fig. 4.20.

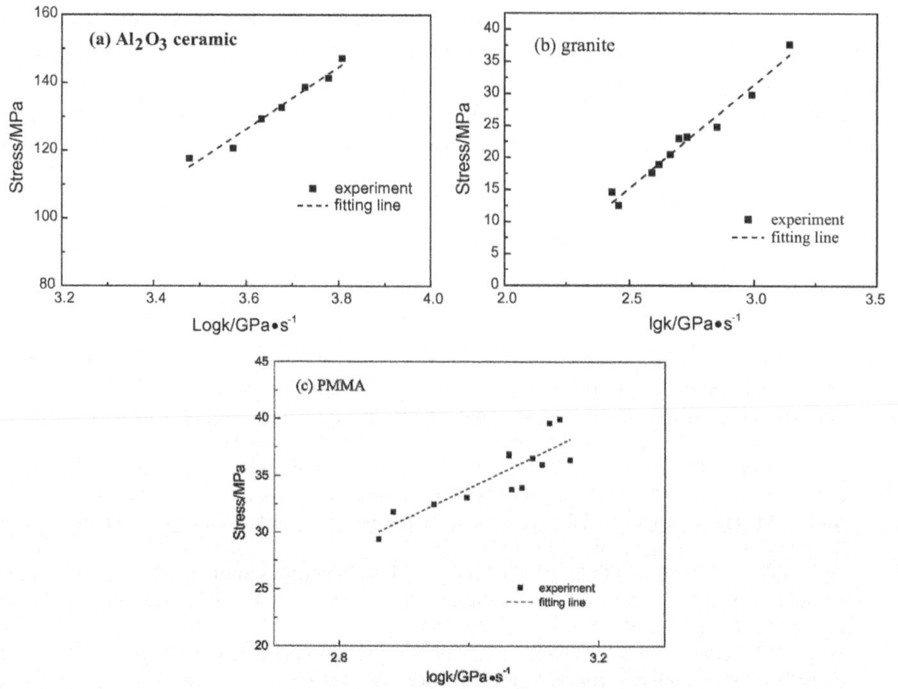

Fig. 4.20 Loading rate related dynamic tensile strengths of the three tested brittle materials

4.7 Concluding Remark and Perspective

The dynamic Brazilian test was performed to study the dynamic tensile properties of three brittle materials: Al_2O_3 ceramic, granite and PMMA. The split Hopkinson pressure bar was used to apply dynamic loading. The DIC was used to conduct a full field deformation analysis of the disc samples using images recorded by a high speed camera. Based on the DIC full field analysis and stress pulse measurement, stress equilibrium is satisfied in the dynamic Brazilian test of brittle materials. In the dynamic Brazilian test, the crack initiates at the disc center, propagates along the loading direction and finally causes the splitting failure of the disc sample. The dynamic Brazilian test is effective in evaluating the tensile properties of brittle materials. The tensile strengths of the Al_2O_3 ceramic, granite and PMMA under various loading rate were determined. Loading rate dependence of the tensile strength was observed for the three brittle materials.

The Kolsky-Hopkinson bar apparatus is acting as a most important dynamic loading device. Using the Hopkinson bar, the standard Brazilian test and its modified versions has been widely used for the research on the dynamic behaviors of brittle materials. The optical full-field technique and high speed photography equipment offer us an important opportunity in dynamic tests. However, it's limited by a lot of technical aspects. Many challenging technique details remain during the images acquisition by using HS and UHS cameras, such as the compromise between the frame rate and the image resolution, the powerful light source, the synchronization of cameras and flash light. The high speed full-field measurement in dynamic test will be a common method with the rapid evolution of the HS or UHS photography techniques.

References

1. SRM (1978) Suggested methods for determining tensile strength of rock materials. Int J Mech Min Sci Geomech Abs 15:99–103
2. ASTM (2004) C496/ C496 M-04: standard test method for splitting tensile strength of cylindrical concrete samples. ASTM International, West Conshohocken, USA
3. ASTM (2008) D 3967-08: standard test method for splitting tensile strength of intact rock core specimens. ASTM International, West Conshohocken, USA
4. Mellor M, Hawkes I (1971) Measurement of tensile strength by diametral compression of discs and annuli. Eng Geol 5:173–225
5. Wang QZ, Jia XM, Kou SQ (2004) The flattened Brazilian disc sample used for testing elastic modulus, tensile strength and fracture toughness of brittle rocks: analytical and numerical results. Int J Rock Mech Min Sci 41:245–253
6. Wang QZ, Xing L (1999) Determination of fracture toughness KIC by using the flattened Brazilian disk sample for rocks. Eng Fract Mech 64(2):193–201
7. Ye JH, Wu FQ, Sun JZ (2009) Estimation of the tensile elastic modulus using Brazilian disc by applying diametrically opposed concentrated loads. Int J Rock Mech Min Sci 46:568–576
8. Ye JH, Wu FQ, Zhang Y, Ji HG (2012) Estimation of the bi-modulus of materials through deformation measurement in a Brazilian disk test. Int J Rock Mech Min Sci 2:122–131
9. Johnstone C, Ruiz C (1995) Dynamic testing of ceramics under tensile stress. Int J Solids Struct 32:2647–2656
10. Zhao J, Li HB (2000) Experimental determination of dynamic tensile properties of a granite. Int J Rock Mech Min Sci 37(5):861–866
11. Proveti JRC, Michot G (2006) The Brazilian test: a tool for measuring the toughness of a material and its brittle to ductile transition. Int J Fract 139:455–460
12. Antonn, JR, Rajendran, AM (1992) Effect of strain rate and size on tensile strength of concrete. In: Proceedings of the topical conference on shock compression of condensed matter. Elsevier, Amsterdam, p 501
13. Dong SM, Xia KW, Huang S (2011) Tubing Yin. Rate dependence of the tensile and flexural strengths of glass–ceramic Macor. J Mater Sci 46:394–399
14. Yan F, Feng XT, Chen R (2012) Dynamic tensile failure of the rock interface between Tuff and Basalt. Rock Mech Rock Eng 45:341–348
15. Zhou YX, Xia K, Li XB (2012) Suggested methods for determining the dynamic strength parameters and mode-I fracture toughness of rock materials. Int J Rock Mech Min Sci 49:105–112
16. Hopkinson B (1914) A method of measuring the pressure produced in the detonation of high explosives or by the impact of bullets. Phil. Trans. Royal Soc. London A213:437–456
17. Kolsky H (1949) An investigation of the mechanical properties of materials at very high rates of loading. Proc Phys Soc Lond B62:676–700
18. Chen Weinong W, Song Bo (2011) Split Hopkinson (Kolsky) bar design, testing and applications. Springer, US

19. Huang S, Chen R, Xia KW (2010) Quantification of dynamic tensile parameters of rocks using a modified Kolsky tension bar apparatus. J Rock Mech Geotech Eng 2(2):162–168
20. Huang S, Xia K, Yan F, Feng X (2010) An experimental study of the rate dependence of tensile strength softening of Longyou sandstone. Rock Mech Rock Eng 43(6):677–683
21. Zhang QB, Zhao J (2014) A review of dynamic experimental techniques and mechanical behaviour of rock materials. Rock Mech Rock Eng 47(4):1411–1478
22. Field JE, Walley SM, Proud WG, Goldrein HT, Siviour CR (2004) Review of experimental techniques for high rate deformation and shock studies. Int J Impact Eng 30:725–775
23. Peters WH, Ranson WF (1982) Digital imaging techniques in experimental stress analysis. Opt Eng 21:427–431
24. Sutton M, Orteu J, Schreier H (2009) Image correlation for shape, motion and deformation measurements: basic concepts, theory and applications. Springer, Berlin
25. Pan B, Qian KM, Xie HM, Asundi A (2009) Two-dimensional digital image correlation for in-plane displacement and strain measurement: a review. Meas Sci Technol 20(6)
26. Liu C (2010) Elastic constants determination and deformation observation using Brazilian disk geometry. Exp Mech 50:1025–1039
27. Chen R, Dai F, Lu F (2013) Flattened Brazilian disc method for determining the dynamic tensile stress-strain curve of low strength brittle solids. Exp Mech 53:1153–1159
28. Grantham SG, Siviour CR, Proud WG (2004) High-strain rate Brazilian testing of an explosive simulant using speckle metrology. Meas Sci Technol 15:1867–1870
29. Zhou ZB, Chen PW, Huang FL (2012) Study on dynamic fracture and mechanical properties of a PBX simulant by using DIC and SHPB method. AIP Conf Proc 1426:665
30. Bhattacharya R, Goulbourne NC (2013) Deformation mechanisms in Mn + 1AXn phase ternary ceramics at high strain rates. In: Dynamic behavior of materials, vol 1. Springer, New York, pp 583–597
31. Zhang QB, Zhao J (2013) Determination of mechanical properties and full-field strain measurements of rock material under dynamic loads. Int J Rock Mech Min Sci 60:423–439
32. Pierron F, Sutton MA, Tiwari V (2011) Ultra high speed dic and virtual fields method analysis of a three point bending impact test on an aluminium bar. Exp Mech 51(4):537–563
33. Martin HS (2005) Elasticity theory, applications, and numerics. Elsevier Butterworth–Heinemann. ISBN 0-12-605811-3
34. Rastogi PK (2000) Photomechanics (Topics in applied physics). Springer, Berlin
35. Rastogi P (2015) Digital optical measurement: techniques and applications. Artech House Publishers, USA
36. Xia Kaiwen, Yao Wei (2015) Dynamic rock tests using split Hopkinson (Kolsky) bar system—a review. J Rock Mech Geotech Eng 7:27–59
37. Vic-2D Reference Manual, Correlated Solutions. www.correlatedsolutions.com
38. Zhou Zhongbin, Chen Pengwan, Huang Fenglei, Liu Siqi (2011) Experimental study on the micromechanical behavior of a PBX simulant using SEM and digital image correlation method. Opt Lasers Eng 49:366–370
39. Reu PL, Miller TJ (2008) The application of high-speed digital image correlation. J Strain Anal Eng Des 43:673–688
40. Pierron F (2013) Benchmarking ultra-high speed cameras for full-field deformation measurement. 21st DYMAT Technical Meeting. UK
41. Rossi M, Pierron F, Forquin P (2014) Assessment of the metrological performance of an in situ storage image sensor ultra-high speed camera for full-field deformation measurements. 2014 Meas Sci Technol 25:025401
42. Photron limited, http://www.photron.com/
43. Chen W, Lu F, Cheng M (2002) Tension and compression tests of two polymers under quasi-static and dynamic loading. Polym Test 21(2):113–121

Chapter 5
Comparative Study of the Dynamic Fracture Toughness Determination of Brittle Materials Using the Kolsky-Hopkinson Bar Machine

Pengwan Chen, Baoqiao Guo and Jingjing Chen

5.1 Introduction

Due to its high strength and hardness, high temperature resistance and low cost, alumina ceramics are widely used in a large range of applications such as armor systems, aerospace industry. In these cases, the ceramic materials inevitably subjected to dynamic loading, its dynamic mechanical properties consequently become the criterion. Fracture toughness is the key parameter in fracture mechanics, which defines a material's resistance to crack propagation for plain strain loading. Measuring this parameter requires knowledge of the specimen geometry and a preset crack within the material. Metals are ductile materials and have traditionally used pre-notch methods to grow a natural crack starter. However, it's difficult to machine a three-point bending (TPB) specimen for brittle materials because of low tensile strength.

The ceramics are typical brittle materials, of which the tensile strength is much lower than the compressive strength. Furthermore, the dynamic fracture toughness is loading rate related. It's even hard to measure the dynamic fracture toughness of ceramic materials. Because of the simplicity of the sample preparation and the experiment process, the Brazilian test has become a standard method and widely used in measuring the dynamic tensile strength of brittle materials such as rock, concrete and ceramic [1–3]. As the most common dynamic impact loading device, the split Hopkinson pressure bar (SHPB) technique [4] is widely used in mechanical study of brittle material [5–8].

P. Chen (✉) · B. Guo · J. Chen
State Key Laboratory of Explosion Science and Technology,
Beijing Institute of Technology, Beijing 100081, China
e-mail: pwchen@bit.edu.cn

R. Othman (ed.), *The Kolsky-Hopkinson Bar Machine*,
https://doi.org/10.1007/978-3-319-71919-1_5

Based on the standard Brazilian test, Wang et al. [9, 10] proposed the flatted Brazilian disc (FBD) test, by which the indirect tensile strength and the fracture toughness of brittle materials could be determined. Chong and Kuruppu also proposed the notched semi-circular bending (NSCB) test to measure the fracture toughness of brittle materials [11].

For the rocks, the dynamic fracture phenomena are frequently encountered in geophysical processes and engineering applications such as the earthquake, projectile penetration, rock blasts [12, 13]. As typical brittle materials, rocks are the mostly studied materials in both static and dynamic tests. Many testing standards on brittle materials mainly come of rocks [14–17].

The dynamic fracture toughness plays an important role in mechanical properties of alumina ceramics. Samborski presented the experimental results of static and dynamic fracture toughness of alumina, and analyzed the influence of porosity fraction on fracture toughness [18]. However, the loading rate effect on the fracture toughness was not investigated. Wang et al. [19] measured dynamic tensile strength of a brittle rock using FBD test, but the time to equilibrium and pattern of stress distribution for different FBD sample was given by finite element analysis. Xia et al. studied the loading rate dependence of fracture toughness on rock [20] and PMMA [21]. PMMA is typical polymer and it shows a non-linear ductile behavior in static case [22]. Due to the strain-rate effect, PMMA presents the brittle behavior from the dynamic deformation and failure in dynamic Brazilian tests [23]. Rittel and Maigre studied the dynamic fracture of PMMA using the so-called compact compression specimen (CCS) configuration [24], the results of dynamic fracture toughness present loading related dependency.

Recently, the digital image correlation (DIC) method has become widely applied in experimental mechanics as a practical and effective tool for full-field deformation measurement [25, 26]. Zhang and Zhao studied the loading rate related dynamic toughness of rocks by NSCB test [27]. The dynamic deformation and failure was also analyzed by the strains fields measured DIC. In the same way, Chen et al. measured the dynamic fracture toughness and failure behavior of explosive mock materials [28]. In the review researches on the dynamic tests on rocks [12, 13], the comprehensive details have been presented on the development of split Hopkinson bar techniques, high rate deformation measurement techniques, the samples geometrical configurations, and etc.

A comparative study using the TPB, NSCB and FBD tests was presented to determine the fracture toughness of a polymer-bonded explosive simulant [29]. The results of three different tests were in good agreement. But the tests were performed only in static, the stain rate effects were not performed.

In this study, the dynamic deformation and failure process, the dynamic fracture toughness on the alumina ceramic and PMMA was experimentally studied by using a SHPB apparatus and high speed DIC technique. The loading rate related dynamic fracture toughness was measured by using three test methods: FBD test, NSCB test and TPB test. The results were analyzed and the testing methods for testing dynamic properties of brittle materials was discussed.

5.2 Testing Principle and Experimental Setup

By using a SHPB apparatus, the dynamic tests were performed on two different materials: alumina ceramic and PMMA. The dynamic fracture toughness at different loading rates was measured by three test methods: the FBD test, the NSCB test and the TPB test. The FBD and NSCB samples are machined from the standard Brazilian disc. The disc dimensions made of alumina ceramic are Φ40 × 18 mm, and those of PMMA are Φ32 × 13 mm. The mechanical principles and the theoretical equations of these three testing methods are therefore introduced.

5.2.1 Flatted Brazilian Disc (FBD) Test

In standard Brazilian disc test, the stress concentration on the loading points may cause the pre-crack. As an alternative, Wang et al. [9, 10] proposed the FBD test. As shown in Fig. 5.1, two parallel planes are processed as the top and bottom surfaces of the disc sample. The pair of loads are applied on the two flat planes. According to Griffith's strength theory, the crack initiates in the center of the FBD sample in condition that the loading angle $2\alpha \geq 20°$, which is the critical prerequisite for the validation of the tensile strength and the fracture toughness. The tensile strength and fracture toughness can also be achieved by this proposed test. The tensile strength of FBD sample is expressed as:

$$\sigma_t = k\frac{2P_t}{\pi DB} \tag{5.1}$$

Fig. 5.1 Schematic of FBD test

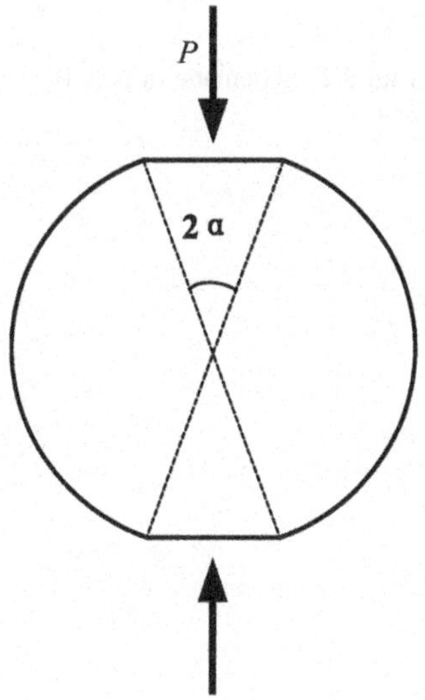

where P_t is the maxi load, D and B are the disc diameter and thickness respectively. The coefficient k, which mainly related to the loading angle α, can be determined by finite element analysis and in the case of standard Brazilian, the loading angle $2\alpha = 0°$, we get $k = 1$.

At the same time, the fracture toughness of FBD sample can be obtained by

$$K_{IC} = \frac{P_{cr}}{B\sqrt{D/2}}\Phi \tag{5.2}$$

where P_{cr} is on the local inflection point after the failure load P_t, which will be demonstrated in the subsequent experimental tests. Φ is a coefficients relevant to load angle. From the finite element analysis on the FBD sample by Wang et al. [10], the crack initiates in the sample center when the flat angle is under the condition $20° \leq 2\alpha \leq 30°$. Two coefficients k and Φ can also be achieved by from the numerical analysis. In this study, the loading angle is $2\alpha = 30°$, so we get $k = 0.9205, \Phi = 0.5895$.

5.2.2 Notched Semi-circular Brazilian (NSCB) Test

Proposed by Chong and Kuruppu [11], the fracture toughness of brittle materials such as rock was studied by the NSCB test. Compared to the TPB test, the NSCB sample is compact. That's great advantage for sample preparation and also the dynamic test with SHPB device. As shown in Fig. 5.2, the NSCB sample with a pre-fabricated crack is supported by two brackets, and the load is applied in the middle of sample. The fracture toughness of NSCB can be determined by

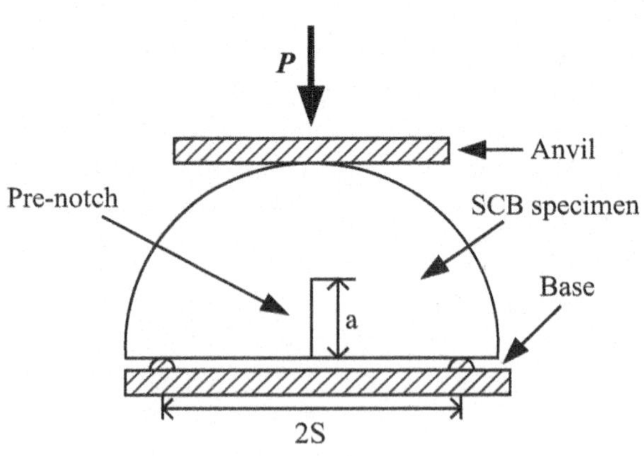

Fig. 5.2 Schematic of NSCB test

$$K_{IC} = \frac{P_Q\sqrt{\pi a}}{DB} Y_K(x) \tag{5.3}$$

$$Y_K(x) = 4.47 + 7.4\frac{a}{D} - 106(\frac{a}{D})^2 + 433.3(\frac{a}{D})^3, \quad \text{where} \quad x = \frac{a}{D}, \frac{2S}{D} = 0.8, \tag{5.4}$$

where P_Q is critical damage load, D and B are the diameter and thickness of sample, α is the pre-fabricated crack length, Y_K is a function of a/D and $2S$ (the bending span). Equations 5.3 and 5.4 are valid only when the conditions are satisfied: $0.25 \leq 2\alpha \leq 0.35$ and $2S/D = 0.8$. The pre-fabricated crack length of NSCB samples made of alumina ceramic is $a = 12$ mm. For the NSCB samples of PMMA, $a = 9.6$ mm.

5.2.3 *Three-Point Bending (TPB) Test*

The three-point bending test is common method to determine the fracture toughness of metal materials, as illustrated in Fig. 5.3. The fracture toughness K_{IC} of the long rectangular sample with a pre-crack can be determined by:

$$K_{IC} = \frac{P_q \cdot S}{B \cdot W^{3/2}} f(x) \tag{5.5}$$

$$f(x) = \frac{3x^{1/2}[1.99 - 2.15x(1-x) - 3.93x + 2.7x^2]}{2(1+x)(1-x)^{3/2}}, \quad x = \frac{a}{W} \tag{5.6}$$

where $B \times W \times S$ are the thickness, height and the length between the two supports, a is the length of pre-crack, and P_Q is the failure load. In the current study, the dimensions of TPB samples both for alumina ceramic and PMMA are $B \times W \times S = 5\ \text{mm} \times 10\ \text{mm} \times 30\ \text{mm}$, and the length of pre-crack is $a = 2$ mm.

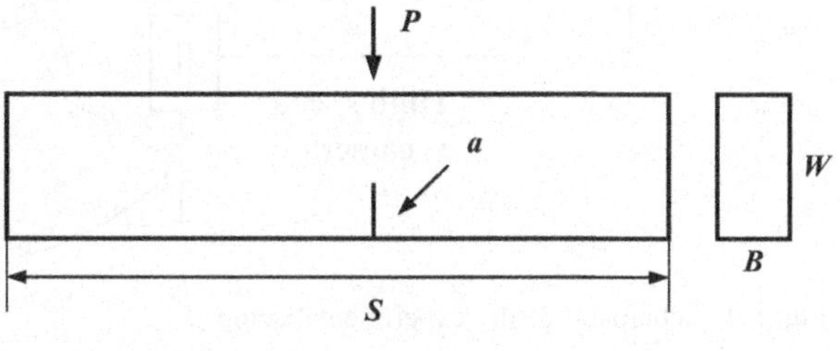

Fig. 5.3 Schematic of TPB test

5.2.4 Sample Preparation and Experimental Setup

The samples made of fine alumina ceramics powder, were prepared by cold compression molding and then sintered under very high temperature. The relative density is about 94.3%. The diameter and thickness of disc-like samples are Φ40 × 16 mm. The FBD and NSCB samples were machined from the original disc samples. The diameters of those two samples are the same with the original disc sample. The flat angle of the flat Brazilian disc is 30°.

In this work, a SHPB system was used as the dynamic loading device. The samples were placed between the incident bar and the transmitted bar. The SHPB apparatus dynamic testing system with the geometrical configuration of three different samples is illustrated in Fig. 5.4. The strain signals from the strain gages attached on the incident and transmitted bars, were recorded by the dynamic strain indicator. a high speed (HS) camera (type Photron Fastcam SA5) was used to measure the dynamic displacement and strain fields.

The HS camera and dynamic strain indicator are simultaneously triggered by the shifting signal generated from the velocity meter. The digital images and stress wave signals are then recorded simultaneously by the high speed camera and dynamic strain indicator. The typical incident, reflected and transmitted pulses recorded in the Brazilian test of the alumina ceramic are shown in Fig. 5.5.

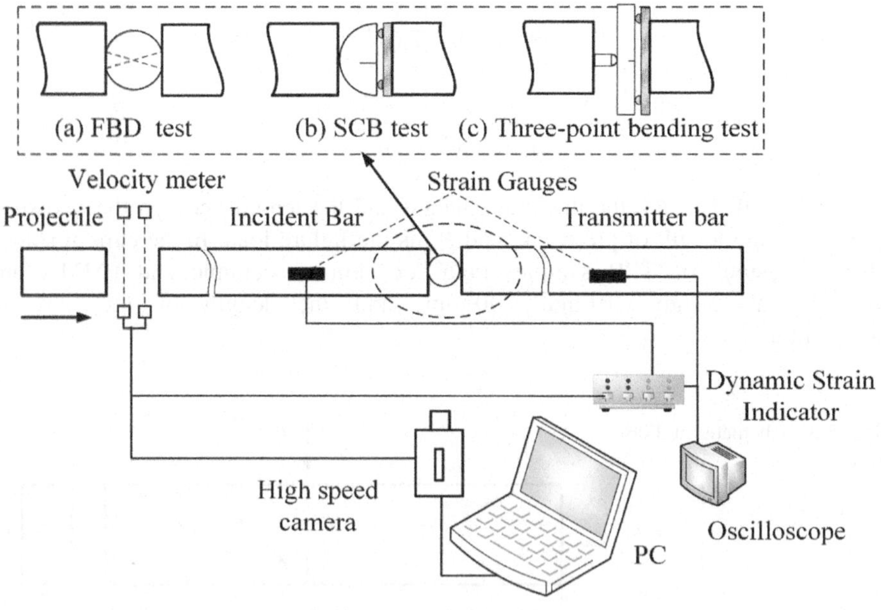

Fig. 5.4 Schematic of the experimental setup

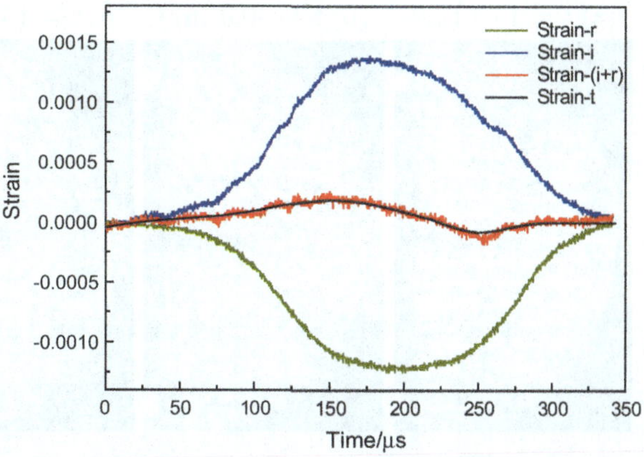

Fig. 5.5 Typical incident, reflected, and transmitted strain pulses of alumina ceramic NSCB sample

5.3 Dynamic Deformation and Failure Analysis

From the pulses measured by the strain gage technique, the stress-strain curve can be deduced. Form the above testing system, the dynamic kinematic fields can be calculated from the images captured by HS camera. This proves a quantitative and evidence information for the analysis on the dynamic deformation and the fracture procedure of the tested materials. In this section, the results only on alumina ceramic will be presented. The deformation and failure mode are found similar on PMMA. The results on PMMA will not be redundantly presented.

The displacement vector fields of NSCB sample at different moments are shown in Fig. 5.6. The frame rate is 75,000 fps, and the image resolution is 320 × 264 pixels. The results show that the local tensile strain concentration appears obviously at pre-fabricated crack tip at 40.0 μs, and the crack propagates till finally the sample is split into two parts, which rotate oppositely around the left point contacted by the incident bar. The Displacement vector fields are illustrated in Fig. 5.7. At 160 μs, it's obvious the two parts are rotating oppositely.

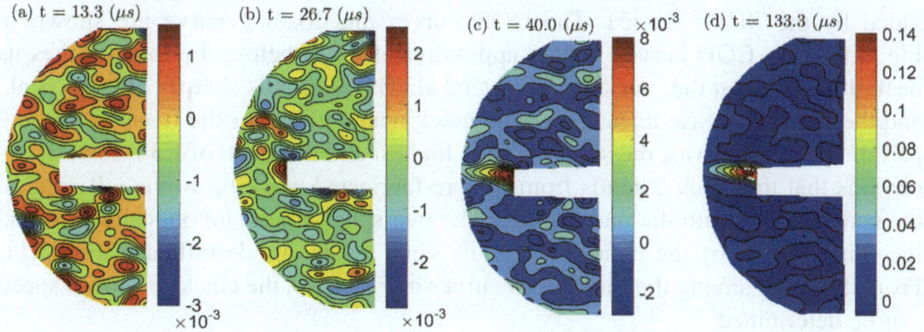

Fig. 5.6 Tensile strain fields of NSCB sample at different moments

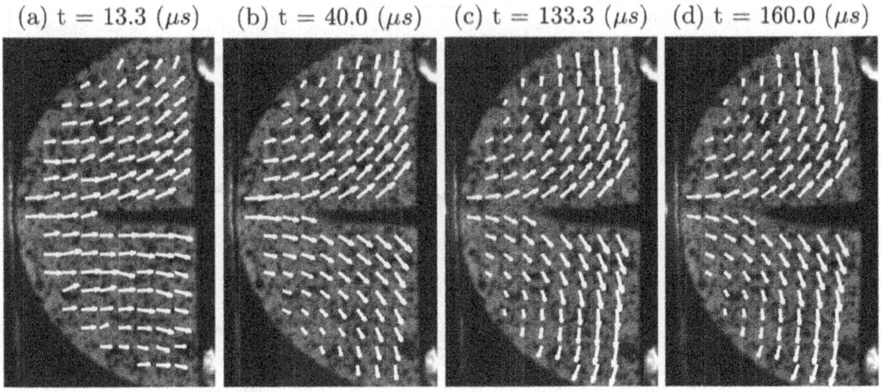

Fig. 5.7 Displacement vector fields of NSCB sample at different moments

From the above NSCB tests, the images on nearly whole surface are recorded and the frame rate is less than 100,000 fps. The deformation fields are calculated by DIC on whole surface of the samples. However, the frame rate is not enough high, only the main crack steps are observed and analyzed. It's difficult to study the crack propagation process of brittle materials from the raw images captured by high-speed camera. In order to observe the details on the crack propagation, the frame rate should be higher, while the image resolution will be decreased. Instead on the whole surface of sample, the deformation fields only on the local area of crack tip are measured. This can improve the displacement resolution. By the combined high-speed DIC-SHPB system, the crack propagation process of alumina ceramic under dynamic loading is therefore studied.

For the NSCB test, the images of resolution 128 $\times$ 128 pixels on the local zone of crack tip were taken at frame rate 210,000 fps. The time interval between two images is 4.76 μs. As shown in Fig. 5.8a, three pairs of points on both sides of the crack extension path are chosen, which are indicated by **E1**, **E2** and **E3**. From the displacement fields measured by DIC technique during the dynamic deformation process, the COD (crack opening displacement) curves in function of time of between each pair of points can be extracted. The dynamic loading curve is provided by the SHPB device. The COD curves and loading curve are shown in Fig. 5.8b. The COD curves remain approximately zero before the load reaches its peak. It shows that the crack hasn't extended. The load falls sharply after its peak, and the CODs increase linearly in rapid crack open phase. For the linear increasing part of the COD curves, the slope of **E1** is highest and the slope of **E3** is lowest. It's obvious that the crack extends from the pre-fabricated crack tip to the left side till the contact point with the incident bar. The two split parts rotate oppositely around the contact point by the incident bar. The same results are described in Fig. 5.11. From the delay among the three COD curves in Fig. 5.8b, the crack extension speed can be determined.

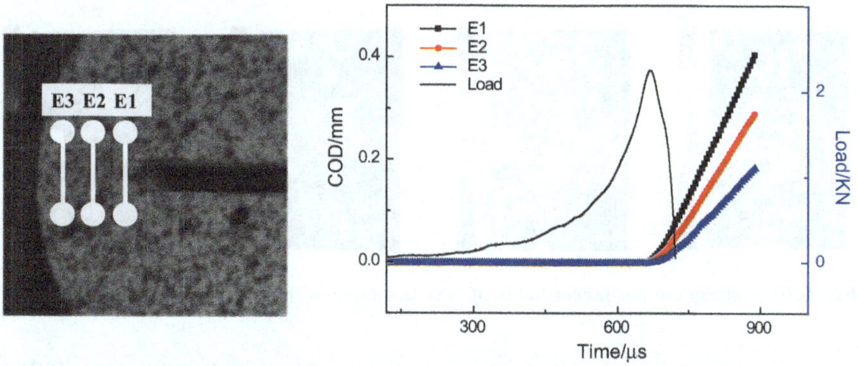

(a) Crack extension in NSCB sample (b) COD curves and loading curve

Fig. 5.8 Crack propagation of NSCB sample under dynamic load, **a** Crack extension in NSCB sample **b** COD curves and loading curve

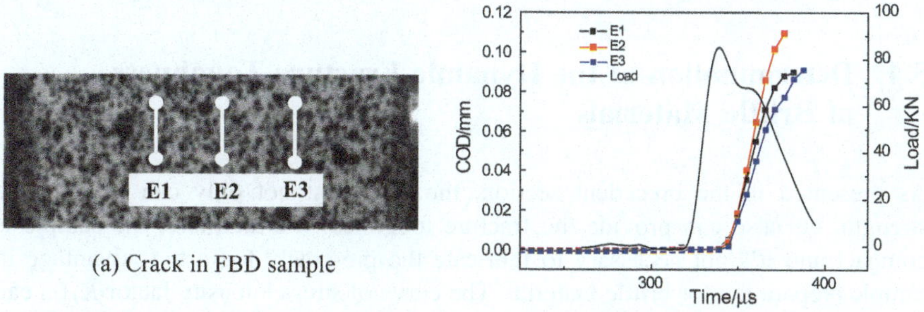

(a) Crack in FBD sample (b) COD curves and loading curve

Fig. 5.9 Crack propagation of FBD sample under dynamic load, **a** crack in FBD sample **b** COD curves and loading curve

In the same way, for alumina ceramic FBD sample, the COD curves on three different locations across the loading diameter are analyzed. As illustrated in Fig. 5.9a, E2 is in the center of the circular sample, E1 and E3 are on its both sides. The frame rate is 100,000 fps. The COD curves as well as the loading curve are shown in Fig. 5.9b. Similar to the NSCB test, the COD curves rise after the maximum loading value, it indicates that the crack begins to extend when the load exceeds the strength of material. The distances of three locations are quite close, there is little different among the COD curves. However, by theoretic analysis, the strain in the center of sample should be maximal. This is confirmed by the COD curves in Fig. 5.9b. The crack initiates in the center (location **E2**) and extends on both sides (location **E1 and E3**) under the dynamic load.

The correlation coefficient presents the matching level between the undeformed and deformed images. The mismatching in local area of the tow images makes the

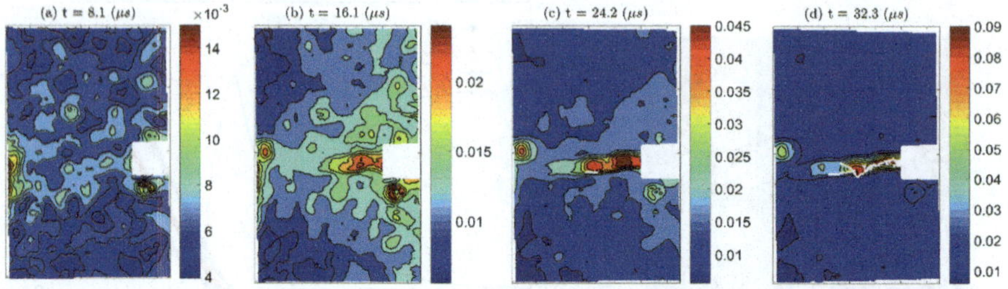

Fig. 5.10 Correlation coefficient fields of NSCB sample at different moments

correlation coefficient map becomes a good indicator of crack and fracture [30]. The correlation coefficient fields of TPB sample at different moments during the dynamic fracture procedure are presented in Fig. 5.10. The image size is 192 × 224 pixels at 124,000 fps. From the contour plots of the correlation coefficient, the crack initiation and propagation can be detected.

5.4 Determination of the Dynamic Fracture Toughness of Brittle Materials

As presented in the precedent section, the FBD test not only can achieve the strength, but also can provide the fracture toughness. Furthermore, the sample is compact and it's not necessary to fabricate the pre-crack. It's a big advantage in sample preparation for brittle material. The curve of stress intensity factor $K_I(t)$ can be obtained according to the Eq. (5.2), as shown in Fig. 5.11. After the collision and compaction process at initial contact, the stress intensity factor $K_I(t)$ curve changes over time with the load increases. There is a linear phase before the peak in the $K_I(t)$ curve, the slope of the linear phase is the dynamic loading rate $\mathrm{d}K_I/\mathrm{d}t$. Because the section of circular samples along the loading direction is not constant, the loading rate instead of the strain rate is presented in this study. The damage occurs when the load reaches the maximum; the cracks appeared in the center of the disc, and growth with the loading increase. And the crack went throughout the disc only when the load down decreased to local inflection load. Therefore, P_{cr} at inflection point represents the load of fracture toughness.

Based on the previous analysis, the crack initiate at the maximum load, and crack penetrate at the local minimum load P_{cr}. The time interval between the maximum load and P_{cr} (Δt in Fig. 5.11) can be determined. As the radius of the disc is known, the crack propagation velocity of alumina ceramic can be determined in the FBD test. The velocity is determined about 2300 ± 200 m/s.

For the NSCB test with the pre-fabricated crack, the fracture toughness of alumina ceramics can be calculated with relevant parameters according to Eqs. (5.3) and (5.4). The dynamic loading rate calculation method of NSCB test is the same

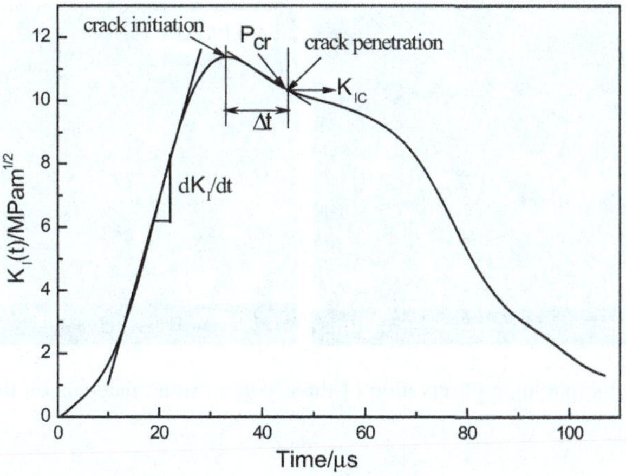

Fig. 5.11 Typical stress intensity factor curve of FBD test

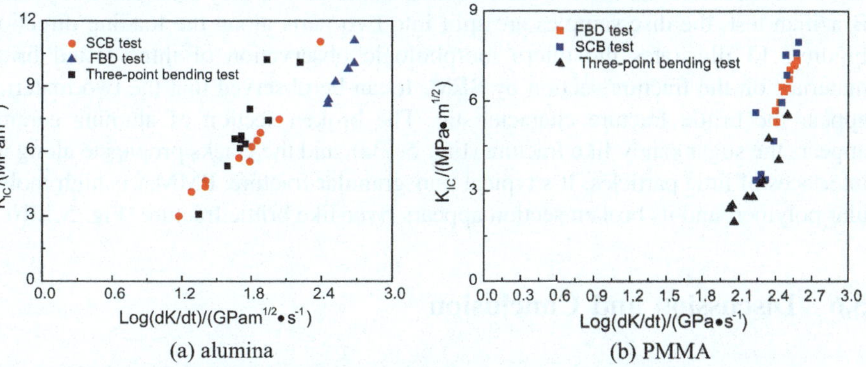

Fig. 5.12 Dynamic fracture toughness in function of strain rate by three test methods

with the FBD test. The relationship between fracture toughness of alumina ceramics with two densities measured by TPB test, FBD test and NSCB test and loading rate is shown in Fig. 5.12a. The fracture toughness measured by three methods are increased with the loading rate, and the three methods have good consistency. The loading rate dependence of alumina ceramics fracture toughness is presented by Kishi [31], it is consistent with the result of in this work. The results of dynamic fracture toughness of PMMA measured by three test methods are presented in Fig. 5.12b. The results are the obvious loading rate related, which is similar as the alumina ceramic. The effectiveness of FBD test to measure fracture toughness of the brittle material was verified by comparing those three methods. The FBD test has certain advantages in the aspect of toughness measurement for brittle material fracture because of sample preparation is simple and it does not need to prefabricate crack in the sample.

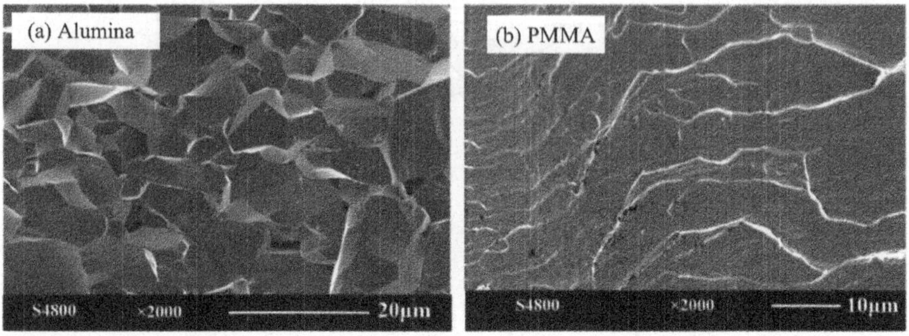

Fig. 5.13 Micro morphologic observation of three tested brittle materials on the fracture section by SEM

5.5 Micro Morphologic Analysis on the Fracture

The microstructures of the two tested brittle materials are very different. In dynamic Brazilian test, the disc samples are split into two parts along the loading direction. Figure 5.13 illustrates the micro morphologic observation of three tested brittle materials on the fracture section by SEM. It can be observed that the two materials appear the brittle fracture characteristic. The broken section of alumina ceramic appears the sugar candy-like fracture (Fig. 5.13a), and the cracks propagate along the interfaces of little particles. It's typical transgranular fracture. PMMA is high molecular polymer, and its broken section appears river-like brittle fracture (Fig. 5.13b).

5.6 Discussion and Conclusion

In this study, we made a comparative study of the dynamic fracture toughness of alumina ceramic by three test methods: FBD test, NSCB test and TPB test. The dynamic fracture toughness of alumina ceramic and PMMA exhibits an obvious rate dependence upon the loading rate. The result shows the reliability of the three methods to determine the dynamic fracture toughness of brittle materials. However, the FBD samples are not necessary to fabricate a pre-crack. In addition, the FBD test can also measure the dynamic tensile strength. Considering these advantage, the FBD test is recommended as an alternative method for measuring the dynamic fracture toughness of brittle materials.

References

1. Johnstone C, Ruiz C (1995) Dynamic testing of ceramics under tensile stress. Int J Solids Struct 32(17/18):2647–2656
2. Zhao J, Li HB (2000) Experimental determination of dynamic tensile properties of a granite. Int J Rock Mech Min Sci 37(5):861–866

3. Rafael J, Proveti C, Michot G (2006) The Brazilian test: a tool for measuring the toughness of a material and its brittle to ductile transition. Int J Fract 139:455–460
4. Chen WW, Song B (2011) Split Hopkinson (Kolsky) bar design, testing and applications. Springer, USA
5. Zhu J, Shisheng H, Wang L (2009) An analysis of stress uniformity for concrete—like specimens during SHPB tests. Int J Impact Eng 36:61–72
6. Wang S, Liu KX (2011) Experimental research on dynamic mechanical properties of PZT ceramic under hydrostatic pressure. Mater Sci Eng, A 528:6463–6468
7. Chang H, Binner J, Higginson R, Myers P, Webb P, King G (2011) High strain rate characteristics of 3-3 metal-ceramic interpenetrating composites. Mater Sci Eng, A 528:2239–2245
8. Dong S, Xia K, Huang S, Yin T (2011) Rate dependence of the tensile and flexural strengths of glass-ceramic Macor. J Mater Sci 46:394–399
9. Wang QZ, Xing L (1999) Determination of fracture toughness KIC by using the flattened Brazilian disk specimen for rocks. Eng Fract Mech 64:193–201
10. Wang QZ, Jia XM, Kou SQ et al (2004) The flattened Brazilian disc specimen used for testing elastic modulus, tensile strength and fracture toughness of brittle rocks: analytical and numerical results. Int J Rock Mech Min Sci 41:245–253
11. Chong KP, Kuruppu MD (1984) New specimen for fracture toughness determination for rock and other materials. Int J Fract 26:59–62
12. Xia K, Yao W (2015) Dynamic rock tests using split Hopkinson (Kolsky) bar system—a review. J Rock Mech Geotech Eng 7:27–59
13. Zhang QB, Zhao J (2014) A review of dynamic experimental techniques and mechanical behaviour of rock materials. Rock Mech Rock Eng 47(4):1411–1478
14. ASTM (2004) C496/C496 M-04, standard test method for splitting tensile strength of cylindrical concrete samples. ASTM International, West Conshohocken, USA
15. ASTM (2008) D 3967-08: standard test method for splitting tensile strength of intact rock core specimens. ASTM International, West Conshohocken, USA
16. ISRM Testing Commission (1988) Suggested methods for determining the fracture toughness of rock. Int J Rock Mech Min Sci Geomech Abstr 25(2):71–96
17. ISRM Testing Commission (1995) Suggested method for determining mode I fracture toughness using cracked chevron notched Brazilian disc (CCNBD) specimens. Int J Rock Mech Min Sci Geomech Abstr 32(1):57–64
18. Samborski S, Sadowskiw T (2010) Dynamic fracture toughness of porous ceramics. J Am Ceram Soc 93(11):3607–3609
19. Wang QZ, Li W, Xie HP (2009) Dynamic split tensile test of flattened Brazilian disc of rock with SHPB setup. Mech Mater 41:252–260
20. Dai F, Xia K, Zheng H, Wang YX (2011) Determination of dynamic rock mode-I fracture parameters using cracked chevron notched semi-circular bend specimen. Eng Fract Mech 78:2633–2644
21. Huang S, Luo S, Xia K (2009) Dynamic fracture initiation toughness and propagation toughness of PMMA. Proceedings of the SEM Annual Conference June 1–4, 2009
22. Liu C (2010) Elastic constants determination and deformation observation using Brazilian disk geometry. Exp Mech 50:1025–1039
23. Chen J, Guo B, Liu H, Liu H, Chen P (2014) Dynamic Brazilian test of brittle materials using the split Hopkinson pressure bar and digital image correlation. Strain 50(6):563–570
24. Rittel D, Maigre H (1996) An investigation of dynamic crack initiation in PMMA. Mech Mater 23:229–239
25. Pan B, Qian KM, Xie HM, Asundi A (2009) Two-dimensional digital image correlation for in-plane displacement and strain measurement: a review. Meas Sci Technol 20(6):062001–062007
26. Sutton M, Orteu J, Schreier H (2009) Image correlation for shape, motion and deformation measurements: basic concepts, theory and applications. Springer, Berlin

27. Zhang QB, Zhao J (2013) Determination of mechanical properties and full-field strain measurements of rock material under dynamic loads. Int J Rock Mech Min Sci 60:423–439
28. C P, Z Zhongbin, Ma S et al (2011) Measurement of dynamic fracture toughness and failure behavior for explosive mock materials. Front Mech Eng 6(3):292–295
29. Zhou Z, Chen P, Duan Z, Huang F (2011) Comparative study of the fracture toughness determination of a polymer-bonded explosive simulant. Eng Fract Mech 78:2991–2997
30. Zhou Z, Chen P, Huang F, Liu S (2011) Experimental study on the micromechanical behavior of a PBX simulant using SEM and digital image correlation method. Opt Lasers Eng 49:366–370
31. Kishi T (1991) Dynamic fracture toughness in ceramics and ceramics matrix composites. Eng Fract Mech 40(415):785–790

Chapter 6
Wave Dispersion in Kolsky-Hopkinson Bar Machine

Ramzi Othman

6.1 Introduction

This chapter is interested in the wave dispersion effects in Kolsky-Hopkinson bar machine. First, Sect. 6.2 introduces how to assess wave dispersion parameters in three-dimensional elastic and viscoelastic bars using analytical analysis. Subsequently, Sect. 6.3 introduces experimental techniques used to measure the wave dispersion.

6.2 Analytical Analysis

6.2.1 Pochhammer-Chree Equation for an Infinite Viscoelastic Rod

The bars used in Kolsky-Hopkinson bar machines are long. Typically, they are 1 to 3-m long and few cm in diameter. The first machines used metallic (high-strength steel) bars. The stress wave propagation in bars can then be approached by studying the wave propagation in infinite elastic bars. Later polymeric bars have been introduced, for which it is more appropriate to study the wave propagation in infinite viscoelastic bars. In this section, the latter case is considered. The viscoelasticity of the bar is taken into consideration through the use of a complex stiffness of the bar. The particular elastic bar situation can be taken into account by considering the stiffness modulus as a real number. The cross-section of the bar is

R. Othman (✉)
Mechanical Engineering Department, Faculty Engineering,
King Abdulaziz University, P.O. Box, 80248, Jeddah 21589, Saudi Arabia
e-mail: rothman1@kau.edu.sa

R. Othman (ed.), *The Kolsky-Hopkinson Bar Machine*,
https://doi.org/10.1007/978-3-319-71919-1_6

here assumed to be circular (Fig. 6.1). A given point P of the bar is then represented by its cylindrical coordinates (r, θ, z). We aim at deriving the displacement and stress components at point P.

As material behavior is assumed to be elastic or viscoelastic, the problem to be solved is linear. Consequently, the displacement field is searched as a superposition of harmonic solutions. To this goal, Fourier transforms are used here. Let f be a function. Its Fourier transform $\tilde{f}$ is defined by:

$$\tilde{f}(\omega) = \int_{-\infty}^{+\infty} f(t)\mathrm{e}^{-i\omega t}dt. \quad (6.1)$$

The original function can then be recovered knowing its Fourier transform as follows:

$$f(t) = \frac{1}{2\pi}\int_{-\infty}^{+\infty} \tilde{f}(\omega)\mathrm{e}^{i\omega t}dt. \quad (6.2)$$

Within this framework, the displacement field $\vec{u}(\vec{X},t)$, the strain tensor $\boldsymbol{\varepsilon}(\vec{X},t)$ and the stress tensor $\boldsymbol{\sigma}(\vec{X},t)$ are recovered in terms of their Fourier transform in the following way:

$$\vec{u}(\vec{X},t) = \frac{1}{2\pi}\int_{-\infty}^{+\infty} \tilde{u}(\vec{X},\omega)\mathrm{e}^{i\omega t}dt, \quad (6.3)$$

$$\boldsymbol{\varepsilon}(\vec{X},t) = \frac{1}{2\pi}\int_{-\infty}^{+\infty} \tilde{\boldsymbol{\varepsilon}}(\vec{X},\omega)\mathrm{e}^{i\omega t}dt, \quad (6.4)$$

and

$$\boldsymbol{\sigma}(\vec{X},t) = \frac{1}{2\pi}\int_{-\infty}^{+\infty} \tilde{\boldsymbol{\sigma}}(\vec{X},\omega)\mathrm{e}^{i\omega t}dt, \quad (6.5)$$

respectively. In the above equations $\vec{X}(r,\theta,z)$ holds for the space position and t for the time instant.

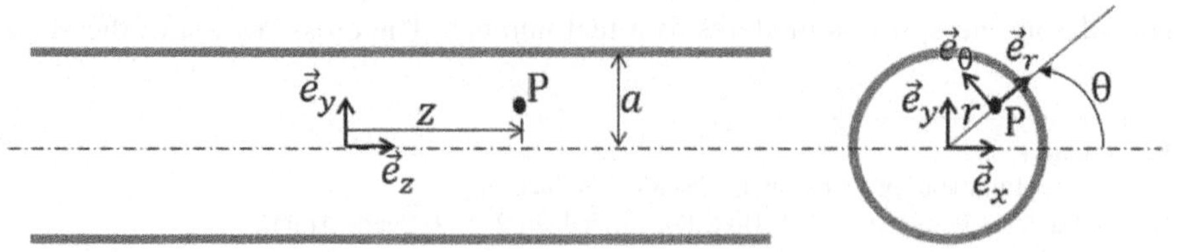

Fig. 6.1 Infinite bar of circular cross-section

Neglecting volume forces, the second Newton's law yields:

$$\overrightarrow{\mathrm{div}}\,\tilde{\boldsymbol{\sigma}} = -\rho\omega^2\tilde{u}, \tag{6.6}$$

where ρ is the bar density and $\overrightarrow{\mathrm{div}}$ is the divergence operator.

As linear visco-elastic waves induce small deformation, the strain tensor can be expressed as:

$$\tilde{\boldsymbol{\varepsilon}} = \frac{1}{2}(\mathbf{grad}\,\tilde{u} + {}^{t}\mathbf{grad}\,\tilde{u}), \tag{6.7}$$

where **grad** is the gradient operator and ^{t} defines the transpose of a tensor.

The Pochhammer-Chree solutions were originally developed for infinite elastic rods [1, 2]. Later they were extended to account for viscoelasticity of polymeric materials by Zhao and Gary [3]. In this section, the equations are derived assuming a viscoelastic behavior of rods. The elastic behavior is only a particular case of the more general viscoelastic case. The viscoelastic constitutive equation is written as follows [4, 5]:

$$\tilde{\boldsymbol{\sigma}}(\omega) = \lambda^*(\omega)tr(\tilde{\boldsymbol{\varepsilon}}(\omega))\boldsymbol{\delta} + 2\mu^*(\omega)\tilde{\boldsymbol{\varepsilon}}(\omega), \tag{6.8}$$

where $\lambda^*(\omega)$ and $\mu^*(\omega)$ are the complex Lamé coefficients, $\boldsymbol{\delta}$ is the Kronecker tensor, and *tr*() is the trace operator. The complex Lamé coefficients are used to account for the viscoelastic behavior. They are given by:

$$\lambda^*(\omega) = \frac{vE^*(\omega)}{(1+v)(1-2v)}, \tag{6.9}$$

and

$$\mu^*(\omega) = \frac{E^*(\omega)}{2(1+v)}, \tag{6.10}$$

where $E^*(\omega)$ is the complex Young's modulus and v is Poisson ratio. $E^*(\omega)$ accounts here for both the storage and loss moduli and reads:

$$E^*(\omega) = E_1(\omega) + i\,E_2(\omega), \tag{6.11}$$

where $E_1(\omega)$ and $E_2(\omega)$ are two frequency-dependent functions and $i^2 = -1$. $E^*(\omega)$ can be experimentally characterized using DMA (dynamic mechanical analysis) or using wave propagation techniques [6–8]. Its frequence dependency can be written in terms of rheological models like the generalized Maxwell model (GMM) (Fig. 6.2).

The Poisson ratio v was considered real and frequency-independent [3]. Actually, the Poisson ratio is complex and frequency-dependent as argued by Pritz [10–12].

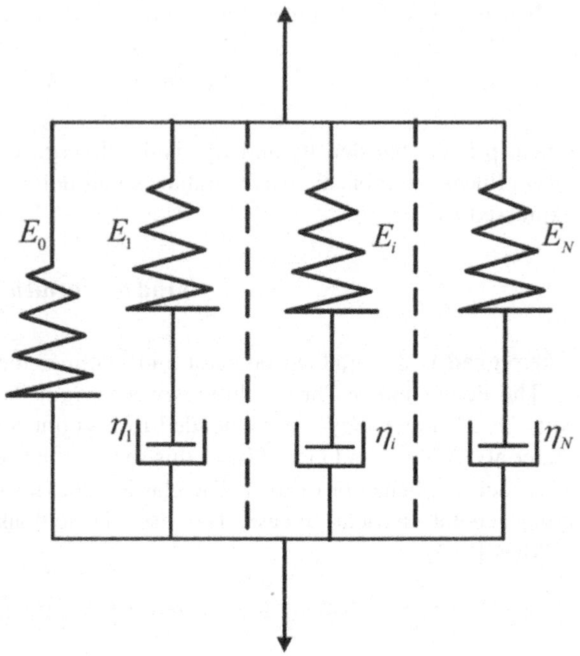

Fig. 6.2 A generalized Maxwell rheological model with N Maxwell units (reproduced from Ref. [9] with permissions from Elsevier)

However, there is a shortage of data about the dependency of this parameter in terms of the frequency as it is highly difficult to assess it. For this reason the frequence dependency of the Poisson ratio is neglected [3, 9].

Substituting Eqs. (6.7) and (6.8) in Eq. (6.6) yields [13–15]:

$$(\lambda^*(\omega) + \mu^*(\omega))\overrightarrow{\text{grad}}\left(\overrightarrow{\text{grad}}.\tilde{u}\right) - \mu^*(\omega)\overrightarrow{\text{grad}} \times \overrightarrow{\text{grad}} \times \tilde{u} = -\rho\omega^2\tilde{u}, \tag{6.12}$$

where $\cdot$ stand for the scalar product and $\times$ writes the cross product.

To solve Eq. (6.12), the displacement field is written in terms of Helmoltz potentials, i.e.,

$$\tilde{u} = \overrightarrow{\text{grad}}\left(\tilde{\Phi}\right) + \overrightarrow{\text{rot}}\left(\tilde{\Psi}\right), \tag{6.13}$$

where $\tilde{\Phi}$ is the scalar Helmoltz potential and $\tilde{\Psi}$ is the vector Helmoltz portential.

Substituting Eqs. (6.13) into (6.12) yields the following two independent equations (6.13–6.15):

$$\text{grad}^2\tilde{\Phi} = -\frac{\omega^2}{c_1^2(\omega)}\tilde{\Phi}, \tag{6.14}$$

and

$$\mathrm{grad}^2\,\widetilde{\Psi} = -\frac{\omega^2}{c_2^2(\omega)}\,\widetilde{\Psi}. \tag{6.15}$$

where $c_1(\omega) = \left(\frac{\lambda^*(\omega)+2\mu^*(\omega)}{\rho}\right)^{\frac{1}{2}}$ and $c_2(\omega) = \left(\frac{\mu^*(\omega)}{\rho}\right)^{\frac{1}{2}}$. $c_1(\omega)$ stands for the complex pressure wave speed in viscoelastic media. Likewise, $c_2(\omega)$ stands for the complex shear wave speed in viscoelastic media.

In this chapter, we only deal with the displacement due to waves propagating along the z-direction. Thus, the Helmoltz potentials are written:

$$\widetilde{\Phi}(\vec{x},\omega) = \varphi(r,\theta,\omega)\mathrm{e}^{i\xi(\omega)z}, \tag{6.16}$$

and

$$\widetilde{\Psi}(\vec{x},\omega) = \begin{pmatrix} \psi_r(r,\theta,\omega) \\ \psi_\theta(r,\theta,\omega) \\ \psi_z(r,\theta,\omega) \end{pmatrix} \mathrm{e}^{i\xi(\omega)z}. \tag{6.17}$$

As we are only interested in longitudinal waves and accounting for axi-symmetry, the tangent displacement u_θ can be neglected. Moreover, the displacement field only depends on the first and third coordinates, i.e., r and z. Within this framework, the Helmoltz potentials can be simplified as follows [16]:

$$\widetilde{\Phi}(\vec{x},\omega) = \varphi(r,\omega)\mathrm{e}^{-i\xi(\omega)z}, \tag{6.18}$$

and

$$\widetilde{\Psi}(\vec{x},\omega) = \begin{pmatrix} 0 \\ \psi_\theta(r,\omega) \\ 0 \end{pmatrix} \mathrm{e}^{-i\xi(\omega)z}. \tag{6.19}$$

Substituting Eqs. (6.18) and (6.19) into Eqs. (6.14) and (6.15), respectively, yields the following two uncoupled differential equations:

$$\frac{\partial^2\varphi}{\partial r^2} + \frac{1}{r}\frac{\partial\varphi}{\partial r} + \alpha^2\varphi = 0, \tag{6.20}$$

and

$$\frac{\partial^2\psi_\theta}{\partial r^2} + \frac{1}{r}\frac{\partial\psi_\theta}{\partial r} + \left(\beta^2 - \frac{1}{r}\right)\psi_\theta = 0, \tag{6.21}$$

where
$\alpha^2 = -\xi^2 + \frac{\rho\omega^2}{\lambda^*(\omega)+2\mu^*(\omega)} = -\xi^2 + \frac{\omega^2}{c_1^2(\omega)}$ and $\beta^2 = -\xi^2 + \frac{\rho\omega^2}{\mu^*(\omega)} = -\xi^2 + \frac{\omega^2}{c_2^2(\omega)}$.

The solutions of Eqs. (6.20) and (6.21) are then expressed in terms of Bessel's functions. Actually, Neumann functions are also solutions of the Eqs. (6.20) and (6.21). However, they go to infinity when r approaches 0. For this reason, Neumann functions are disregarded here [17]. Consequently, the two potential functions φ and ψ_θ read:

$$\varphi(r,\omega) = K_1(\omega)J_0(\alpha(\omega)r), \tag{6.22}$$

and

$$\psi_\theta(r,\omega) = K_2(\omega)J_1(\beta(\omega)r), \tag{6.23}$$

where J_0 and J_1 are zero and first order Bessel's functions and K_1 and K_2 are two frequency-dependent constants.

Considering the new expression of Helmoltz potentials established in Eqs. (6.22) and (6.23), the radial and axial displacements write:

$$u_r(\vec{x},\omega) = u_r(r,z,\omega) = (-\alpha K_1 J_1(\alpha r) + i\beta\xi K_2 J_1(\beta r))\mathrm{e}^{-i\xi z}, \tag{6.24}$$

and

$$u_z(\vec{x},\omega) = u_z(r,z,\omega) = \left(-i\xi K_1 J_0(\alpha r) + \beta^2 K_2 J_0(\beta r)\right)\mathrm{e}^{-i\xi z}, \tag{6.25}$$

respectively. In order to determine the constants K_1 and K_2, the following boundary conditions are considered:

$$\tilde{\sigma}_{rr}(r = a, z, \omega) = 0, \tag{6.26}$$

and

$$\tilde{\sigma}_{rz}(r = a, z, \omega) = 0, \tag{6.27}$$

where $\tilde{\sigma}_{rr}$ is the radial stress component and $\tilde{\sigma}_{rz}$ is a shear stress components. Equations (6.26) and (6.27) write that the outer surface of the rod is free. Using Eqs. (6.7) and (6.8), the two stress components $\tilde{\sigma}_{rr}$ and $\tilde{\sigma}_{rz}$ are expressed in terms of the radial and axial displacements as follows [16]:

$$\tilde{\sigma}_{rr}(r,z,\omega) = \lambda^*(\omega)\left(\frac{\tilde{u}_r}{r} + \frac{\partial \tilde{u}_r}{\partial r} + \frac{\partial \tilde{u}_z}{\partial z}\right) + 2\mu^*(\omega)\frac{\partial \tilde{u}_r}{\partial r} \tag{6.28}$$

and

$$\tilde{\sigma}_{rz}(r,z,\omega) = \mu^*(\omega)\left(\frac{\partial \tilde{u}_r}{\partial r} + \frac{\partial \tilde{u}_z}{\partial z}\right). \tag{6.29}$$

Developing Eqs. (6.28) and (6.29) by substituting the expression of the radial and axial displacement of Eqs. (6.24) and (6.25) and then considering the boundary conditions of Eqs. (6.26) and (6.27) yield the following system of two algebraic equations in terms of the constants K_1 and K_2:

$$K_1\left(-\frac{(\beta^2-\xi^2)}{2}J_0(\alpha a)+\frac{\alpha}{a}J_1(\alpha a)\right)+K_2\left(i\beta^2\xi J_0(\beta a)+\frac{\alpha}{a}J_1(\beta a)\right)=0, \quad (6.30)$$

and

$$K_1(2i\alpha\xi J_1(\alpha a))-K_2(\beta(\beta^2-\xi^2)J_1(\beta a))=0. \quad (6.31)$$

For non-trivial solutions, the determinant of the above system should be zero. This leads to the so-called Pochhammer-Chree wave dispersion equation which was first established in the case of elastic bars by Pochhammer [1] and Chree [2] and later generalized to the case of viscoelastic bars by Zhao and Gary [3]:

$$f(\xi)=\frac{2\alpha}{a}\frac{(\beta^2+\xi^2)}{2}J_1(\alpha a)J_1(\beta a)-(\beta^2-\xi^2)J_0(\alpha a)J_1(\beta a)-4\xi^2\alpha\beta J_1(\alpha a)J_0(\beta a)=0, \quad (6.32)$$

6.2.2 *Solutions of the Pochhammer-Chree Equation*

Bancroft [18] numerically solved the Pochhammer-Chree equation established in Eq. (6.32) in the case of elastic bars. He mainly argued that the longitudinal wave velocity can be written as the wave velocity at infinite wavelengths/low frequencies multiplied by a function that dependent on two parameters, which are the Poisson's ratio and the diameter-to-wavelength ratio. Davies [19] showed that this equation has multiple solutions and he calculated the three first solutions. Each of these solutions corresponds to a wave propagation mode [16]. These propagation modes were later observed experimentally in Refs. [20–22].

Zhao and Gary [3] have solved the Pochhammer equation in the case of viscoelastic bars using the iterative formula of Newton:

$$\xi_{n+1}=\xi_n+\frac{f(\xi_n)}{f'(\xi_n)}. \quad (6.33)$$

The derivative f' was derived knowing the expression of f given in Eq. (6.32). Coquin [5] considered the following approximations:

$$\frac{|Im(\lambda^*(\omega))|}{|Re(\lambda^*(\omega))|} \ll 1, \tag{6.34}$$

and

$$\frac{|Im(\mu^*(\omega))|}{|Re(\mu^*(\omega))|} \ll 1. \tag{6.35}$$

Within this framework and using a Voigt rheological model, he succeeded to calculate the first three propagation modes in a viscoelastic rod. Dealing also with viscoelastic rods, Othman [23] numerically solved the Pochhammer-Chree using the regula falsi method (also called the false position method). This method is quite similar to the iterative formula of Newton, except that the derivative is calculated numerically.

Alternative approximate solutions for the wave dispersion relation have been also developed. Love's equation accounts for the radial inertia [24]. The kinetic energy of the bar is written as the sum of the axial and the radial kinetic energies. The radial kinetic energy is deduced from the axial kinetic energy using Poisson' ratio. The Love's approximation was developed for elastic bars. Recently, this approximation is extended to viscoelastic bars [25, 26]. Anderson [27] derived higher-order approximations for the elastic propagation of longitudinal waves in bars. Following Mindlin and McNiven [28], he represented the axial and radial displacements in terms of Jacobi polynomials. He showed that the four-mode equation gives good approximation of the Pochhammer-Chree solution (Fig. 6.3). Żak and Krawczuk [29] enriched the displacement field so to fulfill the traction-free boundary condition on the outer surface of the rod.

Recently, Othman [30] proposed a phenomenological fraction equation that involves the Poisson's ratio as a parameter. The proposed new equation approximates well the wave dispersion relation obtained by the Pochhammer-Chree equation for Poisson's ratio ranging from 0.25 to 0.4 and for wavelengths four times higher than the radius of the bar. This roughly corresponds to frequencies up to 75 kHz. The phenomenological fractional equation is then highly useful for processing signals obtained in Kolsky-Hopkinson bar experiments.

6.2.3 *Simplification of the Pochhammer-Chree Solution*

6.2.3.1 Multi-mode Propagation

The above cited works [5, 19–23] agree that the Pochhammer-Chree solution have multiple solutions ξ_m, where m denotes the solution order. Consequently, the displacement and then strain is the superposition of the contributions of all modes. Namely, the axial strain can be written as:

$$\tilde{\varepsilon}_{zz}(r,z,\omega) = \sum_{m=1}^{\infty} A_m^{\pm}(r,\omega)\mathrm{e}^{\mp i\xi_m(\omega)z}, \tag{6.36}$$

where A_m^+ and A_m^- is the amplitudes of the two waves of the mth mode and that propagate in opposite directions. ξ_m is the wave dispersion relation/solution of the mth mode. In the most general case, ξ_m is a complex number and can then be split in real and imaginary parts:

$$\xi_m(\omega) = k_m(\omega) + i\alpha_m(\omega) = \frac{\omega}{c_m(\omega)} + i\alpha_m(\omega), \tag{6.37}$$

where $k_m(\omega)$, $c_m(\omega)$ and $\alpha_m(\omega)$ are the wave number, the wave velocity and the damping coefficient of the mth mode.

In the case of the elastic rods, the first solution of the Pochhammer-Chree equation (giving the lowest wave velocity) is real for all frequencies. When excited, the first mode then propagates on the whole frequency domain without any damping. The wave dispersion solution ξ_m of the higher modes ($m > 1$) is only real for frequencies higher than a critical frequency $f_m^{(c)}$ [16]. For the frequencies lower than $f_m^{(c)}$ the real part of ξ_m:

$$Re(\xi_m(\omega)) = k_m(\omega) = 0, \quad \text{when} \quad -2\pi f_m^{(c)} \le \omega \le 2\pi f_m^{(c)}. \tag{6.38}$$

This means that higher modes do not propagate at low frequencies and only the first mode does. It is then possible to assume that there is only one excited mode at low frequencies, which highly simplifies the equation. In this case, the axial strain is reduced to the contribution of only the first mode:

$$\tilde{\varepsilon}_{zz}(r,z,\omega) = A_1^+(r,\omega)\mathrm{e}^{-i\xi_1(\omega)z} + A_1^-(r,\omega)\mathrm{e}^{i\xi_1(\omega)z}. \tag{6.39}$$

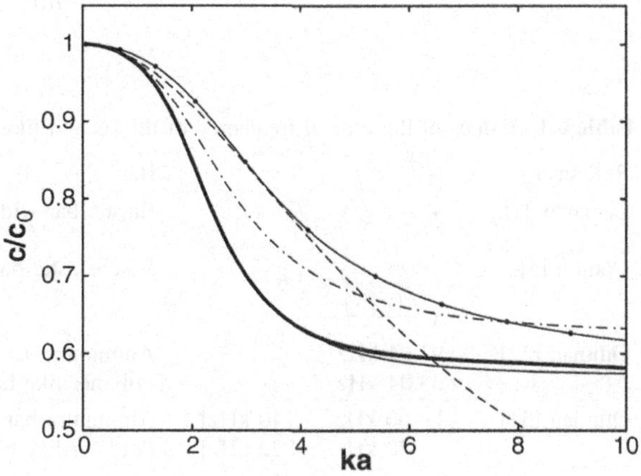

Fig. 6.3 Dispersion curves for an aluminium bar: Love, dashed line; two-mode, solid line with dot markers; three-mode, dash-dot line; four-mode, thick solid line; Pochhammer, thin solid line (reproduced from Ref. [27] with permissions from Elsevier)

We recall that $A_1^+(r, \omega)$ is the amplitude of the wave that propagates in positive z direction and $A_1^-(r, \omega)$ is the amplitude of the wave that propagates in negative z direction.

The simplified Eq. (6.39) can only be used for frequencies lower than the critical frequency of the second mode $f_2^{(c)}$. Several values of this critical frequency have been reported in the literature depending on the material and diameter of the bar. They are synthesized in Table 6.1. The second mode's critical frequency is mainly higher than 90 kHz for elastic metallic bars and 30 kHz for viscoelastic polymeric bars, if the bar radius does not exceed 20 mm. Thus, for non-large diameter Kolsky-Hopkinson bars, it is reasonable to assume that only the first-mode propagates as soon as the impact pulses do not include frequencies higher than 90 kHz in the case of metallic bars and 30 kHz in the case of polymeric bars. Indeed, it is hard to have impact pulses involving frequencies higher than 50 kHz (metallic bars) and 20 kHz (polymeric bars) in typical Kolsky-Hopkinson bars tests at strain rates lower than 5000/s. However, this would not be the case at very high strains (strain rates higher than 5000/s) where the critical frequency of the second mode and the cut-off frequency of the strain-gauge should be compared to the maximum excited frequency by the impact load [31].

6.2.3.2 Variation Within the Cross-Section

In Kolsky-Hopkinson bar machine experiments, strain measurements are recorded on the outer surface of the bar using strain gauges. This means that, it is only possible to have a direct knowledge of the strain at ($r = a$). However, the displacement field obtained by solving the Pochhammer-Chree solution is depending not only on the axial coordinate z but also on the radial coordinate r [32–35]. Tyas and Watson [36, 37] and Tyas and Pope [38] wrote the Fourier transform of the axial displacement $\tilde{u}_z$ of a mode m as :

$$\tilde{u}_{z,m}(r, z, \omega) = \xi_m f A \frac{J_1(ha)}{ha} e^{i\xi_m}, \tag{6.40}$$

Table 6.1 Values of the critical frequency of the second model

References	$f_2^{(c)}$	Bar
Zemanek [16]	$2\pi f_2^{(c)} \approx \frac{3.6}{a}\sqrt{\frac{\mu}{\rho}}$	Elastic bar with Poisson ratio $\nu = 0.3317$
Coquin [5]	$\frac{2\pi f_2^{(c)} a}{\sqrt{\frac{\mu^*\left(2\pi f_2^{(c)} a\right)}{\rho}}} \approx 3.6$	Viscoelastic bar with Poisson ratio $\nu = 0.36$
Othman [23]	91.94 kHz 33.04 kHz	Aluminum bar (a = 20 mm) Polymer-like bar (a = 20 mm)
Othman [31]	[~60 kHz, ~740 kHz] [~20 kHz, ~220 kHz]	Aluminum bar (a ranging in [2 mm, 40 mm]) Polymer-like bar (a ranging in [2 mm, 40 mm])

where A is the wave's magnitude, $f = \frac{ha\,J_0(hr)}{J_1(ha)} + \frac{1-SZ}{Z-1}\frac{\kappa a\,J_0(\kappa r)}{J_1(\kappa a)}$, $S = \frac{1-2\nu}{1-\nu}$, $Z = (1+\nu)\frac{\rho}{E^*(\omega)}\left(\frac{\omega}{\xi_m}\right)^2$, $h = -i\xi_m\sqrt{SZ-1}$ and $\kappa = -i\xi_m\sqrt{2Z-1}$. Similar expressions can be deduced for strain and stress. Yet, for higher mode ($m \geq 2$), the displacement, strain and stress are highly sensitive to the radial coordinate (Fig. 6.4). Considering the first mode, Tyas and Watson [36] plotted the strain over the cross-section and showed that the strain is flat at low frequencies and its dependency to the radial position increases as the frequency increases. Assuming that the displacement is constant over the cross-section can lead to errors as high as 24 and 64% if the bar radius-to-wavelength ratio is equal to 0.2 and 0.3, respectively [37]. Tyas and Ozdemir [39] advised against assuming uniform distribution of axial strain over the cross-section for bar radius-to-wavelengths ratios higher than 0.28–0.32.

Idesman and Mates [40] carried out finite element analysis for elastic wave propagation in rods. They noticed that the non-uniform radial distribution of the axial particle velocity is important not far from the impact end. However, the particle velocity is more uniform over the cross-section as the distance from the impact end increases. This can be interpreted by the fact that near the impact end higher modes contribute to particle velocity (and of course to strain and stress). As the waves propagate away from the impact end, the higher modes vanish and the particular velocity is only due the first Pochhammer-Chree mode, which has less dependency on the radial coordinate. Moreover, Idesman and Mates [40] reported that the radial non-uniformity is more important for the short times of impact pulse

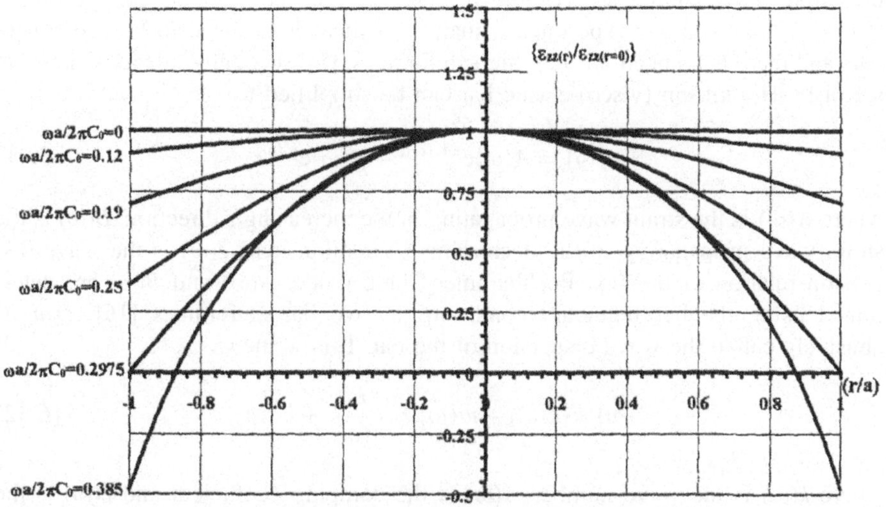

Fig. 6.4 Mode 1 Pochhammer-Chree distribution of the axial strain over the cross-section of an elastic bar (ν = 0.29) (reproduced from Ref. [36] with permissions from Springer)

than it is for the long times. This is due to the fact that radial coordinate dependency is more important at high frequencies than it is at low frequencies.

Though in Kolsky-Hopkinson bar experiments short wavelengths as short as 3 or 5 times the bar radius can be present, the effect radial coordinate dependency is less important as most of the wave impact energy is concentrated at longer wavelengths (longer than 10 times the radius). The effect of radial coordinate dependency should increase as the bar radius increases. Studying the over-cross-section distribution of stress and particle velocity in large Kolsky-Hopkinson bars, Merle and Zhao [41] reported errors of 12% for 100-mm-diameter bars whereas less than 5% of error observed for 60-mm-diameter bars.

6.2.3.3 Finite-Length Bars

The Pochhammer-Chree equation and approximations was established for an infinite bar. Skalak [42] and Folk et al. [43] studied the response of semi-infinite bars to a pressure impulse. Malkov [44] derived the exact short-time using Sobolev-Smirov method. Folk et al. [43] used a double transform method. They showed that the exact solution can be expressed in terms of Fourier integrals, whose integrands have similar form as Pochhammer-Chree waves. They noticed that these integrals can hardly be evaluated by simple techniques. Using asymptotic solutions, they solved for large distances of propagation. It is then possible to assume that modes of propagation of infinite bars are not altered by the boundary conditions of finite-length bars [45]. Provided that the bar length of the bar is much higher than its radius, the Pochhammer-Chree modes of infinite-length can be used for a finite-length bar [19]. Following the discussions of Sects. 6.2.3.1 and 6.2.3.2, only the first Pochhammer-Chree mode has to be considered for most of Kolsky-Hopkinson bar experiments; mainly, if there is no need to large-diameter bars and there is no need for very high strain rates. Consequently, the axial strain in a Kolsky-Hopkinson (visco)-elastic bar can be simplified to:

$$\tilde{\varepsilon}(z,\omega) = A(\omega)e^{-i\xi(\omega)z} + B(\omega)e^{i\xi(\omega)z}. \tag{6.41}$$

where $A(\omega)$ is the strain wave propagating in the increasing z direction, $B(\omega)$ is the strain wave propagating in the decreasing z direction, and $\xi(\omega)$ is the wave dispersion relation of the first Pochhammer-Chree mode. $A(\omega)$ and $B(\omega)$ are determined using the bar boundary conditions at two bar extremities [16]. $\xi(\omega)$ is abusively called the wave dispersion of the bar. It is written

$$\xi(\omega) = k(\omega) + i\alpha(\omega) = \frac{\omega}{c(\omega)} + i\alpha(\omega), \tag{6.42}$$

where $k(\omega)$ is the wave number, $\alpha(\omega)$ is the damping coefficient and $c(\omega)$ is the wave velocity. All these three parameters are frequency-dependent. The bar wave dispersion relation is also written

$$\xi(\omega) = \frac{\omega}{c^*(\omega)}, \tag{6.43}$$

where $c^*(\omega)$ is the complex wave velocity.

Starting from Eq. (6.41), it is simple to derive the axial stress, the axial particle velocity and the axial displacement. Namely,

$$\tilde{\sigma}(z,\omega) = E^*(\omega)\left(A(\omega)\mathrm{e}^{-i\xi(\omega)z} + B(r,\omega)\mathrm{e}^{i\xi(\omega)z}\right), \tag{6.44}$$

$$\tilde{v}(z,\omega) = \omega\frac{-A(\omega)\mathrm{e}^{-i\xi(\omega)z} + B(r,\omega)\mathrm{e}^{i\xi(\omega)z}}{\xi(\omega)}, \tag{6.45}$$

and

$$\tilde{u}(z,\omega) = i\frac{A(\omega)\mathrm{e}^{-i\xi(\omega)z} - B(r,\omega)\mathrm{e}^{i\xi(\omega)z}}{\xi(\omega)}, \tag{6.46}$$

respectively.

In the following Sect. 6.3, we will be dealing with the experimental techniques proposed in the literature to measure the wave dispersion of elastic and viscoelastic bars.

6.3 Experimental Analysis

Several experimental techniques have been developed to analyse the wave dispersion in long elastic and visco-elastic rods. These techniques are based on the classical Kolsky-Hopkinson bar rig. However, each bar is studied separately. Thus, the experimental setup mainly consists in a long (visco)-elastic rod striked by a projectile at one end, the other end being kept free. The projectile can be a short rod or a spherical ball. In classical Kolsky-Hopkinson bar tests, the length of the striker rod is maximized, but kept lower than half the incident bar length, in order to increase the test duration and then to increase deformation of the specimen. In wave dispersion experiments, the seak is for the opposite. Indeed, the striker rod length is minimized and sometimes replaced by a spherical ball in order to widen the spectrum of the incident pulse. The shorter the wave pulse, the larger is the frequency for which we can measure the wave velocity and the wave damping. Sometimes, the experimental techniques are developed only to measure the wave dispersion relation (velocity and damping) [46, 47]. However, most of techniques are developed aiming at measuring the complex Young's modulus of polymeric materials [48, 49]. In this chapter, we will review all of these techniques. However, we will disregard the step of calculating the Young's modulus knowing the wave dispersion relation.

The experimental methods characterizing the wave dispersion in rods can be split into two groups. The first group of methods isolates the first transient waves, i.e., the first or second incident waves and/or the first and second reflected waves. This group will be referred as «short-duration signals» techniques. The second group of methods records strain or velocity measurements in one or several cross-section of the rod and allows multiple round-trips of the wave. The measurements are recorded for longer duration. This second group will be referred as «long-duration signals» techniques. However, in both groups the wave pulse is short to achieve large frequency range.

6.3.1 Techniques Using Short-Duration Signals

In this type of techniques, the rod is impact by a ball or a short striker (Figs. 6.5 and 6.6). The second end of the bar is kept free. This impact induces a short incident compressive wave which travels along the bar. When the wave reaches the opposite side, it is reflected back as a tensile wave. Unless damped, the wave does infinite number of round-trips in the rod. The techniques using short-duration signals focus on the first round-trip in the wave. They assume that it is possible to isolate the first incident compressive wave in one or two positions of the rod and or the first reflected tensile wave (Fig. 6.5).

Let $\theta(t)$ be the strain at the impact side of the rod which is generated by the impact of the projectile (ball or short rod) at that side, and let $\tilde{\theta}(\omega)$ be its Fourier transform. This wave will propagate along the rod. The first incident and reflected waves will be recorded by one strain gauge after some delays. Thus, the two waves are superimposed within the same signal recorded by the gauge. However, the incident is considered short enough to end up before the reflected wave arrives to the strain gauge cross-section. In this case the two waves can be separated and

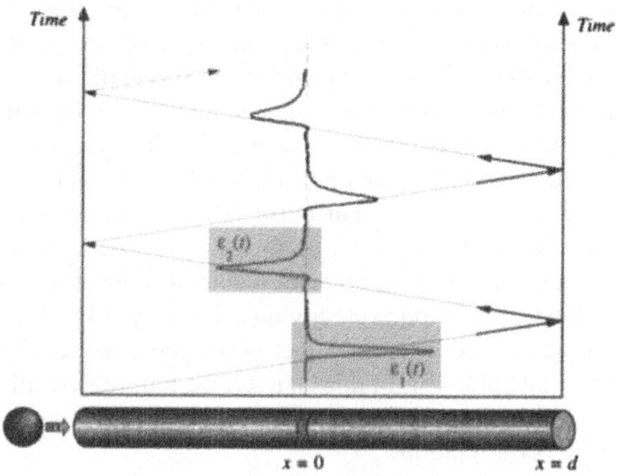

Fig. 6.5 A bar impacted by a ball and the corresponding strain signal in one cross-section (reproduced from Ref. [46] with permissions from Springer)

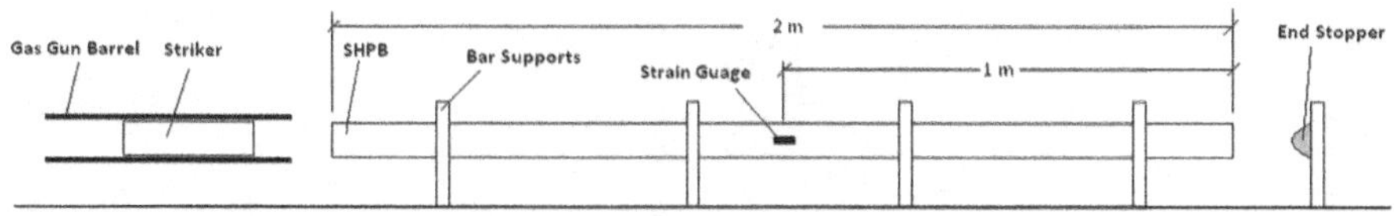

Fig. 6.6 A rod impacted by a short striker bar (reproduced from Ref. [48] with permissions from Springer)

isolated as shown in Fig. 6.5. Let $\varepsilon_{inc}(t)$ and $\varepsilon_{ref}(t)$ be the incident and reflected waves, respectively, and let $\tilde{\varepsilon}_{inc}(\omega)$ and $\tilde{\varepsilon}_{ref}(\omega)$ be their Fourier transforms. Using the fact that the right side of the bar is free, the compressive incident wave is reflected, at this bar end, as a tensile wave, with the same magnitude. Consequently, the Fourier transform of the incident and reflected waves read:

$$\tilde{\varepsilon}_{inc}(\omega) = \tilde{\theta}(\omega)\mathrm{e}^{-i\xi(\omega)d}, \tag{6.47}$$

and,

$$\tilde{\varepsilon}_{ref}(\omega) = -\tilde{\theta}(\omega)\mathrm{e}^{-i\xi(\omega)(2L-d)}, \tag{6.48}$$

respectively, where d is the distance separating the gauge station from the impact end and L the bar length. Dividing Eqs. (6.48) by (6.47) yields:

$$\mathrm{e}^{-2i\xi(\omega)(L-d)} = -\frac{\tilde{\varepsilon}_{ref}(\omega)}{\tilde{\varepsilon}_{inc}(\omega)}, \tag{6.49}$$

Considering the expression of the wave dispersion given in Eq. (6.49), the wave damping and the wave number are given by:

$$\alpha(\omega) = \mathrm{e}^{-2i\xi(\omega)(L-d)} = \frac{1}{2(L-d)}\left|\frac{\tilde{\varepsilon}_{ref}(\omega)}{\tilde{\varepsilon}_{inc}(\omega)}\right|, \tag{6.50}$$

and

$$k(\omega) = \frac{\arg(\tilde{\varepsilon}_{inc}(\omega)) - \arg(\tilde{\varepsilon}_{ref}(\omega)) + 2n\pi}{2(L-d)}, \tag{6.51}$$

respectively, where $\arg(x)$ is the argument of the a complex number x and n is an integer. Figure 6.7 depicts the wave phase velocity and the wave damping coefficient for two PMMA bars as obtained by Bacon [46].

This approach has been considered by several authors (see for Ex. Refs. [6, 46, 50–54]). It is mainly used for its simplicity. However, it has some drawbacks. First, the product $2(L - d)k(\omega)$ is only known modulo 2π. Some information is lost when considering the phase or the arg operation. Thus a correction have to be considered

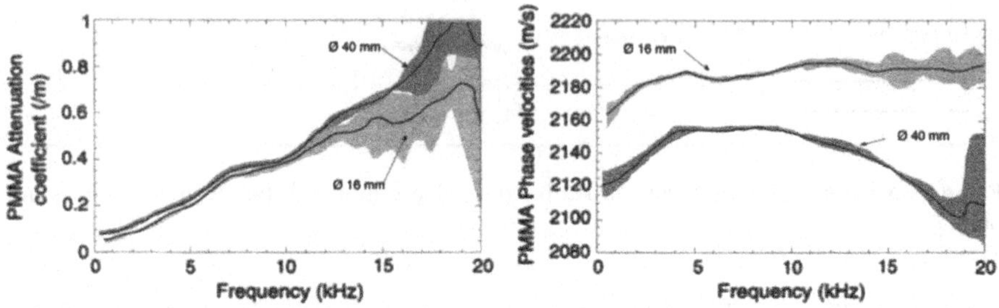

Fig. 6.7 The damping coefficient and phase velocity of two PMMA bars (reproduced from Ref. [46] with permissions from Springer)

by adding $2n\pi$. At very low frequencies n is fixed to 0. While going to the higher frequencies, any drop in $\arg(\tilde{\varepsilon}_{inc}(\omega)) - \arg(\tilde{\varepsilon}_{ref}(\omega))$ by 2π corresponds to an increase of n by 1. However, there is no drop in $\arg(\tilde{\varepsilon}_{inc}(\omega)) - \arg(\tilde{\varepsilon}_{ref}(\omega))$ which is exactly equal to 2π because of noise. Thus, n is incremented by 1 for any significant drop in $\arg(\tilde{\varepsilon}_{inc}(\omega)) - \arg(\tilde{\varepsilon}_{ref}(\omega))$. This can lead to substantial errors. Second, this technique is sensitive to noise. Indeed, the calculation of the damping is based on Eq. (6.50) which uses a ratio of the spectrum of the two waves. As soon as the magnitude of the spectrum of the incident wave vanishes, the noise will be highly amplified. Moreover, the calculation of the wave number or wave velocity is based on Eq. (6.51) which is based on the difference in phase between Fourier spectrums of incident and reflected waves. For the frequencies where the two waves have the same phase, the measured wave number corresponds rather to the difference in phase between the noises recorded on the two waves. Third, this technique is clearly sensitive to the measurement of the distance $(L - d)$.

6.3.2 Techniques Using Long-Duration Signals

Alternative solutions have been proposed in order to take advantage of the multiple wave reflections within rods [6, 8, 55, 56]. The experimental rig is similar to the one used in Sect. 6.3.1 and depicted in Figs. 6.5 and 6.6, except that sometimes multiple strain gauge stations are cemented on the rod.

Lundberg and Blanc [6] have used two strain gauge stations on the rod. Thus, two strain measurement are recorded at two cross-section z_1 and z_2. As in Sect. 6.3.1, one end is kept free. Let $\tilde{\varepsilon}_1(\omega)$ and $\tilde{\varepsilon}_2(\omega)$ be the Fourier transforms of the two strain measurements. Their ratio is denoted $\psi(\omega)$ and writes:

$$\psi(\omega) = \rho(\omega)e^{i\varphi(\omega)} = \frac{\tilde{\varepsilon}_1(\omega)}{\tilde{\varepsilon}_2(\omega)}. \tag{6.52}$$

Assuming $z = 0$ at the free end, the ratio between the two strain measurements simply reads:

$$\psi(\omega) = \frac{e^{-i\xi(\omega)z_1} - e^{i\xi(\omega)z_1}}{e^{-i\xi(\omega)z_2} - e^{i\xi(\omega)z_2}}. \tag{6.53}$$

Lundberg and Blanc [6] also assumed that $z_2 = 2z_1 = -2b$, where b is a distance. Separating the real and imaginary parts of Eq. (6.53) and solving for $\rho(\omega)$ and $\varphi(\omega)$ yields:

$$\rho(\omega) = 2\sqrt{\cosh^2(b\,\alpha\,(\omega)) - \sin^2(b\,k\,(\omega))}, \tag{6.54}$$

and

$$\varphi(\omega) = \tan^{-1}(-\tanh(b\,\alpha\,(\omega))\tan(b\,k\,(\omega))) + m\pi, \tag{6.55}$$

where $m = 0, 1, \ldots$ is an integer that is chosen such as $\varphi(\omega = 0)$ and $\varphi(\omega)$ is an increasing, continuous function. Solving Eqs. (6.54) and (6.55) for $\alpha(\omega)$ and $k(\omega)$ gives:

$$k(\omega) = \frac{s\varsigma + n\pi}{b}, \tag{6.56}$$

and

$$\alpha(\omega) = -\frac{\sinh^{-1}\left(\sqrt{C - A}\right)}{b}, \tag{6.57}$$

where $C = \sqrt{A^2 + B^2}$, $A = \frac{1-(\rho(\omega)/2)^2}{2}$, $B = \frac{\rho(\omega)\sin\varphi(\omega)}{2}$, $\varsigma = \sin^{-1}\left(\sqrt{C + A}\right)$, $s = \mp 1$, and $n = 0, 1, \ldots$ is an integer which is chosen such as $k(\omega = 0)$ and $k(\omega)$ is an increasing, continuous function. An alternative solution have been proposed by Lundberg and Ödenn [8] which uses three strain measurements in order to overcome the free-end condition.

This type of methodology has the advantage of not assuming any time limitation on the strain measurements. However, it has the same disadvantages as the techniques using short-duration signals. Mainly, this methodology is sensitive to noise in the frequency ranges where the spectrums of the strain measurements vanish.

In order to measure wave dispersion in rods it is indispensable to solve an algebraic system of three equations. Indeed, the problem has three unknowns: the complex wave dispersion $\xi(\omega)$ and the two waves propagating in opposite directions $A(\omega)$ and $B(\omega)$. This explains why Lundberg and Ödenn [8] have recorded three strain measurements on the rod. Lundberg and Blanc [6] have recorded only two strain measurements on the rod but they used a free-end condition which is equivalent to a strain measurement. Even the short-duration signals method, presented in Sect. 6.3.1, can not go round the three minimum equations. It is true that this technique records one strain measurements but it is equivalent to two separate measurements as the incident and the reflected waves are separated. It also assumes a free-end condition which is equivalent to a third strain measurement.

When no noise is recorded, it is perfect to have only three strain measurements or two strain measurements and one boundary condition. However, noise is inescapable. It is intrinsically recorded together with the strain measurements. In this case, using redundant measurements is necessary [57, 58]. Hillström et al. [55] considered that it is possible to recorded N strain measurements with $N > 3$ in order to ensure redundancy. Let $\hat{\varepsilon}_1(\omega), \ldots, \hat{\varepsilon}_N(\omega)$ be the Fourier transforms of the N strain measurements and let $z_1, \ldots, z_N$ the coordinates of the cross-section of the strain measurements. Each strain can be written as the theoretical strain $\tilde{\varepsilon}_j(\omega)$, plus an experimental noise $n_j(\omega)$:

$$\hat{\varepsilon}_j(\omega) = \tilde{\varepsilon}_j(\omega) + n_j(\omega). \tag{6.58}$$

The theoretical strain is the contribution of the two waves propagating in opposite direction using Eq. (6.41) yields to the following matrix equation:

$$M(\omega)P(\omega) = E(\omega), \tag{6.59}$$

where

$$M(\omega) = \begin{bmatrix} e^{-i\xi(\omega)z_1} & e^{i\xi(\omega)z_1} \\ \vdots & \vdots \\ e^{-i\xi(\omega)z_N} & e^{i\xi(\omega)z_1} \end{bmatrix}, P(\omega) = \begin{bmatrix} A(\omega) & B(\omega) \end{bmatrix} \text{ and } E(\omega) = \begin{bmatrix} \tilde{\varepsilon}_1(\omega) \\ \vdots \\ \tilde{\varepsilon}_N(\omega) \end{bmatrix}$$

Hillström et al. [55] computed the wave dispersion relation $\xi(\omega)$ using the least-square method. Actually, both $M(\omega)$ and $P(\omega)$ are unknowns. The problem is solved in two steps by solving successively two optimization problems. The first optimization problem is solved analytically to express $P(\omega)$ and a second one to determine $M(\omega)$ or $\xi(\omega)$. The cost function $f_1(P)$ of the first optimization problem is defined as the relative difference between the theoretical strain vector given by Eq. (6.59) and the experimental strain vector

$$\widehat{E}(\omega) = \begin{bmatrix} \hat{\varepsilon}_1(\omega) \\ \vdots \\ \hat{\varepsilon}_N(\omega) \end{bmatrix}. \tag{6.60}$$

Namely,

$$f_1(P(\omega)) = \frac{\left\| M(\omega)P(\omega) - \widehat{E}(\omega) \right\|}{\left\| \widehat{E}(\omega) \right\|}. \tag{6.61}$$

where $\|\cdot\|$ is the Eucleadian norm. This cost function is minimized with respect to $P(\omega)$ using the Moore-Penrose pseudo-inverse matrix. More precisely,

$$P(\omega) = (M^*(\omega)M(\omega))^{-1}M(\omega)\widehat{E}(\omega). \tag{6.62}$$

where M^* is the adjoint (conjugate and transpose) matrix of M. The cost function of the second optimization problem is obtained by substituting the expression of $P(\omega)$ expressed by Eq. (6.62) in the expression of first cost function of Eq. (6.61). Put in other words,

$$f_2(P(\omega)) = \frac{\left\| M(\omega)(M^*(\omega)M(\omega))^{-1}M(\omega)\widehat{E}(\omega) - \widehat{E}(\omega)\right\|}{\left\|\widehat{E}(\omega)\right\|}. \tag{6.63}$$

The minimization of the cost function in Eq. (6.63) can be undertaken for each frequency. The optimizing parameters are $\alpha(\omega)$ and $k(\omega)$. This is the non-parametric method. The minimization of the cost function $f_2(P(\omega))$ can also be considered just one time for all frequencies. However, a viscoelastic model for the complex Young's modulus of the polymer has to be assumed. The optimizing parameters are then the constants of the viscoelastic model. This yields a parametric identification [59].

Both techniques, the parametric and non-parametric identifications, have the main advantage of being strong against noise. Its stability against noise is increased as the number of measured strain signals N increases. Hillström et al. [55] used five strain measurements and one boundary conditions ($N = 6$) to measure the wave dispersion in polypropylene and polymethylmethacrylate rods (Fig. 6.8). Mossberg et al. [60] have undertaken a statistical analysis of the non-parametric method and confirmed its accuracy.

The main motivation of the works [55, 59, 60] is to reduce the effect of noise. To this aim, they mainly used redundant measurements. With the same motivation, it is possible to exploit the bar resonances [61–65] as at these frequencies the signal-to-noise (SNR) ratio is the highest. Thus, the information given by measurements in the frequency range around the resonances is the most accurate.

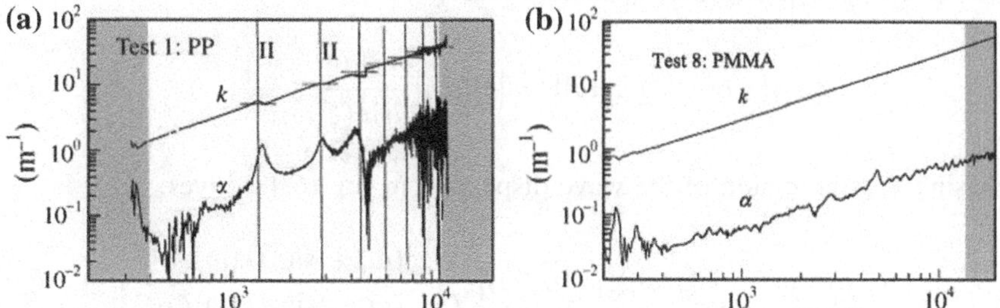

Fig. 6.8 The damping coefficient and phase velocity of two PMMA bars (reproduced from Ref. [55] with permissions from Springer)

Othman et al. [56] took advantage of the rod resonances to develop a method measuring the wave phase velocity and wave damping. The experimental set-up is similar to the one used by Bacon [46] and Butt et al. [48, 49] and presented in Figs. 6.5 and 6.6, except that the strain measurements are recorded for long duration without any time limitation. The resonance-based method does not need any separation of waves propagating in opposite directions, indeed. The rod can be elastic or viscoelastic. It is impacted at one side by a projectile that can be a ball or a short rod. The opposite side of the rod is assumed free. This side is considered as the origin ($z = 0$). Considering Eq. (6.41), the boundary condition at the origin yields:

$$B(\omega) = -A(\omega). \tag{6.64}$$

recall that $A(\omega)$ is the Fourier transform of the successives waves propagating in positive z direction (successives incident waves) and $B(\omega)$ is the Fourier transform of the successives waves propagating in negative z direction (successives reflected waves). Let $\widetilde{\Theta}(\omega) = \tilde{\varepsilon}(z = -L, \omega)$ be Fourier transform of the strain at the impact side, where L the length of the rod. Using one more time Eq. (6.41) leads to:

$$A(\omega) = \frac{\widetilde{\Theta}(\omega)}{e^{i\xi(\omega)L} - e^{-i\xi(\omega)L}}. \tag{6.65}$$

Now substituting Eqs. (6.64) and (6.65) in Eq. (6.41) yields:

$$\tilde{\varepsilon}(z, \omega) = -\widetilde{\Theta}(\omega)\frac{e^{-i\xi(\omega)z} - e^{i\xi(\omega)z}}{e^{-i\xi(\omega)L} - e^{i\xi(\omega)L}}. \tag{6.66}$$

or

$$\tilde{\varepsilon}(z, \omega) = -\widetilde{\Theta}(\omega)\frac{\sin(\xi(\omega)z)}{\sin(\xi(\omega)L)}. \tag{6.67}$$

Equation (6.67) gives the general expression of the Fourier transform of the strain that can be measured at any cross-section of the rod, assuming that one rod side is free. The spectrum of this strain reads:

$$S(z, \omega) = |\tilde{\varepsilon}(z, \omega)|^2 = \left|\widetilde{\Theta}(\omega)\right|^2 \frac{|\sin(\xi(\omega)z)|^2}{|\sin(\xi(\omega)L)|^2}. \tag{6.68}$$

Using the expression of the wave dispersion in Eq. (6.43) gives:

$$S(z, \omega) = |\tilde{\varepsilon}(z, \omega)|^2 = \left|\widetilde{\Theta}(\omega)\right|^2 \frac{z^2\alpha^2(\omega) + \sin^2(k(\omega)z)}{L^2\alpha^2(\omega) + \sin^2(k(\omega)L)}. \tag{6.69}$$

Othman et al. [56] neglected the damping effect in a first step. Thus, the bar resonances occur for the angular frequencies ω_p such as:

$$k(\omega_p) = \frac{\omega_p}{c(\omega_p)} = \frac{p\pi}{L}. \tag{6.70}$$

where p is an integer representing the order of the resonance. $p = 1$ for the first resonance, etc. Equation (6.70) gives a measurement method for the wave phase velocity. More precisely, it is possible to calculate the wave velocity knowing the resonance position (resonance frequency), the resonance order and the rod length, as follows:

$$c(\omega_p) = \frac{\omega_p L}{p\pi}. \tag{6.71}$$

However, it is important to ensure that the numerator of Eqs. (6.58) and (6.69) does not vanish in ω_p, i.e., $\sin(\xi(\omega_p)z)$ should be different from zero. This can be obtained by choosing the gauge position such as $|z/L|$ is not a fraction. Actually, when $\sin(\xi(\omega_p)z)$ vanishes, the height of the resonance is highly reduced and the signal-to-noise ratio is also decreased. Othman et al. [56] chose to cement the strain gauge in the middle of the bar, i.e., $|z/L| \approx 1/2$. This cancels all resonances with even p. However, this highly improves the signal-to-noise ratio for resonances with odd p.

Equation (6.71) gives the wave velocity knowing the resonance position. In order to get the wave damping, Othman et al. [56] assumed that the damping is almost constant in the neighbourhood of a resonance, i.e., $\alpha(\omega) \approx \alpha(\omega_p) \equiv \alpha_p$. Then, they made a development of $\sin(\xi(\omega_p)L)$ to the first order of $(\omega - \omega_p)$. In this case, Eq. (6.69) yields:

$$S(z,\omega) = |\tilde{\varepsilon}(z,\omega)|^2 = \left|\tilde{\Theta}(\omega)\right|^2 \frac{z^2\alpha_p^2 + \sin^2(k(\omega_p)z)}{L^2\alpha_p^2 + L^2(\omega - \omega_p)^2\left(\left.\frac{\partial k}{\partial \omega}\right|_{\omega=\omega_p}\right)^2}. \tag{6.72}$$

Subsequently, they considered the values of the $S(z, \omega)$ at half height. This gives:

$$\alpha_p = -\delta\omega_p \left.\frac{\partial k}{\partial \omega}\right|_{\omega=\omega_p} = -\delta\omega_p c_g(\omega_p). \tag{6.73}$$

where $\delta\omega_p$ is half the bandwidth at the half height and $c_g(\omega_p)$ is the group velocity.

The disadvantage of this method is that Eq. (6.73) assumes that there is an infinite wave round-trips within the bar. However, the strain gauge signals are recorded for a finite time. Thus they include only effects of a finite number of round-trips within the bar. Consequently, the wave damping is overestimated when calculated by Eq. (6.73). Othman et al. [56] proposed a correction method that accounts for the number of wave round-trip within the bar, M_p, which is approximated as:

$$M_p \approx \left[\frac{T \times c(\omega_p)}{2L}\right], \tag{6.74}$$

where T the strain measurement duration.

Othman [23] used this resonance method to measure the wave dispersion and damping in an aluminum bar and a Nylon bar (see Figs. 6.9 and 6.10, respectively).

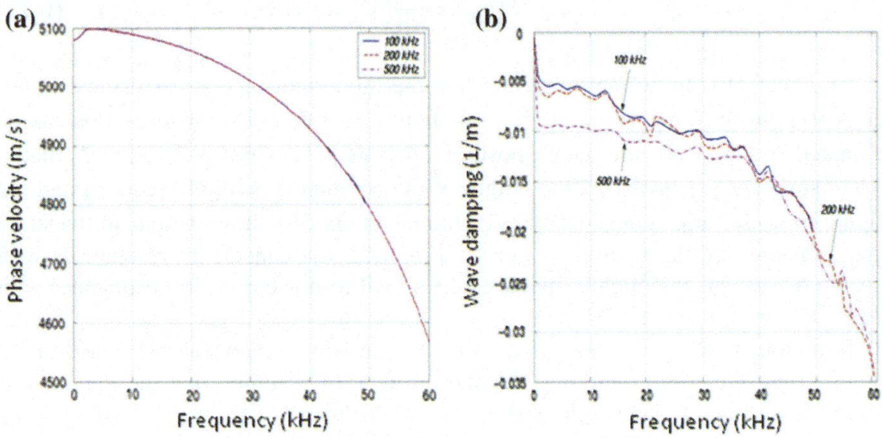

Fig. 6.9 The phase velocity (**a**) and damping (**b**) of an aluminum bar of 40 mm in diameter (reproduced from Ref. [23])

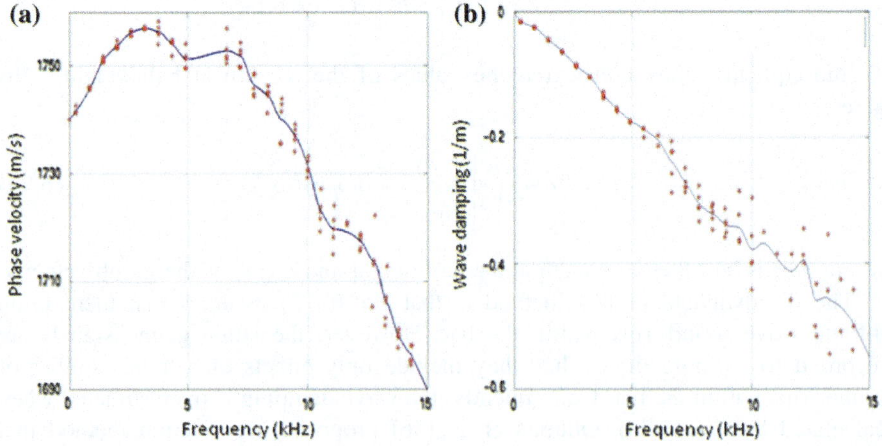

Fig. 6.10 The phase velocity (**a**) and damping (**b**) of a Nylon bar of 40 mm in diameter (reproduced from Ref. [23])

It is also possible to improve the experimental measurements of wave dispersion, obtained by any of the above methods, using the Kramers-Kronig equation writing causality [66, 67]. This works in three steps, first the complex Young's modulus is deduced from the wave dispersion measurements. Subsequently, Kramers-Kronig equations are applied to the real and imaginary parts of the complex Young's modulus. A complex Young's modulus is then obtained respecting the physical principle of causality. Finally, the wave dispersion can be calculated using the new complex Young's modulus.

References

1. Pochhammer L (1876) Uber die Fortpflanzungsgeschwindigkeinten kleiner Schwingungen in einem unbergrenzten isotropen Kreiszylinder. J die Reine Angew Math 81:324–336
2. Chree C (1889) The equations of an isotropic elastic solid in polar and cylindrical co-ordinates, their solutions and applications. Cambridge Phil Soc Trans 14:250–369
3. Zhao H, Gary G (1995) A three dimensional analytical solution of longitudinal wave propagation in an infinite linear viscoelastic cylindrical bar. Application to experimental techniques. J Mech Phys Solids 43:1335–1348
4. Bland DR (1960) The theory of linear viscoelasticity. Oxford University Press, Oxford
5. Coquin GA (1964) Attenuation of guided waves in isotropic viscoelastic materials. J Acoust Soc Am 36:1074–1080
6. Lundberg B, Blanc RH (1988) Determination of mechanical material properties from the two point response of an impacted linearly viscoelastic rod specimen. J Sound Vib 126:97–108
7. Lundberg B, Carlsson J, Sundin KG (1990) Analysis of elastic waves in non-uniform rods from two-point strain measurement. J Sound Vib 137:483–493
8. Lundberg B, Ödeen S (1993) In situ determination of the complex modulus from strain measurements on an impacted structure. J Sound Vib 167:413–419
9. Lu Q, Wang ZJ (2016) Studies of the propagation of viscoelastic spherical divergent stress waves based on the generalized Maxwell model. J Sound Vib 371:183–196
10. Pritz T (1998) Frequency dependences of complex moduli and complex Poisson's ratio of real solid materials. J Sound Vib 214:83–104
11. Pritz T (2000) Measurement methods of complex Poisson's ratio of viscoelastic materials. Appl Acoust 60:279–292
12. Pritz T (2000) The Poisson's loss factor of solid viscoelastic materials. J Sound Vib 306: 790–802
13. Kolsky H (1963) Stress waves in solids. Clarendon Press, Oxford
14. Achenbach JD (1978) Wave propagation in elastic solids. North-Holland Publ. Co, Amsterdam
15. Graff KF (1991) Wave motion in elastic solids. Dover, New York
16. Zemanek J (1972) An experimental and theoretical investigation of elastic wave propagation in a cylinder. J Acoust Soc Am 51:265–283
17. Redwood M (1960) Mechanical waveguides, the propagation of acoustic and ultrasonic waves in fluids and solids with boundaries. Pergamon Press, Paris
18. Bancroft D (1941) The velocity of longitudinal waves in cylindrical bar. Phys Rev 59:588–593
19. Davies RM (1948) A critical study of Hopkinson pressure bar. Phil Trans Roy Soc A240:375–457
20. Curtis CW (1954) Second mode vibration of the Pochhammer-Chree frequency equation. J Appl Phys 25:928

21. Fox G, Curtis CW (1958) Elastic strain produced by sudden application of pressure to one end of a cylindrical bar. II. Experimental observation. J Acoust Soc Am 30:559–563
22. Mikolowitz J, Nisewanger CR (1957) The propagation of compressional waves in a dispersive elastic rod. J Appl Mech 24:240–244
23. Othman R (2002) Extension du champ d'application du système des barres de Hopkinson aux essais à moyennes vitesses de deformation. Ph.D. Thesis, Ecole Polytechnique, France
24. Love AE (1927) A treatise on the mathematical theory of elasticity, 4th edn. Dover Publications, New York
25. Aleyaasin M, Harrigan JJ (2010) Wave dispersion and attenuation in viscoelastic polymeric bars: Analysing the effect of lateral inertia. Int J Mech Sci 52:754–757
26. Ahonsi B, Harrigan JJ, Aleyaasin M (2012) On the propagation coefficient of longitudinal stress waves in viscoelastic bars. Int J Impact Eng 45:39–51
27. Anderson SP (2006) Higher-order rod approximations for the propagation of longitudinal stress waves in elastic bars. J Sound Vib 290:290–308
28. Mindlin RD, McNiven HD (1960) Axially symmetric waves in elastic rods. ASME J Appl Mech 27:145–151
29. Żak A, Krawczuk M (2010) Assessment of rod behaviour theories used in spectral finite element modeling. J Sound Vib 329:2099–2113
30. Othman R (2017) A fractional equation to approximate wave dispersion relation in elastic rods. Strain (in press)
31. Othman R (2012) Cut-off frequencies induced by the length of strain gauges measuring impact events. Strain 48:16–20
32. Bertholf LD (1967) Numerical solution for two-dimensional elastic wave propagation in finite bars. J Appl Mech 34:724–734
33. Bertholf LD, Karnes J (1975) Two-dimensional analysis of the split Hopkinson pressure bar system. J Mech Phys Solids 23:1–19
34. Kennedy LW, Jones OE (1969) Longitudinal wave propagation in a circular bar loaded suddenly by a radially distributed end stress. J Appl Mech 36:470–478
35. Boström A (2000) On wave equations for elastic rods. J Appl Math Mech (ZAMM) 4:245–251
36. Tyas A, Watson AJ (2000) Experimental evidence of Pochhammer-Chree strain variations in elastic cylinders. Exp Mech 40:331–337
37. Tyas A, Watson AJ (2001) An investigation of frequency domain dispersion correction of pressure signals. Int J Impact Eng 25:87–101
38. Tyas A, Pope DJ (2005) Full correction of first-mode Pochhammer-Chree dispersion effects in experimental pressure bar signals. Meas Sci Technol 16:642
39. Tyas A, Ozdemir Z (2014) On backward dispersion correction of Hopkinson pressure bar signals. Philos Trans Roy Soc A 372:20130291
40. Idesman AV, Mates SP (2014) Accurate finite element simulation and experimental study of elastic wave propagation in a long cylinder under impact loading. Int J Impact Eng 71:1–16
41. Merle R, Zhao H (1980) On the errors associated with the use of large diameter SHPB, correction for radially non-uniform distribution of stress and particle velocity in SHPB testing. Int J Impact Eng 32:1964–1980
42. Skalak R (1957) Longitudinal impact of a semi-infinite circular elastic bar. J Appl Mech 24:59–64
43. Folk R, Fox G, Shook CA, Curtis CW (1958) Elastic strain produced by sudden application of pressure to one end of a cylindrical bar. I. Theory. J Acoust Soc Am 30:552–558
44. Malkov MA (2013) Longitudinal impact of semi-infinite elastic bars: Plane short time solution. J Appl Mech 80:#011023 (5pages)
45. McNiven HD, Perry DC (1962) Axially symmetric waves in finite elastic rods. J Acoust Soc Am 34:433–437
46. Bacon C (1998) An experimental method considering dispersion and attenuation in a viscoelastic bar. Exp Mech 38:242–250

47. Theodorakopoulos ID (2015) Determination of the useful frequency for the calculation of the attenuation coefficient and phase velocity in a viscoelastic Hopkinosn pressure bar. Int Appl Mech 51:360–368
48. Butt HSU, Xue P (2014) Determination of the wave propagation coefficient of viscoelastic SHPB: significance for characterization of cellular materials. Int J Impact Eng 74:83–91
49. Butt HSU, Xue P, Jiang TZ, Wang B (2015) Parametric identification for material of viscoelastic SHPB from wave propagation data incorporating geometrical effects. Int J Mech Sci 91:46–54
50. Gorham DA, Wu XJ (1996) An empirical method for correcting dispersion in pressure bar measurements of impact stress. Meas Sci Technol 7:1227–1232
51. Yew EH, Chen CS (1978) Experimental study of dispersive waves in beam and rod using FFT. J Appl Mech 45:940–942
52. Cheng ZQ, Crandall JR, Pilkey WD (1998) Wave dispersion and attenuation in viscoelastic split Hopkinson pressure bar. Shock Vib 5:307–316
53. Humen V, Potesil A (1993) Pulse method used to identify material properties in linear viscoelastic media. Int J Impact Eng 13:85–98
54. Lee CKB, Crawford RC (1993) A new method for analyzing dispersed bar gauge data. Meas Sci Technol 4:931–937
55. Hillström L, Mossberg M, Lundberg B (2000) Identification of complex modulus from measured strains on an axially impacted bar using least squares. J Sound Vib 230:689–707
56. Othman R, Blanc RH, Bussac MN, Collet P, Gary G (2002) Identification de la relation de la dispersion dans les barres. Comptes Rendus de Mécanique 330:849–856
57. Bussac MN, Collet P, Gary G, Othman R (2002) An optimisation method for separating and rebuilding one-dimensional dispersive waves from multi-point measurements. Application to elastic or viscoelastic bars. J Mech Phys Solids 50:321–349
58. Othman R, Bussac MN, Collet P, Gary G (2001) Séparation et reconstruction des ondes dans les barres élastiques et viscoélastiques à partir de mesures redondantes. Comptes Rendus de l'Académie des Sciences Série IIb 329:369–376
59. Mossberg M, Hillström L, Abrahamsson L (2001) Parametric identification of viscoelastic materials from time and frequency data. Inverse Probl Eng 9:645–670
60. Mossberg M, Hillström L, Söderström T (2001) Non-parametric identification of viscoelastic materials from wave propagation experiments. Atutomatica 37:511–521
61. Madigosky WM, Lee GF (1983) Improved resonance technique for materials characterization. J Acoust Soc Am 73:1374–1377
62. Jimeno-Fernandez I, Überall H, Madigosky WM, Fiorito RB (1992) Resonance decomposition for vibratory response of a viscoelastic rod. J Acoust Soc Am 91:2030–2033
63. Garrett SL (1990) Resonant acoustic determination of elastic moduli. J Acoust Soc Am 88:210–221
64. Guo Q, Brown DA (2000) Determination of the dynamic elastic moduli and internal friction using thin rods. J Acoust Soc Am 108:167–174
65. Cheng S (2000) Resonant frequency method for the measurement and uncertainty analysis of acoustic and elastic properties. Ultrasonics 38:206–211
66. Collet P, Gary G, Lundberg B, Mohr D (2012) Complex modulus estimation respecting causality: application to viscoelastic bars. In EPJ Web of conferences, vol 26, p 01012 (6 pages). (https://doi.org/10.1051/epjconf/20122601012)
67. Collet P, Gary G, Lundberg B (2012) Noise-Corrected Estimation of Complex Modulus in Accord With Causality and Thermodynamics: Application to an Impact Test. J Appl Mech 80: paper 011018 (7 pages). (https://doi.org/10.1115/1.4007081)

Chapter 7
Wave Separation Techniques

Ramzi Othman

7.1 Introduction

Wave separation techniques are methods that use mathematical tools to process at least two measurements in order to separate the effects of two or more different types of waves. They are first introduced to the Kolsky/Hopkinson bar machine by Lundberg and Henchoz [1] and Yanagihara [2]. The use of wave separation techniques overcomes the limitation of the test duration caused by the waves' overlap in strain gauge cross-sections. The wave separation techniques are either based on the time expressions of waves or their frequency expressions. The subsequent Sect. 7.2 will first discuss the wave separation techniques that are based on wave solutions that are written in the time domain. Later, Sect. 7.3 will discuss the wave separation techniques which are written in the frequency domain.

7.2 Wave Separation in Time Domain

The wave separation techniques in the time domain were developed in the framework of the elementary wave theory of thin elastic rods. Namely, the displacement of any cross-section of the bar is written as a superposition of two waves propagating at the same constant speed but in opposite directions:

$$u(x,t) = f(t - x/c) + g(t + x/c), \tag{7.1}$$

R. Othman (✉)
Mechanical Engineering Department, Faculty Engineering,
King Abdulaziz University, P.O. Box 80248, Jeddah 21589, Saudi Arabia
e-mail: rothman1@kau.edu.sa

R. Othman (ed.), *The Kolsky-Hopkinson Bar Machine*,
https://doi.org/10.1007/978-3-319-71919-1_7

where x is the position of the cross-section, t the time, f is the displacement due to the wave propagating in the positive x direction (mostly called the incident wave), g is the displacement due to the wave propagating in the negative x direction (mostly called the reflected wave) and c is the sound speed.

Lundberg and Henchoz [1] and Yanagihara [2] have independently proposed to indirectly measure the force and velocity at the end of an elastic bar using two strain measurements recorded at two different cross-sections, far from the bar ends (Fig. 7.1). An impact load is applied on either side of the bar. The solution proposed in Refs. [1, 2] works with any boundary condition on the opposite end of the bar. Without any loose of generality, the wave separation method finds the strain and velocity at a cross-section ($x = 0$) knowing two strain measurement $\varepsilon_1(t)$ and $\varepsilon_2(t)$ at two cross-section x_1 and x_2.

Lundberg and Henchoz [1] and Yanagihara [2] assumed that the bar is thin and elastic. They independently came up with the following solutions:

$$\begin{aligned}\varepsilon_0(t) = \varepsilon_0(t-2T) + \varepsilon_1(t+T_1) - \varepsilon_1(t+T_1-2T_2)\\ + \varepsilon_2(t-T_2) - \varepsilon_1(t-T_2+2T_1),\end{aligned} \tag{7.2}$$

and

$$\begin{aligned}\frac{v_0(t)}{c} = \frac{v_0(t-2T)}{c} - \varepsilon_1(t+T_1) - \varepsilon_1(t+T_1-2T_2)\\ + \varepsilon_2(t-T_2) + \varepsilon_1(t-T_2+2T_1),\end{aligned} \tag{7.3}$$

where, $\varepsilon_0(t)$ and $v_0(t)$ are the strain and velocity at ($x = 0$), respectively, c is the sound speed in the bar, $T_1 = x_1/c$ is the wave propagation time from ($x = 0$) till the first strain gauge, $T_2 = x_2/c$ is the wave propagation time from ($x = 0$) till the second strain gauge and $T = T_2 - T_1$ is the wave propagation time between the two strain strain gauges. Without loss of generality, $T_1 < T_2$.

Equations (7.2) and (7.3) can be proved using the characteristics diagram depicted in Fig. 7.2. It is also based on the solution of one-dimensional propagation of elastic waves in thin rods given by Eq. (7.1). Considering this equation the strain is also written as a contribution of two waves propagating in opposite directions:

$$\varepsilon(x,t) = \left(-f'\left(t-\frac{x}{c}\right) + g'(t+x/c)\right)/c, \tag{7.4}$$

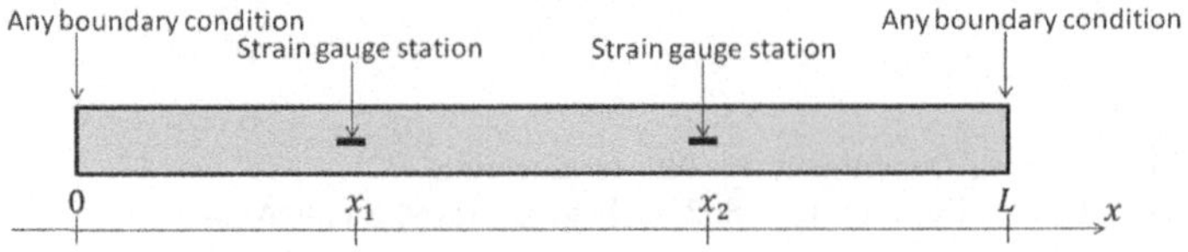

Fig. 7.1 Schematic of an elastic bar with two strain gauge stations cemented on it

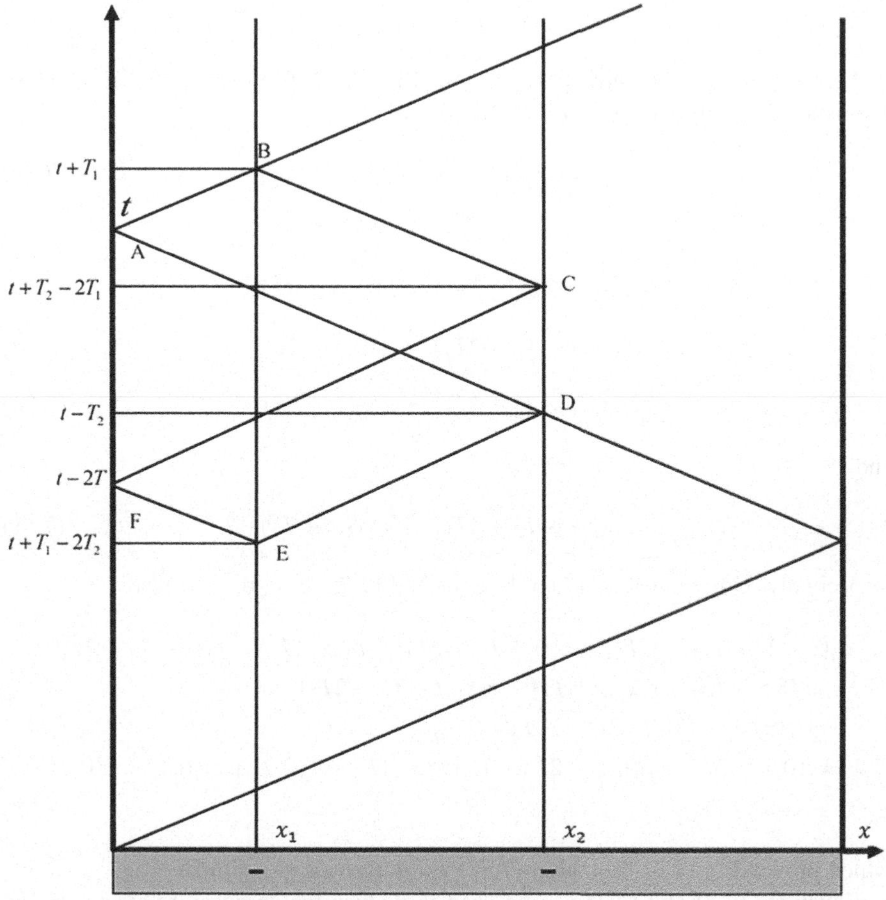

Fig. 7.2 Characteristics diagram for waves propagating in elastic thin bars

This equation can be simplified to:

$$\varepsilon(x,t) = \varepsilon_i(x,t) + \varepsilon_r(x,t), \tag{7.5}$$

Applying the last equation to cross-sections x_1 and x_2 and to times $t+T_1$, $t-T_2, t+T_1-2T_2$ and $t-T_2+2T_1$ yields:

$$\varepsilon(x_1, t+T_1) = \varepsilon_i(x_1, t+T_1) + \varepsilon_r(x_1, t+T_1), \tag{7.6a}$$

$$\varepsilon(x_1, t+T_1-2T_2) = \varepsilon_i(x_1, t+T_1-2T_2) + \varepsilon_r(x_1, t+T_1-2T_2), \tag{7.6b}$$

$$\varepsilon(x_2, t-T_2) = \varepsilon_i(x_2, t-T_2) + \varepsilon_r(x_2, t-T_2), \tag{7.6c}$$

and

$$\varepsilon(x_2, t - T_2 + 2T_1) = \varepsilon_i(x_2, t - T_2 + 2T_1) + \varepsilon_r(x_2, t - T_2 + 2T_1). \qquad (7.6d)$$

Considering the segments [AB], [AD], [ED], [EF], [CF] and [CB] of the characteristics diagram depicted in Fig. 7.2 gives:

$$\varepsilon_i(x_1, t + T_1) = \varepsilon_i(0, t), \qquad (7.7a)$$

$$\varepsilon_r(x_2, t - T_2) = \varepsilon_r(0, t), \qquad (7.7b)$$

$$\varepsilon_i(x_1, t + T_1 - 2T_2) = \varepsilon_i(x_2, t - T_2), \qquad (7.7c)$$

$$\varepsilon_r(x_1, t + T_1 - 2T_2) = \varepsilon_r(0, t - 2T), \qquad (7.7d)$$

$$\varepsilon_i(x_2, t - T_2 + 2T_1) = \varepsilon_i(0, t - 2T), \qquad (7.7e)$$

and

$$\varepsilon_r(x_1, t - T_2 + 2T_1) = \varepsilon_r(x_2, t + T_1), \qquad (7.7f)$$

respectively. Combing Eqs. (7.6a, b, c, d) and (7.7a, b, c, d, e, f) leads:

$$\begin{aligned} &\varepsilon_0(t - 2T) + \varepsilon_1(t + T_1) - \varepsilon_1(t + T_1 - 2T_2) + \varepsilon_2(t - T_2) - \varepsilon_1(t - T_2 + 2T_1) \\ &= \varepsilon(0, t - 2T) + \varepsilon(x_1, t + T_1) - \varepsilon(x_1, t + T_1 - 2T_2) \\ &\quad + \varepsilon(x_2, t - T_2) - \varepsilon(x_2, t - T_2 + 2T_1) \\ &= \varepsilon(0, t - 2T) - \varepsilon_i(0, t - 2T) - \varepsilon_r(0, t - 2T) + \varepsilon_i(0, t) + \varepsilon_r(0, t) = \varepsilon(0, t), \end{aligned} \qquad (7.8)$$

which proves Eq. (7.2). Equation (7.3) can be proved in a similar way.

Equations (7.2) and (7.3) can be used to deduce the strain and velocity at any cross-section of the bar knowing the strain at two different cross-sections. It works in an iterative way. Put in other words, the strain and velocity should be known for a period of time of length $2T$. This is a minor difficulty as the strain and velocity can be assumed equal to 0 for the period of time $[-2T, 0]$, i.e., just before the test. Once this assumption is considered, Eqs. (7.2) and (7.3) can be used one time to get the strain and velocity for time period $[0, 2T]$, a second time to get the strain and velocity for time period $[2T, 4T]$, etc. Adachi et al. [3] used this wave separation technique to process two strain signals recorded on an elastic bar to measure the force in an impact pendulum machine.

As analysed by Lundberg and Henchoz [1], this wave separation technique has some drawbacks. Mainly,

- It does not take into account the wave dispersion in bars as it is based on the one-dimensional wave propagation theory in elastic rods.

- This wave separation technique is sensitive to the noise recorded on the strain gauges.
- It is also sensitive to errors in measuring the gauge positions.

It is also possible to add that this technique is sensitive to errors in measuring the wave speed in the bar. Actually, a misknowledge of the gauge positions or the wave speed leads to errors in calculating the shifting times T_1, T_2 and T. It is worth noting that these drawbacks are not specific to this wave separation technique. They are common with all wave separation techniques using two measurements. However, the major drawback of this technique is that it does not account for the wave dispersion effects. Mainly, it assumes that the wave pulses propagate along the bar without any change of shape or loose of energy.

Park and Zhou [4] used a similar iterative approach, except that the two strain measurements are replaced by one strain measurement and one boundary condition. This solution has the advantage of reducing the number of recorded signals. However, it has the drawback that the assumed boundary conditions are not realistic in intermediate strain rate testing. Actually, Park and Zhou [4] assumed that impact end of the input bar is traction-free after one-round trip of the wave. Moreover, they assumed one end of the output bar is also traction free. This highly limit the maximum deformation of the specimen and the test duration. Actually, the Park and Zhou technique can only double the test duration and the maximum deformation of the specimen. Yet, Park and Zhou's solution increases the test duration by 100%. But, it is insufficient for testing materials in the intermediate strain rate range. Moreover, a 100% increase of the test duration can be alternatively achieved without any use of wave separation technique as proposed by Gary and Mohr [5]. Indeed, the strain gauge station of the input bar is placed near the impact end. Subsequently, the stress, strain and strain rate in the specimen are deduced only from the incident and transmitted waves. However, Gary and Mohr' solution assumes the deformation of the specimen is homogeneous.

Meng and Li [6] were also interested in Lundberg-Henchoz-Yanangihara solution. In order to overcome the drawback related to the wave dispersion effects, they proposed to reduce the distance between the two strain gauges recording the two strain measurements. This reduces the shifting times T_1, T_2 and T. As a consequence, the effects of wave dispersion are reduced. However, by reducing the time T the number of iterations needed to process signals using Eqs. (7.2) and (7.3) is increased. Consequently, the errors related to the noise or the position of gauges, that are discussed above, are more amplified than when longer distance is considered between the two gauges.

In order to overcome the problem of noise amplification, Jacquelin and Hamelin [7] proposed a time-domain wave separation technique using three strain measurements and one velocity measurement. Actually, the wave separation problem is an ill-conditioned mathematical problem [8]. Thus, redundant measurements are necessary to reduce noise amplification [9, 10]. Jacquelin and Hamelin [7] assumed that the third strain gauge position and the velocity measurement position are the

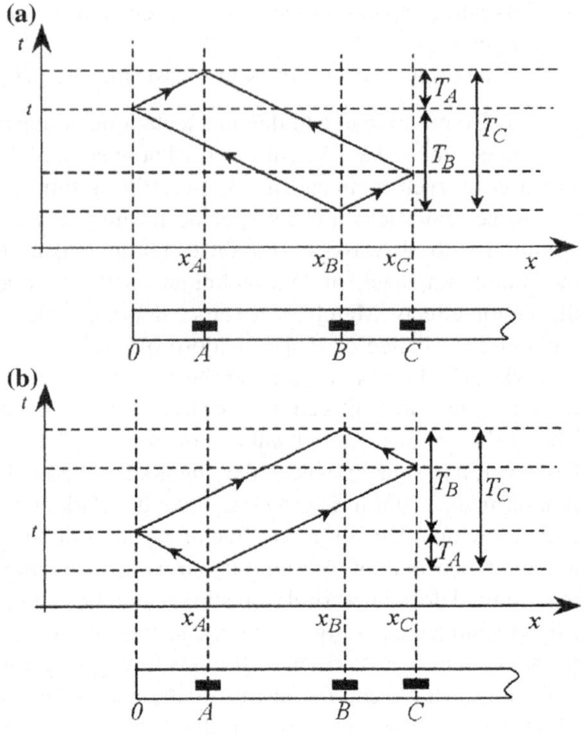

Fig. 7.3 Characteristics diagram for waves propagating in elastic thin bars. (Reproduced from Ref. [7] with permissions from Elsevier) (to match notations of the text we consider $x_1 = x_A$, $x_2 = x_B$, $x_3 = x_C$, $T_1 = T_A$, $T_2 = T_B$ and $T_3 = T_C$)

same and such as $x_3 = x_1 + x_2$. Considering the characteristics diagram shown in Fig. 7.3 it is possible to show that:

$$\varepsilon_0(t) = \varepsilon(x_1, t + T_1) + \varepsilon(x_2, t - T_2) - \varepsilon(x_3, t + T_1 - T_2), \tag{7.9}$$

and

$$\frac{v_0(t)}{c} = -\varepsilon(x_1, t + T_1) + \varepsilon(x_2, t - T_2) + \frac{v(x_3, t + T_1 - T_2)}{c}. \tag{7.10}$$

Using the same characteristics diagram of Fig. 7.3, it is also possible to prove that:

$$\varepsilon_0(t) = \varepsilon(x_1, t - T_1) + \varepsilon(x_2, t + T_2) - \varepsilon(x_3, t + T_2 - T_1), \tag{7.11}$$

and

$$\frac{v_0(t)}{c} = \varepsilon(x_1, t - T_1) - \varepsilon(x_2, t + T_2) + \frac{v(x_3, t + T_2 - T_1)}{c}. \tag{7.12}$$

Equations (7.9)–(7.12) can be combined to obtain the following symmetric equations:

$$\begin{aligned}\varepsilon_0(t) = &\frac{1}{2}(\varepsilon(x_1, t+T_1) + \varepsilon(x_1, t-T_1) + \varepsilon(x_2, t-T_2) \\ &+ \varepsilon(x_2, t+T_2) - \varepsilon(x_3, t+T_1-T_2) - \varepsilon(x_3, t+T_2-T_1)),\end{aligned} \tag{7.13}$$

and

$$\begin{aligned}\frac{v_0(t)}{c} = &\frac{1}{2}(-\varepsilon(x_1, t+T_1) + \varepsilon(x_1, t-T_1) + \varepsilon(x_2, t-T_2) - \varepsilon(x_2, t+T_2) \\ &+ \frac{v(x_3, t+T_1-T_2)}{c} + \frac{v(x_3, t+T_2-T_1)}{c}\Bigg).\end{aligned} \tag{7.14}$$

In order to obtain the force in a cross-section ($x = 0$), and only at this cross-section, it is possible to use either Eqs. (7.9), (7.11) or (7.13). Likewise, it is possible to use either Eqs. (7.10), (7.12) or (7.14) to deduce the velocity at the cross-section ($x = 0$), and only at this cross-section.

The wave separation technique proposed by Jacquelin and Hamelin is stable against noise [11]. Moreover, it can be easily extended to take into account wave dispersion effects as will be exposed in the coming Sect. 7.3.2. However, it has two main drawbacks. First, once the strain gauges are cemented on the bar, it is possible to calculate the strain and velocity at only one cross-section, which is ($x = 0$). Second, it assumes that a strain measurement and a velocity measurement can be recorded at the same cross-section $x_3 = x_1 + x_2$. Assume that this is possible and affordable from a technical point of view, there is much easier processing technique to identify the strain and velocity at ($x = 0$). This solution was proposed by Casem et al. [12]. They used single-point measurements. Let be x_0 the cross-section where a strain and a velocity measurements are recorded. The strain can be written:

$$c\varepsilon(x_0, t) = -f'\left(t - \frac{x_0}{c}\right) + g'(t + x_0/c), \tag{7.15}$$

whereas the velocity reads:

$$v(x_0, t) = f'\left(t - \frac{x_0}{c}\right) + g'(t + x_0/c). \tag{7.16}$$

It is possible to separate the effects of the two waves f and g, which are propagating in opposite directions, by adding and subtracting Eqs. (7.15) and (7.16). Namely,

$$f'\left(t - \frac{x_0}{c}\right) = \frac{v(x_0, t) - c\varepsilon(x_0, t)}{2}, \tag{7.17}$$

and

$$g'\left(t+\frac{x_0}{c}\right)=\frac{v(x_0,t)+c\varepsilon(x_0,t)}{2}. \quad (7.18)$$

As f' and g' are determined, it is then possible to calculate the force and the velocity at any cross-section of the bar. More precisely, the velocity at any cross-section x is calculated as:

$$v(x,t)=f'\left(t-t'-\frac{x_0}{c}\right)+g'\left(t+t'+\frac{x_0}{c}\right), \quad (7.19)$$

where $t'=x_0/c$. Likewise, the force at any cross-section x is calculated as:

$$F(x,t)=\frac{EA}{c}\left(f'\left(t-t'-\frac{x_0}{c}\right)+g'\left(t+t'+\frac{x_0}{c}\right)\right), \quad (7.20)$$

where E and A are the Young's modulus and the cross-sectional area of the bar, respectively. We want to point out that $f'\left(t-t'-\frac{x_0}{c}\right)$ and $g'\left(t+t'+\frac{x_0}{c}\right)$ are a shift in the time domain of the velocity waves $f'\left(t-\frac{x_0}{c}\right)$ and $g'\left(t+\frac{x_0}{c}\right)$, respectively.

Equations (7.19) and (7.20) close the solution proposed by Casem et al. [12]. This wave separation technique is strong against noise [11], mainly, because the associated mathematical problem is not ill-posed. Moreover, it has the advantage that it can be easily extended to account for the wave dispersion effects in elastic or viscoelastic bars [12]. The only drawback is that it needs a velocity measurement at the same cross-section as the strain measurement. A velocity measurement needs more advanced technology [13] and rare who instrumented a Kolsky-Hopkinson bar set-up with such technology.

7.3 Wave Separation in the Frequency Domain

Zhao and Gary [14, 15] are the first to deal with wave dispersion effects in wave separation problem. They proposed an iterative methodology based on the pioneer solution of Lundberg-Henchoz-Yanigihara [1, 2]. Zhao and Gary [14, 15] mainly used the frequency domain to shift the measurements from one cross-section to nother. Several works have then followed and wrote the wave separation problem completely in the frequency domain. These will be called the "direct solutions". The methodology proposed by Zhao and Gary [14, 15] is classified as an "iterative solution".

7.3.1 The Iterative Solution of Zhao and Gary

The main motivation of Zhao and Gary to develop a new wave separation technique was to take into consideration the wave dispersion effects. The recorded two strain measurements on a bar which can be either elastic or viscoelastic. The diameter of the bar is of some tens of millimeters. Thus the one-dimensional propagation of waves is ensured. However, waves can distort while propagating because of 3D geometrical effects. The Zhao and Gary technique is based on the fact the in the beginning of the test, the strain gauge measures only the effects of the incident wave that propagates forwards (rightwards in Fig. 7.1) as the incident wave is created at one end of the bar (left end) and needs a certain period of time to propagate along the bar and then reflects back when it reaches the second end (right end).

Let R_1 and R_2 are time when the incident (rightwards) wave arrives at the first and second, respectively, strain gauge cross-sections. Obviously,

$$\varepsilon_i(x_1,t) = \varepsilon_1(t), \quad \text{if } t < R_1 \tag{7.21}$$

and

$$\varepsilon_i(x_2,t) = \varepsilon_2(t), \quad \text{if } t < R_2 \tag{7.22}$$

where $\varepsilon_1(t)$ and $\varepsilon_2(t)$ are the strains recorded at the first and second, respectively, strain gauges.

The time is divided in intervals $I_k = [R_1 + 2(k-2)T, R_1 + 2(k-1)T]$ and $J_k = [R_2 + 2(k-1)T, R_2 + 2kT]$, where $T = (x_2 - x_1)/c$ the time needed by the wave to go from one gauge to another and k is an integer different from 0. The two strain measurements are sliced in $\varepsilon^{(k)}(x_1,t)$ and $\varepsilon^{(k)}(x,t)$, such as:

$$\varepsilon^{(k)}(x_1,t) = \varepsilon(x_1,t) \times 1_{I_k}, \tag{7.23}$$

and

$$\varepsilon^{(k)}(x_2,t) = \varepsilon(x_2,t) \times 1_{J_k}, \tag{7.24}$$

where 1_{I_k} and 1_{J_k} are the characteristic functions of the intervals I_k and J_k, respectively.

Each strain measurement is the superposition of the effects of the forward (incident) and backward (reflected) waves. Therefore,

$$\varepsilon^{(k)}(x_1,t) = \varepsilon_i^{(k)}(x_1,t) + \varepsilon_r^{(k)}(x_1,t), \tag{7.25}$$

and

$$\varepsilon^{(k)}(x_2,t) = \varepsilon_i^{(k)}(x_2,t) + \varepsilon_r^{(k)}(x_2,t). \tag{7.26}$$

The goal of the wave separation technique proposed by Zhao and Gary [14, 15] is to calculate separately the effects of the incident and reflected waves $\varepsilon_i^{(k)}(x_1,t)$, $\varepsilon_r^{(k)}(x_1,t)$, $\varepsilon_i^{(k)}(x_2,t)$ and $\varepsilon_r^{(k)}(x_2,t)$ knowing the two recorded strain measurements.

In order to account for wave dispersion effects, it is impossible to avoid the use of Fourier transform. More precisely, the Fourier transform of the participation of the incident wave to the second strain measurement can be expressed in terms of the Fourier transform of the participation of the incident wave to the first strain measurement:

$$\tilde{\varepsilon}_i^{(k)}(x_2,\omega) = \mathrm{e}^{-i\xi(\omega)\Delta}\tilde{\varepsilon}_i^{(k-1)}(x_1,\omega), \tag{7.27}$$

where $\tilde{f}$ is the Fourier transform of f, and $\Delta = x_2 - x_1$ is the distance between the two strain gauges. Likewise,

$$\tilde{\varepsilon}_r^{(k)}(x_1,\omega) = \mathrm{e}^{-i\xi(\omega)\Delta}\tilde{\varepsilon}_r^{(k)}(x_2,\omega). \tag{7.28}$$

The above Eqs. (7.27) and (7.28) are approximate equations. They are only exact if there in no wave dispersion as the intervals I_k and J_k are built assuming a constant wave velocity c, which is the wave velocity of non dispersive waves. However, Eqs. (7.27) and (7.28) write the shifting of waves from one strain gauge to another that accounts for wave dispersion effects as an exact wave dispersion relation $\xi(\omega)$ can be used.

Let's assume that $\varepsilon_i^{(k)}(x_1,t)$ is known for a given k, $\varepsilon_i^{(k+1)}(x_2,t)$ can be deduced from Eq. (7.27) as:

$$\varepsilon_i^{(k+1)}(x_2,t) = \int_{-\infty}^{+\infty} \mathrm{e}^{-i(\xi(\omega)\Delta-\omega t)}\tilde{\varepsilon}_i^{(k)}(x_1,\omega)d\omega. \tag{7.29}$$

Equation (7.29) means that once the incident wave at the first strain gauge cross-section is known on an interval I_k, the incident wave at the second strain gauge cross-section can be known on the interval J_{k+1} by using direct and inverse Fourier transforms. As $\varepsilon_i^{(k+1)}(x_2,t)$ is now known, it is easy to deduce $\varepsilon_r^{(k+1)}(x_2,t)$ the reflected wave at the second strain gauge cross-section on the interval J_{k+1}. More precisely, Eq. (7.26) yields:

$$\varepsilon_r^{(k+1)}(x_2,t) = \varepsilon^{(k+1)}(x_2,t) - \varepsilon_i^{(k+1)}(x_2,t). \tag{7.30}$$

Now $\varepsilon_r^{(k+1)}(x_2,t)$ is known. Thus $\varepsilon_r^{(k+1)}(x_1,t)$ can be calculated using Eq. (7.28) and, direct and inverse Fourier transforms:

$$\varepsilon_r^{(k+1)}(x_1,t) = \int_{-\infty}^{+\infty} e^{-i(\xi(\omega)\Delta-\omega t)}\tilde{\varepsilon}_r^{(k+1)}(x_2,\omega)d\omega. \tag{7.31}$$

Once $\varepsilon_r^{(k+1)}(z_1,t)$ is known, it easy to calculate $\varepsilon_i^{(k+1)}(z_1,t)$ using Eq. (7.25):

$$\varepsilon_i^{(k+1)}(x_1,t) = \varepsilon^{(k+1)}(x_1,t) - \varepsilon_r^{(k+1)}(x_1,t). \tag{7.32}$$

Thus starting from $\varepsilon_i^{(k)}(z_1,t)$, it is possible to calculate $\varepsilon_i^{(k+1)}(z_1,t)$. Moreover, it is possible to deduce $\varepsilon_i^{(k+1)}(x_2,t)$, $\varepsilon_r^{(k+1)}(x_2,t)$ and $\varepsilon_r^{(k+1)}(x_1,t)$.

The wave separation procedure proposed by Zhao and Gary [14, 15] can be synthesized in the following scheme:

$$\left|\begin{array}{l}\varepsilon_i^{(0)}(x_1,t) = \varepsilon^{(0)}(x_1,t)\\ \left\|\begin{array}{l}\varepsilon_i^{(k+1)}(x_2,t) = \int_{-\infty}^{+\infty} e^{-i(\xi(\omega)\Delta-\omega t)}\tilde{\varepsilon}_i^{(k)}(x_1,\omega)d\omega\\ \varepsilon_r^{(k+1)}(x_2,t) = \varepsilon^{(k+1)}(x_2,t) - \varepsilon_i^{(k+1)}(x_2,t), \quad k = 0\ldots N\\ \varepsilon_r^{(k+1)}(x_1,t) = \int_{-\infty}^{+\infty} e^{-i(\xi(\omega)\Delta-\omega t)}\tilde{\varepsilon}_r^{(k+1)}(x_2,\omega)d\omega\\ \varepsilon_i^{(k+1)}(x_1,t) = \varepsilon^{(k+1)}(x_1,t) - \varepsilon_r^{(k+1)}(x_1,t)\end{array}\right.\end{array}\right. \tag{7.33}$$

This wave separation solution partially accounts for wave dispersion as the cutting of intervals I_k and J_k uses a non dispersive wave velocity c. The errors due to this approximation should cumulate and increase with the number of iteration k. However, Zhao and Gary [15] have showed that this methodology can work for a test duration as long as 30 ms, which is more than 60 times the test duration of a

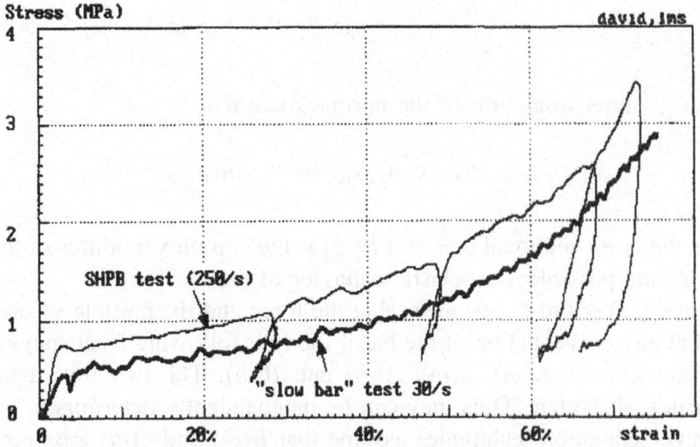

Fig. 7.4 Strain rate effects on foam at medium strain rates. (Reproduced from Ref. [15] with permissions from Elsevier)

classical Kolsky-Hopkinson bar machine. It is then possible to achieve large strains at intermediate strain rates (Fig. 7.4).

7.3.2 Direct Solutions

The advantage of the direct solutions, using Fourier transform, is that they can work with either elastic or viscoelastic bars. They also can take into account the wave dispersion effects.

Let's consider the most general case of a straight uniform viscoelastic bar with possible 3D geometrical effects. In order to consider the wave dispersion effects, it is more appropriate to work in the Fourier domain. The Fourier transform of strain in a cross-section x is written:

$$\tilde{\varepsilon}(x,\omega) = A(\omega)\mathrm{e}^{-i\xi(\omega)x} + B(\omega)\mathrm{e}^{i\xi(\omega)x}. \tag{7.34}$$

where $A(\omega)$ and $B(\omega)$ are the waves propagating in opposite directions and $\xi(\omega)$ is the wave dispersion relation that can account for both material and geometrical effects. The above Eq. (7.34) assumes one-dimensional wave propagation within the bar and takes into account the effects of the wave dispersion. The objective of the wave separation techniques is to recover the two waves $A(\omega)$ and $B(\omega)$ knowing some strain and/or velocity measurements on some cross-sections of the bar.

Knowing the expression of the strain's Fourier transform at one cross-section, it is possible to derive the Fourier transforms of the particle velocity and the normal force at the same cross-section. Namely, the Fourier transform of the particle velocity reads:

$$\tilde{v}(x,\omega) = \frac{\omega}{\xi(\omega)}\left(-A(\omega)\mathrm{e}^{-i\xi(\omega)x} + B(\omega)\mathrm{e}^{i\xi(\omega)x}\right), \tag{7.35}$$

whereas the Fourier transform of the normal force reads:

$$\tilde{N}(x,\omega) = E(\omega)S\left(A(\omega)\mathrm{e}^{-i\xi(\omega)x} + B(\omega)\mathrm{e}^{i\xi(\omega)x}\right), \tag{7.36}$$

where S is the cross-sectional area and $E(\omega)$ is the complex modulus of the bar that accounts for the possible viscoelastic behavior of the bar [16].

Equations (7.35) and (7.36) mean that the force and the particle velocity can be recovered at any cross-section of the bar if the four following frequency-dependent functions are known: $E(\omega)$, $\xi(\omega)$, $A(\omega)$ and $B(\omega)$. The two first functions are characteristics of the bar. Thus they can be independently determined.

The wave separation techniques assume that $E(\omega)$ and $\xi(\omega)$ are known. Their objective is to determine $A(\omega)$ and $B(\omega)$, the two waves propagating in opposite directions, from a reduced number of measurements on the bar.

Since we have two unknowns $A(\omega)$ and $B(\omega)$, for each frequency, the establishment of two equations should be sufficient to solve the problem. To do this, several authors proposed to record two measurements on one or two cross-sections of the bar [12, 15, 17, 18]. The only condition here is that the two measurements should lead to two independent mathematical equations.

Casem et al. [12] used one strain measurement and one velocity measurement at the same cross-section. Let x_0 be the coordinate of the cross-section where the two measurements are recorded. Knowing the expressions of strain and velocity given in Eqs. (7.34) and (7.35), the two waves, $A(\omega)$ and $B(\omega)$ can simply be derived as:

$$A(\omega) = e^{i\xi(\omega)x_0} \frac{\tilde{\varepsilon}(x_0, \omega) - \frac{\tilde{v}(x_0,\omega)}{c(\omega)}}{2}, \tag{7.37}$$

and

$$B(\omega) = e^{-i\xi(\omega)x_0} \frac{\tilde{\varepsilon}(x_0, \omega) + \frac{\tilde{v}(x_0,\omega)}{c(\omega)}}{2}, \tag{7.38}$$

where $c(\omega) = \frac{\omega}{\xi(\omega)}$ is a complex wave velocity. Equations (7.37) and (7.38) show that it is quite simple to solve the wave separation problem, i.e., to determine the two waves propagating in opposite directions using one strain and one velocity measurements, both are recorded in the same cross-section. This wave separation solution has two main advantages. First, the mathematical problem is well-posed. The Eqs. (7.37) and (7.38) are valid and stable for any frequency. The amplification of noise is then highly reduced [11]. Second, Casem et al. [12] wave separation can work either in the frequency domain (Eqs. 7.37 and 7.38) or in the time domain (Eqs. 7.19 and 7.20 of Sect. 7.1). However, the main disadvantage is to afford a velocity measurement and to precisely adjust it at the same cross-section as the strain measurement.

Because of this last disadvantage, several works have argued for the use of two strain measurements which are recorded in two different cross-sections [15, 17, 18]. Two equations can then be derived from Eq. (7.34):

$$\tilde{\varepsilon}_1(\omega) = \tilde{\varepsilon}(x_1, \omega) = A(\omega)e^{-i\xi(\omega)x_1} + B(\omega)e^{i\xi(\omega)x_1}, \tag{7.39}$$

and

$$\tilde{\varepsilon}_2(\omega) = \tilde{\varepsilon}(x_2, \omega) = A(\omega)e^{-i\xi(\omega)x_2} + B(\omega)e^{i\xi(\omega)x_2}. \tag{7.40}$$

where $\tilde{\varepsilon}_1(\omega)$ and $\tilde{\varepsilon}_2(\omega)$ are the Fourier transforms of the measured strain at the two cross-sections x_1 and x_2. If these two equations are independent, we can simply calculate the two waves, as first suggested by Zhao and Gary [15]. More precisely, solving Eqs. (7.39) and (7.40) yields:

$$A(\omega) = \frac{\tilde{\varepsilon}_2(\omega)\mathrm{e}^{i\xi(\omega)x_1} - \tilde{\varepsilon}_1(\omega)\mathrm{e}^{i\xi(\omega)x_2}}{\mathrm{e}^{-i\xi(\omega)\Delta} - \mathrm{e}^{i\xi(\omega)\Delta}}, \tag{7.41}$$

and

$$B(\omega) = \frac{\tilde{\varepsilon}_1(\omega)\mathrm{e}^{-i\xi(\omega)x_2} - \tilde{\varepsilon}_2(\omega)\mathrm{e}^{-i\xi(\omega)x_1}}{\mathrm{e}^{-i\xi(\omega)\Delta} - \mathrm{e}^{i\xi(\omega)\Delta}}, \tag{7.42}$$

where $\Delta = x_2 - x_1$.

Equations (7.41) and (7.42) could have been a very nice solution to the wave separation in elastic and viscoelastic bars, except that their denominator can vanish. In the case of elastic bars, the determinant of Eqs. (7.41) and (7.42) equals zero for a set of resonant frequencies defined by:

$$\Omega_e = \{\omega_n; \xi(\omega_n)\Delta = n\pi, \quad \forall n \in IN\}. \tag{7.43}$$

In the case of a viscoelastic bar, the set of resonant frequencies is theoretically reduced to:

$$\Omega_v = \{0\}. \tag{7.44}$$

However, the denominator has a low value for all the other frequencies in Ω_e.

Bacon [17] split the frequency domain in two sets: Ω_e and $IR - \Omega_e$. Since the problem with Eqs. (7.41) and (7.42) concerns only the frequencies that are in Ω_e, he applied these equations for the frequencies that are in $IR - \Omega_e$, i.e., for frequencies that do not belong to Ω_e. For the frequencies in Ω_e, Bacon [17] proceeded differently. Instead of solving for $A(\omega)$ and $B(\omega)$, he dealt with:

$$P(\omega) = A(\omega)\mathrm{e}^{-i\xi(\omega)x_1}, \tag{7.45}$$

and:

$$Q(\omega) = B(\omega)\mathrm{e}^{i\xi(\omega)x_1}. \tag{7.46}$$

Solving Eqs. (7.39) and (7.40) yields:

$$P(\omega) = \frac{\tilde{\varepsilon}_1(\omega) - \tilde{\varepsilon}_2(\omega)\mathrm{e}^{-i\xi(\omega)\Delta}}{1 - e^{-2i\xi(\omega)\Delta}}, \tag{7.47}$$

and

$$Q(\omega) = \frac{\tilde{\varepsilon}_1(\omega) - \tilde{\varepsilon}_2(\omega)\mathrm{e}^{i\xi(\omega)\Delta}}{1 - \mathrm{e}^{-2i\xi(\omega)\Delta}}. \tag{7.48}$$

Equation (7.47) is equivalent to:

$$P(\omega) = \tilde{\varepsilon}_1(\omega) - \tilde{\varepsilon}_2(\omega)\mathrm{e}^{-i\xi(\omega)\Delta} + \mathrm{e}^{-2i\xi(\omega)\Delta}P(\omega). \tag{7.49}$$

Likewise, Eq. (7.48) is equivalent to:

$$Q(\omega) = \tilde{\varepsilon}_1(\omega) - \tilde{\varepsilon}_2(\omega)\mathrm{e}^{i\xi(\omega)\Delta} + \mathrm{e}^{-2i\xi(\omega)\Delta}Q(\omega). \tag{7.50}$$

Equations (7.39) and (7.40) are then solved, for the frequencies in Ω_e, in the time domain using an iterative procedure.

In line with this work, Zhao and Lok [18] dealt separately with the resonant frequencies in Ω_e. On one hand, they also used Eqs. (7.41) and (7.42) outside Ω_e. On the other hand, Zhao and Lok [18] argued that $A(\omega)$ and $B(\omega)$ are undefined fractions of type 0/0 for the frequencies in Ω_e. Consequently, they proposed to calculate $A(\omega)$ and $B(\omega)$ using the Hopital's rule. The wave separation technique proposed in Ref. [17] calculates the two elementary waves for the resonant frequencies by an iterative method. Therefore, it is time consuming and can amplify experimental noise. On the contrary, the method proposed in Ref. [18] is a direct frequency method. Nevertheless, it uses numerical differentiation of measured signals. This amplifies errors mainly at high frequencies. The maximum test duration presented in Ref. [17] is 6 ms and in Ref. [18] is 2 ms.

To solve the system of the two linear Eqs. (7.41) and (7.42), Bussac et al. [9, 10] proposed a different approach. Mainly, they integrated the Fourier transforms in the lower half-plane of the frequency complex space. This numerical tool is possible because the Fourier transforms of the measured strains $\tilde{\varepsilon}_1(\omega)$ and $\tilde{\varepsilon}_2(\omega)$, and the wave dispersion relation $\xi(\omega)$ are analytical bounded functions in the lower frequency half-plane [19].

Equations (7.41) and (7.42), write an exact mathematical solution for the two elementary waves $A(\omega)$ and $B(\omega)$. However, the denominator vanishes for frequencies in Ω_e. Bussac et al. [9, 10] argued for an integration of the Fourier transforms in the lower half-plane of the complex frequency space. This means that Eqs. (7.41) and (7.42) are solved for a complex frequency $(\omega - i\eta)$ and not for the real frequency (ω). η is an artificial numerical constant damping added to the signal in order to escape from the poles in real frequency axis. The main difficulty here is how to get the wave dispersion for the complex frequencies. Bussac et al. [9, 10] used the Cauchy-Riemann approximation which yields [19]:

$$\xi(\omega - i\eta) = \xi(\omega) - i\eta\frac{\partial\xi}{\partial\eta} = \xi(\omega) + \omega\frac{\partial\xi}{\partial\omega} \equiv \Xi(\omega). \tag{7.51}$$

The complete algorithm is presented in Fig. 7.5. Bussac et al. [10] and Othman [19] showed that their solution is equivalent in the time domain to the solution presented by Lundberg and Henchoz [1] and Yanagihara [2], in the case of non-dispersive waves.

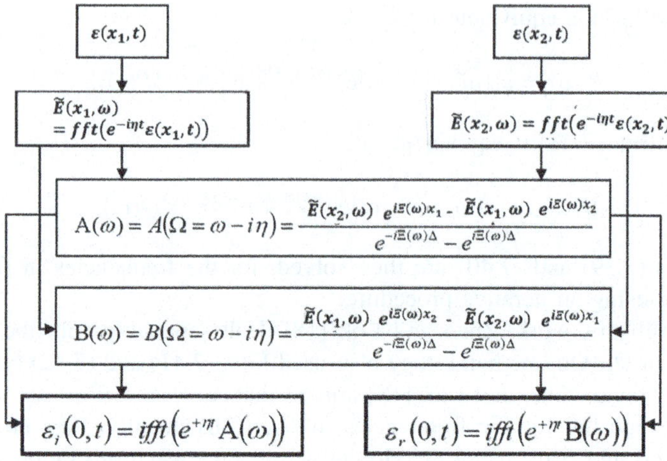

Fig. 7.5 Two-strain measurements wave separation technique using integration of Fourier transforms in the lower half-plance of the complex frequency space

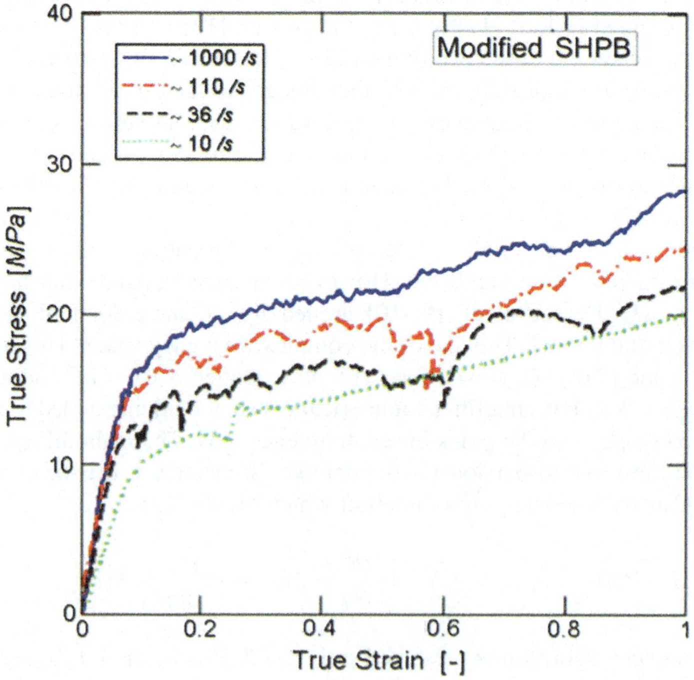

Fig. 7.6 Stress-stress curves on polyurea in the range of 10–1000/s using a modified Hopkinson bar. (Reproduced from Ref. [20] with permissions from Elsevier)

The wave separation methods presented in Refs. [17, 18] and the first method presented in Refs. [9, 10] use only two-strain measurements. This would be enough if the measured strains were perfectly free from noise, which is impossible in actual experimental situations. Besides, these methods, as deconvolution techniques, are an ill-posed problem. Consequently, initial noise is amplified when rebuilding the force and the particle velocity. Jacquelin and Hamelin [8] showed that initial noise is highly amplified by the recovering process in the neighbourhood of the resonant frequencies. It was then suggested in Refs. [7, 9, 10] the use of redundant measurements.

Jacquelin and Hamelin [7] presented a 3-strain method. This technique is already presented in Sect. 7.1 within a time domain approach. It can easily be used within a frequency domain approach in order to take into account the wave dispersion. Simply, one can write the Fourier transforms of Eqs. (7.13) and (7.14). The advantages and disadvantages of the frequency domain solution are similar to the advantages and disadvantages of the time domain solution except that the frequency domain solution accounts for the wave dispersion effects.

Bussac et al. [9, 10] proposed the analysis of the measured strains, and eventually velocities, by the maximum of likelihood method. They assumed that the experimental noises are white, Gaussian and two-by-two independent. Mainly, their solution works with K strain measurements and M velocity measurements. Each strain or velocity measurement is assumed to be the sum of the real physical strain or velocity and an experimental Gaussian noise. More precisely, the measured strain at the cross-section x_k reads:

$$\hat{\varepsilon}_k(t) = \varepsilon(x_k, t) + n_{\varepsilon k}. \tag{7.52}$$

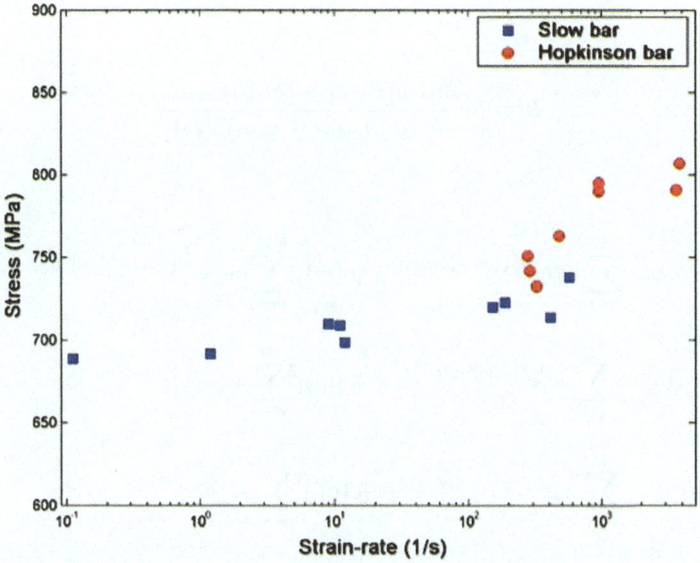

Fig. 7.7 Strain rate sensitivity of an aluminum alloy using a modified Hopkinson bar and a classical Hopkinson bar. (Reproduced from Ref. [22] with permissions from Springer)

where $\varepsilon(x_k,t)$ the real physical strain and $n_{\varepsilon k}$ is the experimental noise recorded together with the strain measurement at cross-section x_j and. Likewise, the measured velocity at the cross-section x_m reads:

$$\hat{v}_j(t) = v(x_m,t) + n_{vm}. \tag{7.53}$$

where $v(x_m,t)$ the real physical velocity and n_{vm} is the experimental noise recorded together with the velocity measurement at cross-section x_m. The elementary functions $A(\omega)$ and $B(\omega)$ are then derived in order to minimize the difference between experimental and theoretical signals. Therefore, $A(\omega)$ and $B(\omega)$ are the solution of an optimization problem. Its cost function is written:

$$\begin{aligned} C =& \frac{1}{2\pi}\int_{-\infty}^{+\infty}\sum_{k=1}^{K} w_k^2\left|\tilde{\varepsilon}_k(\omega) - A(\omega)e^{-i\xi(\omega)x_k} - B(\omega)e^{i\xi(\omega)x_k}\right|^2 d\omega \\ &+ \frac{1}{2\pi}\int_{-\infty}^{+\infty} w_m^2 \sum_{m=1}^{M}\left|\tilde{v}_m(\omega) + c(\omega)A(\omega)e^{-i\xi(\omega)x_m} - c(\omega)B(\omega)e^{i\xi(\omega)x_m}\right|^2 d\omega \end{aligned} \tag{7.54}$$

Recall that $c(\omega) = \frac{\omega}{\xi(\omega)}$.

Minimising the cost function C yields:

$$A(\omega) = \frac{h_2(\omega)E_1(\omega) - g(\omega)E_2(\omega)}{h_1(\omega)h_2(\omega) - g(\omega)\bar{g}(\omega)}, \tag{7.55}$$

and

$$B(\omega) = \frac{h_1(\omega)E_2(\omega) - \bar{g}(\omega)E_1(\omega)}{h_1(\omega)h_2(\omega) - g(\omega)\bar{g}(\omega)}, \tag{7.56}$$

where

$$h_1(\omega) = \sum_{k=1}^{K} w_k^2 e^{-i\left(\xi(\omega)-\bar{\xi}(\omega)\right)x_k} + |c(\omega)|^2\sum_{m=1}^{M} w_m^2 e^{-i\left(\xi(\omega)-\bar{\xi}(\omega)\right)x_m}, \tag{7.57}$$

$$h_2(\omega) = \sum_{k=1}^{K} w_k^2 e^{i\left(\xi(\omega)-\bar{\xi}(\omega)\right)x_k} + |c(\omega)|^2\sum_{m=1}^{M} w_m^2 e^{i\left(\xi(\omega)-\bar{\xi}(\omega)\right)x_m}, \tag{7.58}$$

$$g(\omega) = \sum_{k=1}^{K} w_k^2 e^{i\left(\xi(\omega)+\bar{\xi}(\omega)\right)x_k} + |c(\omega)|^2\sum_{m=1}^{M} w_m^2 e^{i\left(\xi(\omega)+\bar{\xi}(\omega)\right)x_m}, \tag{7.59}$$

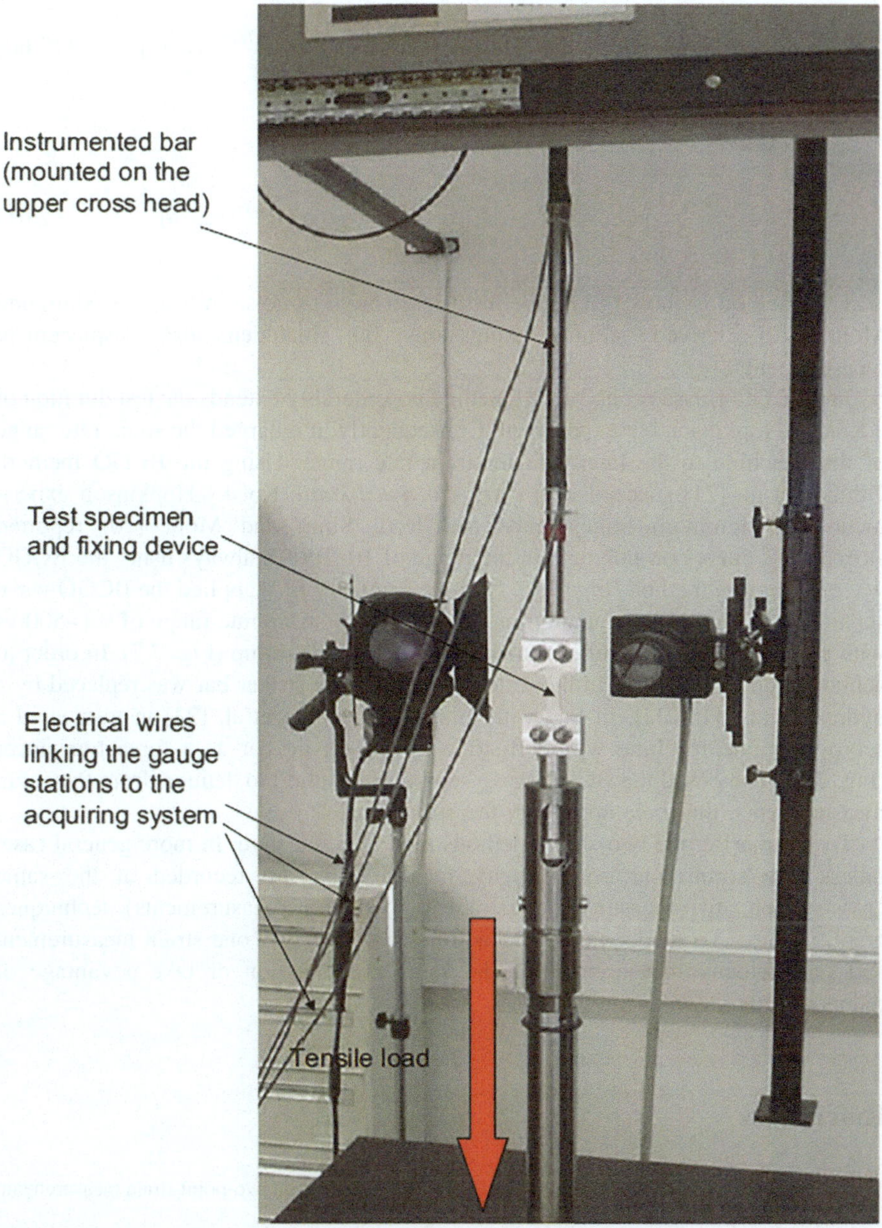

Fig. 7.8 Modified servo-hydraulic machine using an instrumented bar. (Reproduced from Ref. [23] with permissions from Elsevier)

$$E_1(\omega) = \sum_{k=1}^{K} w_k^2 e^{i\bar{\xi}(\omega)x_k}\tilde{\varepsilon}_k(\omega) + c(\omega)\sum_{m=1}^{M} w_m^2 e^{i\bar{\xi}(\omega)x_m}\tilde{v}_m(\omega), \tag{7.60}$$

and

$$E_2(\omega) = \sum_{k=1}^{K} w_k^2 e^{-i\bar{\xi}(\omega)x_k}\tilde{\varepsilon}_k(\omega) + c(\omega)\sum_{m=1}^{M} w_m^2 e^{-i\bar{\xi}(\omega)x_m}\tilde{v}_m(\omega). \tag{7.61}$$

This method is shown to be the most insensitive to noise in Ref. [7]. Shim and Mohr [20] derived similar expressions but included also displacement measurements.

The BCGO Bussac et al. [9, 10] method considerably extends the test duration of a Kolsky-Hopkinson bar experiment. Consequently, it enlarged the strain rate range of this machine to the intermediate strain rate range. Using the BCGO method, Othman et al. [21] succeeded to carry out *quasi-static* Kolsky-Hopkinson experiments on aluminum honeycombs and lead. Shim and Mohr [20] reported stress-stress curves on polyurea in the range of 10–1000/s always using the BCGO wave separation method (Fig. 7.6). Othman and Gary [22] applied the BCGO wave separation method to test an aluminum alloy in the strain rate range of 0.1–5000/s with a classical and a modified Kolsky-Hopkinson bar set-up (Fig. 7.7). In order to achieve high deformation in the tested specimens the striker bar was replaced by a hydraulic jack [15, 22]. In line with this idea, Othman et al. [23] instrumented a servo-hydraulic machine with a 3-strain gauge elastic bar as a force transducer (Fig. 7.8). They used the BCGO wave separation method to deduce, from the strain measurements, the force applied to the specimens.

To summarise, the two-strain methods are noise sensitive. In more general case, unless one strain and one velocity measurement are recorded at the same cross-section, two-measurement (strain or velocity measurements) techniques amplify noise. To reduce noise effects, one can either use one strain measurement and one velocity measurement at the same cross-section or take advantage of redundant measurements.

References

1. Lundberg B, Henchoz A (1977) Analysis of elastic waves from two-point strain measurement. Exp Mech 17:213–217
2. Yanagihara N (1978) New measuring method of impact force. Bull Jpn Soc Mech Eng 21:1085–1087
3. Adachi T, Kataoka T, Higuchi M (2015) Predicting impact shear strength of phenolic resin adhesive blended with nitrile rubber. Int J Adhes Adhes 56:53–60
4. Park SW, Zhou M (1999) Separation of elastic waves in split Hopkinson bars using one-point strain measurements. Exp Mech 39:287–294

5. Gary G, Mohr D (2013) Modified Kolsky formulas for an increased measurement duration of SHPB systems. Exp Mech 53:713–717
6. Meng H, Li QM (2003) An SHPB set-up with reduced time-shift and pressure bar length. Int J Impact Eng 28:677–696
7. Jacquelin E, Hamelin P (2003) Force recovered from three recorded strains. Int J Solids Struct 40:73–87
8. Jacquelin E, Hamelin P (2001) Block-bar device for energy absorption analysis. Mech Syst Sig Process 15:603–617
9. Othman R, Bussac MN, Collet P, Gary G (2001) Separation et reconstruction des ondes dans les barres elastiques et viscoelastiques a partir de mesures Redondantes. C R Acad Sci Ser IIb 329:369–376
10. Bussac MN, Collet P, Gary G, Othman R (2002) An optimisation method for separating and rebuilding one-dimensional dispersive waves from multi-point measurements. Application to elastic or viscoelastic bars. J Mech Phys Solids 50:321–350
11. Othman R (2014) Comparison of three methods to separate waves in the processing of long-time Hopkinson bar experiments. Int J Mech Eng. Technol 5:114–119
12. Casem DT, Fourney W, Chang P (2003) Wave separation in viscoelastic pressure bar using single point measurements of strain and velocity. Polym Test 22:155–164
13. Casem DT, Zellner MB (2013) Kolsky bar wave separation using a photon doppler velocimeter. Exp Mech 53:1467–1473
14. Zhao H, Gary G (1994) A new method of wave separation for application to dynamic testing. C R Acad Sci Paris 319:987–992
15. Zhao H, Gary G (1997) A new method for the separation of waves. Application to the SHPB technique for an unlimited measuring duration. J Mech Phys Solids 45:1185–1202
16. Othman R (2014) On the use of complex Young's modulus while processing polymeric Kolsky-Hopkinson bars's experiments. Int J Impact Eng 73:123–134
17. Bacon C (1999) Separating waves propagating in an elastic or viscoelastic Hopkinson pressure bar with three-dimensional effects. Int J Impact Eng 22:55–69
18. Zhao PJ, Lok TS (2002) A new method for separating longitudinal waves in a large diameter Hopkinson bar. J Sound Vib 257:119–130
19. Othman R (2002) Extension du champ d'application du systeme des barres de Hopkinson aux essais a moyennes vitesses de deformation. Ph.D. thesis, Ecole Polytechnique, France
20. Shim J, Dirk M (2009) Using split Hopkinson pressure bars to perform large strain compression tests on polyurea at low, intermediate and high strain rates. Int J Impact Eng 36:1116–1127
21. Othman R, Bussac MN, Collet P, Gary G (2003) Testing with SHPB from quasi-static to dynamic strain rates. J Phys IV 110:397–402
22. Othman R, Gary G (2007) Testing aluminum alloy from quasi-static to dynamic strain-rates with a modified split Hopkinson bar method. Exp Mech 47:295–299
23. Othman R, Guegan P, Challita G, Pasco F, Lebreton D (2009) A modified servo-hydraulic machine for testing at intermediate strain rates. Int J Impact Eng 36:460–467

Chapter 8
Inertial and Frictional Effects in Dynamic Compression Testing

Clive R. Siviour and Stephen M. Walley

8.1 Introduction to Inertia in Dynamic Compression Tests

An inevitable requirement of high strain rate testing is rapid acceleration of at least part of the specimen. As well as inducing waves that propagate in the specimen, at least during the initial stages of loading, this rapid acceleration gives rise to inertial stresses that affect the force measurements at the specimen-bar interfaces.

The specimen stress measured during a Hopkinson bar experiment may be thought of as the sum of stress components:

$$\sigma_T = \sigma_M + \sigma_I + \sigma_F + \sigma_{other}, \tag{8.1}$$

where σ_M is the material strength (which is desired to be measured), σ_I an enhancement owing to inertia, σ_F an enhancement owing to friction and σ_{other} the stress owing other effects that may be present. The experimenter must therefore design specimens that minimise the effects of friction and inertia, so that σ_I and σ_F are an acceptably small fraction of σ_M, although some authors have proposed that the data be corrected after the experiment had been performed.

In this section of the chapter, we will mainly be concerned with the effects of macro-inertia on the observed specimen response, which is to say the evaluation of

---

C. R. Siviour
Department of Engineering Science, The University of Oxford,
Parks Road, Oxford OX1 3PJ, UK

S. M. Walley (✉)
SMF Group, The Cavendish Laboratory, The University of Cambridge,
J.J. Thomson Avenue, Cambridge CB3 0HE, UK
e-mail: smw14@cam.ac.uk

R. Othman (ed.), *The Kolsky-Hopkinson Bar Machine*,
https://doi.org/10.1007/978-3-319-71919-1_8

σ_I and its minimisation. Many materials also experience *micro* inertia effects, e.g. microbuckling in foams and other structured materials; here, these will be regarded as part of the material response, σ_M, and will not be considered further. The related phenomenon of macro-inertial effects on larger scale structures (e.g. honeycombs) loaded dynamically is also outside the scope of this chapter.

A number of different approaches to deriving values for σ_I will be discussed below. Broadly speaking, the magnitude of this stress depends on the square of the specimen dimensions, and, of course, the specimen density. There is usually a term associated with the longitudinal and radial accelerations of the specimen as the strain rate increases during the early stages of a Hopkinson bar experiment. These are accompanied by contributions associated with advection of the specimen material to larger radii as the specimen continues to deform at constant strain rate. These lead to terms proportional to $\ddot{\varepsilon}$ and $\dot{\varepsilon}^2$, where ε is the true strain in the specimen. Further, because the output bar is not rigid, there is an additional stress proportional to the acceleration $\dot{v}$ of the interface between the specimen and output bar.

In addition to affecting the measured stresses, specimen inertia causes a radial confinement on the specimen. Intuitively, one would expect this stress to be largest along an axis down the centre of the specimen, and it must be zero on the free surfaces. This is indeed the case. By changing the stress state within the specimen, this may lead to changes in specimen behaviour. For brittle materials, the radial stress may inhibit the growth of cracks, and therefore has the potential to increase the observed strength of the specimen, discussed further below. For ductile materials, because the radial stress increases the pressure at the centre of the specimen, and hence decreases the deviatoric stress for a given axial load, the observed yield stress will typically increase. Hence, whilst correcting experimental data for the effects of inertia may seem attractive, in practice this is not easy to achieve because in addition to causing changes in σ_I, σ_M may also be affected.

8.2 Calculations of the Effect of Inertia

From the earliest days of split-Hopkinson bar testing, the importance of inertia was recognised. A common approach is to consider the change in kinetic energy of the specimen, and therefore the work that must be done to provide this energy. The other approach utilised is to calculate the stress distribution in the specimen from its acceleration field. The kinetic energy approach more simply provides the additional forces at the bar-specimen interfaces, which are in practice the forces that are measured. However, the stress-based approach has the advantage of giving a deeper understanding of how inertia might affect the specimen response. The history of both approaches will be considered here. Whilst it may seem unnecessary to consider equations that have now been superseded, it is instructive to understand the different assumptions made and how these affected the final result. Throughout this chapter, the notation, and sign conventions used have been modified from the original papers so that the expressions can be written in a common format (note, in particular, that some references refer to compressive stress as negative, but also express some of these stresses as pressures, which are positive in compression).

8.2.1 Energy Based Approaches

Owing to his concerns about the time it takes to achieve force equilibrium in a dynamic experiment, Kolsky [1] used specimens with a short length to diameter ratio. He was therefore careful to consider the effects of friction and radial inertia in his experiments. Assuming that the stress is axially uniform, he considered the kinetic energy, K, of a cylindrical specimen expanding uniformly, and derived the following equation

$$K = \frac{1}{4} Ah\rho v_r^2, \tag{8.2}$$

where A and h are the cross-sectional area and height of the specimen, respectively, ρ its density, and v_r is the radial velocity of a point on the circumference. This radial velocity can in turn can be calculated from the radius, a, and strain rate, $\dot{\varepsilon}$, as

$$v_r = \nu a\dot{\varepsilon}. \tag{8.3}$$

where ν is the ratio of radial to axial strain, which for elastic deformation is the Poisson ratio. For plastic deformation ν is typically 0.5.

Equations (8.2) and (8.3) may be used to calculate a rate of change of kinetic energy, K, given by

$$\begin{aligned} \frac{\mathrm{d}K}{\mathrm{d}\varepsilon} &= \frac{\mathrm{d}K}{\mathrm{d}t}\frac{\mathrm{d}t}{\mathrm{d}\varepsilon} = \frac{1}{\dot{\varepsilon}}\frac{\mathrm{d}}{\mathrm{d}t}\left(\frac{Ah}{4}\rho\nu^2 a^2\dot{\varepsilon}^2\right) \\ &= \frac{1}{\dot{\varepsilon}}\frac{Ah}{4}\rho\nu^2 a^2 2\dot{\varepsilon}\ddot{\varepsilon} = \frac{Ah}{2}\rho\nu^2 a^2\ddot{\varepsilon} \end{aligned} \tag{8.4}$$

Thus since the work done per unit volume due to an applied stress is $\mathrm{d}W = \sigma\mathrm{d}\varepsilon$, the inertial stress is given by:

$$\sigma_I = \frac{1}{2}\nu^2 a^2\rho\ddot{\varepsilon}, \tag{8.5}$$

where the convention is adopted (and will continue through this chapter) that compressive stresses and strain rates take a positive sign.

The next attempt to derive the additional stress owing to inertia was made by Davies and Hunter [2]. They again used an energy based approach, but this time also considered the effects of the axial acceleration of the specimen, hence allowing a net axial force and including the rigid body motion of the specimen caused by the possible difference in pressure between the two specimen bar interfaces. A detailed derivation is given in their paper, and the final result, for the stress σ_2 measured by the output bar (as is usually the case) is that

$$\sigma_2 = \sigma_M - \rho\left(\frac{h^2}{6} - \frac{\nu^2 a^2}{2}\right)\ddot{\varepsilon}, \tag{8.6}$$

giving the opportunity, for small strains, to minimise the effects of inertia by ensuring that $h = \sqrt{3}\,\nu a$.

Samanta [3] extended this analysis by considering the material rate of change, 'i.e. the rate-of-change of a quantity that is defined with reference to specific particles of the moving continuum', the convective part of the material derivative. Again, considering the work done on the specimen by the external forces, represented by the stresses σ_1 and σ_2 measured on the input and output bars respectively, and how this work changes the internal and kinetic energy of the specimen, he derived

$$\begin{aligned}\frac{1}{2}(\sigma_1 + \sigma_2) = \sigma_M + \rho\left(\frac{h^2}{12} + \frac{a^2}{8}\right)\ddot{\varepsilon} \\ + \rho\left(\frac{a^2}{16} - \frac{h^2}{12}\right)\dot{\varepsilon}^2.\end{aligned} \tag{8.7}$$

Here, the material is assumed to be incompressible, so that $\nu = 0.5$, and this result is therefore consistent with that of Davies and Hunter, but with an additional term in $\dot{\varepsilon}$. Effectively, this term recognises that, even when the strain rate is constant, particles must accelerate as the specimen expands, because the radius at which they are located increases. Note that the sign change for the term in $a^2\ddot{\varepsilon}$ is a result of writing the expression in terms of the mean stress at the two specimen-bar interfaces, rather than just the output bar stress.

Bertholf and Karnes [4] produced a (claimed) corrected version of this analysis, and explicitly presented the result for inertia effects as experienced in the output bar:

$$\sigma_2 = \sigma_M - \rho\left(\frac{h^2}{6} - \frac{a^2}{8}\right)\ddot{\varepsilon} + \rho\left(\frac{a^2}{16} - \frac{h^2}{3}\right)\dot{\varepsilon}^2. \tag{8.8}$$

They performed a two dimensional numerical analysis demonstrating the oscillations in the stress-strain data resulting from the specimen inertia, and also that these oscillations could be reduced through appropriate pulse shaping to increase the rise time of the input pulse, and thereby reduce $\ddot{\varepsilon}$.

Malinowski and Klepaczko [5] derived a very similar equation, with small differences in the numerical constants, and also noted that, whilst the equations above imply an optimal aspect ratio to minimise the effects of inertia when the friction coefficient is zero, a similar equation could be derived for a non-zero friction coefficient, giving the optimal ratio as

$$\left.\frac{h}{a}\right|_{opt} = 2\left[\frac{\mu\sigma_z}{2\rho a^2(\dot{\varepsilon}^2 + \ddot{\varepsilon})}\right]^{1/3} \tag{8.9}$$

Possibly the most complete of the energy based approaches is that due to Gorham [6], who extended Samanta's analysis to the case where both ends of the specimen move. The results of his analysis were

$$\sigma_1 = \sigma_M + \rho\ddot{\varepsilon}\left(\frac{a^2}{8} + \frac{h^2}{3}\right) + \rho\dot{\varepsilon}^2\left(\frac{a^2}{16} - \frac{h^2}{3}\right) + \frac{\rho h\dot{v}}{2}, \tag{8.10}$$

$$\frac{1}{2}(\sigma_1 + \sigma_2) = \sigma_M + \rho\ddot{\varepsilon}\left(\frac{a^2}{8} + \frac{h^2}{12}\right) + \rho\dot{\varepsilon}^2\left(\frac{a^2}{16} - \frac{h^2}{12}\right), \tag{8.11}$$

$$\sigma_2 = \sigma_M + \rho\ddot{\varepsilon}\left(\frac{a^2}{8} - \frac{h^2}{6}\right) + \rho\dot{\varepsilon}^2\left(\frac{a^2}{16} + \frac{h^2}{6}\right) - \frac{\rho h\dot{v}}{2}. \tag{8.12}$$

To the authors' knowledge, this is the most complete of the energy-based analyses, and is the one which should be used as a guide to specimen design. Recently Sen et al. [7] extended this analysis to non-cylindrical specimens, and found the same result for the inertia, but with *a*/8 replaced by *J*/4*A*, where *J* is the second moment of the area, and *A* the specimen cross-sectional area.

8.2.2 Stress-Based Approaches

Whilst Haddow [8] produced a stress-analysis based solution for high speed compression of a thin disc, considering both friction and inertia, the first analysis directly applicable to the split-Hopkinson bar was that of Dharan and Hauser [9], who specifically considered the velocity, acceleration and therefore stress field required to satisfy the dynamic equilibrium equation in the specimen. This allowed calculation of the radial stress as

$$\sigma_r = \frac{\rho}{4l_0(1-\varepsilon)}\left[\frac{3v_x^2}{2l_0(1-\varepsilon)} + \dot{v}_x\right]\left[\frac{a_0^2}{(1-\varepsilon)} - r^2\right], \tag{8.13}$$

where the specimen was assumed to be incompressible, σ_r is the radial component of the stress field, l_0 and a_0 the original length and diameter of the specimen, v_x the axial velocity and r the radial location in the specimen. A clear advantage of this approach is therefore that the radial *distribution* of inertial stress is explicitly produced. In order to evaluate the effect on Hopkinson bar measurements, this equation was simplified by assuming that the stress everywhere was the maximum stress (which in fact occurs at $r = 0$), rather than integrating over the specimen surface. It then remains to convert this radial stress into an axial stress, in this case assuming a von Mises yield surface

$$\sigma = \sigma_x - \sigma_r \quad (\sigma_r = \sigma_\theta, \nu = 0.5), \tag{8.14}$$

where σ represents the flow stress in an ideal compression experiment and σ_x is the measured stress (equivalent to σ_I or σ_2 above). This gave, using notation prevalent in this chapter

$$\begin{aligned} \sigma_1 &= \sigma_M + \frac{3}{8}\rho\left(\frac{a_0}{l_0}\right)^2 \frac{v_x^2}{(1-\varepsilon)^3} \\ &= \sigma_M + \frac{3}{8}\rho v_x^2\left(\frac{a}{l}\right)^2 = \sigma_M + \frac{3}{8}\rho a^2 \dot{\varepsilon}^2. \end{aligned} \tag{8.15}$$

Given the assumption that the stress everywhere was the maximum, this is pleasingly similar to Kolsky's result, when expressed for an incompressible material.

The next consideration of this effect was by Forrestal et al. [10], who derived equations for an elastic-brittle material, assumed to obey Hooke's law, and also assumed (although this was not necessary), that the stress field was uniform along the specimen length. One advantage of this approach is that all three components of stress (r, ϑ, z) can be given as a function of radius in the specimen, where σ_z is the axial stress in the specimen and σ_r provides the confining stress most relevant to brittle specimens. The three stresses derived *for an elastic material* are, when re-expressed in the notation of this chapter:

$$\sigma_Z = \sigma_M + \frac{\nu(3-2\nu)}{4(1-\nu)}\left[a^2 - \frac{2r^2}{(3-2\nu)}\right]\rho\ddot{\varepsilon}, \tag{8.16}$$

$$\sigma_r = \frac{\nu(3-2\nu)}{8(1-\nu)}\left[a^2 - r^2\right]\rho\ddot{\varepsilon}, \tag{8.17}$$

$$\sigma_\theta = \frac{\nu(3-2\nu)}{8(1-\nu)}\left[a^2 - \frac{(1+2\nu)}{(3-2\nu)}r^2\right]\rho\ddot{\varepsilon}, \tag{8.18}$$

These authors showed that, by integrating over the specimen surface, these equations match those derived by Kolsky (Eq. 8.2). A clear implication for specimen design is that σ_I can be reduced by using an annular specimen, a cylinder with an appropriately sized hole cut axially down the centre, as this removes the region in which $\sigma_z(r, \vartheta)$ is smallest [11], whilst the radial inertia contribution to σ_z is [12]:

$$\sigma_I = \left\{ \rho \left(\frac{a^2 + b^2 - r^2}{4} \right) - \frac{3\rho a^2 b^2}{8E} \left[\rho \ln\left(\frac{a}{r}\right) \lambda^2 + \frac{2E}{3a^2} \right] \cos \lambda t \right\} \ddot{\varepsilon} \tag{8.19}$$

where t is time, b is the internal radius of the specimen and

$$\lambda = \sqrt{\frac{2E(b^2 - a^2)}{3\rho \ln\left(\frac{b}{a}\right) b^2 a^2}}. \tag{8.20}$$

This analysis was extended by Zhang et al. [13], who calculated expressions for all three principal components of stress in an elastic annular specimen, as functions of specimen dimensions and strain rate.

A weakness in Eqs. (8.16–8.18) is that the convection term is again neglected. Including convection gives the following expression (due to Warren and Forrestal [14]) for the pressure as a function of radial location in an incompressible material in which engineering stress and strain are used rather than true stress and strain

$$p = \frac{\rho}{4(1 - \varepsilon)^2} \left[\frac{3\dot{\varepsilon}^2}{2(1 - \varepsilon)} + \ddot{\varepsilon} \right] \left(a_0^2 - r_0^2 \right), \tag{8.21}$$

which, when averaged over the specimen surface, gives

$$\sigma_I = \frac{3\rho a_0^2}{16(1 - \varepsilon)^3} \dot{\varepsilon}^2 + \frac{\rho a_0^2}{8(1 - \varepsilon)^2} \ddot{\varepsilon}. \tag{8.22}$$

These expressions are given in terms of the original specimen radius a_0 and radial location, r_0, but may be modified to the current specimen radius as follows

$$\sigma_I = \frac{3\rho a^2}{16(1 - \varepsilon)^2} \dot{\varepsilon}^2 + \frac{\rho a^2}{8(1 - \varepsilon)} \ddot{\varepsilon}. \tag{8.23}$$

It is instructive to compare this equation to the relevant terms in Gorham's analysis. Here, the $(1 - \varepsilon)$ terms can be attributed to the use of engineering strain by Warren and Forrestal [14] whereas Gorham used true strain [6]. The numerical factor of 3/16, instead of 1/16 as observed in the energy-based analyses can be understood by reference to earlier steps in the analyses of Samanta [3] and Gorham [6], in which 3/16 appears in the radial terms, but is partly cancelled when the inhomogeneity along the specimen length is considered. This was recently demonstrated by Hockly and Siviour [15], who showed that by combining Forrestal's equation for the radial dependence of the stress, and then considering a non-uniform axial distribution, a result could be derived that is the same as the Gorham equation.

8.2.3 Comparison of Approaches

Although it has been shown that the energy-based and stress-based approaches yield equivalent results, it is worth remarking on their application. In particular, the energy-based approach more readily accounts for the inertia-induced differences in stress at the two specimen-bar interfaces. If, as is usually the case, it is desired to perform the dynamic experiment under conditions of static equilibrium, the inertial contribution to this stress difference must be considered alongside the contribution from wave oscillation in the specimen. Hence, it is valuable to have an estimate of the difference induced by these inertial stresses. Conversely, the stress analysis based approach, in giving the radial distribution of the inertial stress, demonstrates the increased confinement at the centre of the specimen compared to the periphery. This difference is particularly important for brittle specimens, where a large amount of cracking is often observed at large radii within a specimen, even if the centre remains intact. Furthermore, stress analysis allows a direct consideration of the effects of confinement on yielding, through comparison to an appropriate yield surface and has also suggested novel specimen designs (discussed further below) to minimise the effects of inertia by means other than simply making the specimen smaller.

8.3 Inertia Calculations and Effects in Experimental Data

It is instructive to calculate some typical values of inertial stress using the equations above. Because the stresses scale linearly with specimen density, these calculations will be performed using a density of 1000 kg/m^3, so that they can easily be converted for different materials. The plot in Fig. 8.1 shows the magnitudes of the first two terms in Eq. (8.12) (Gorham's analysis) evaluated for a 10 mm long and 10 mm diameter specimen as a function of strain rate, assuming that the rise time of the input pulse is 10 μs, so that $\ddot{\varepsilon} = 10^5 \times \dot{\varepsilon}$. Again, these values are chosen so that the curves shown can be easily scaled to other specimen dimensions. The term in $\dot{v}$ is more difficult to define as it depends directly on the stress supported by the specimen, being

$$\dot{v} = \frac{\sigma}{\sqrt{E_b \rho_b}} \frac{A_s}{A_b}, \tag{8.24}$$

for an elastic bar of area A_b, density ρ_b and Young's modulus E_b.

Considering Warren and Forrestal's analysis given in Eq. (8.21), Fig. 8.2 shows the pressure as a function of radius for an incompressible material with the same specimen dimensions at a number of different strain rates.

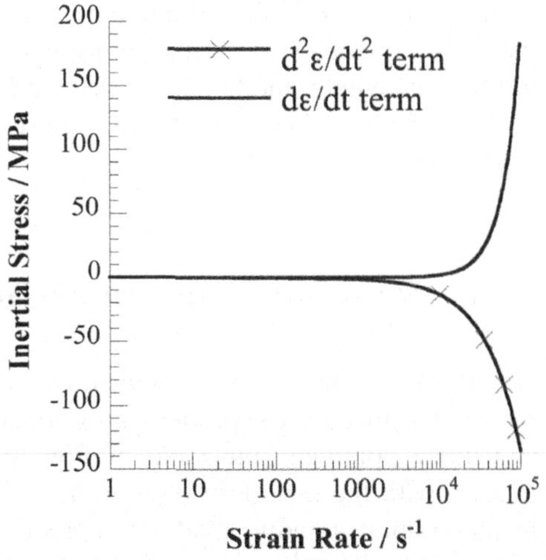

Fig. 8.1 Plots of the dependence of strain rate of two terms in Eq. (8.12) for a 'standard' sized specimen

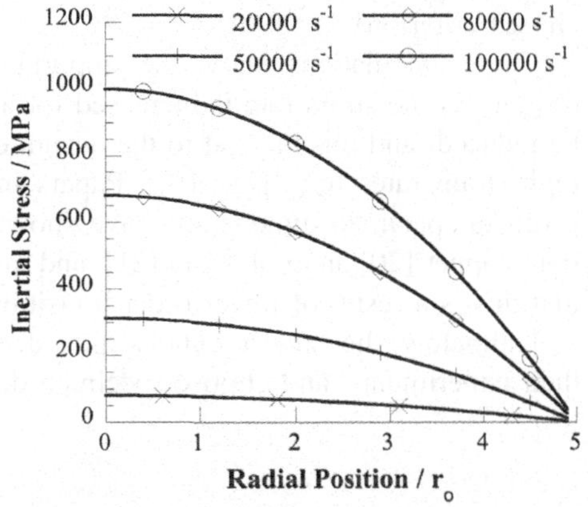

Fig. 8.2 Plots of the dependence on strain rate and radial position of the inertially-induced pressure calculated from Eq. (8.21) for a 'standard' specimen

8.4 Implications for Specimen Design

The above equations have clear implications for specimen design when performing Hopkinson bar experiments on all materials, and, in particular, they place limits on the size of specimen that can be used for any given material density and strain rate. Whilst some authors (e.g. Malinowski and Klepaczko [16]) have proposed using these equations to correct data obtained from Hopkinson bar experiments, there are weaknesses to this approach. Firstly, the required experimental data (specimen dimensions, strain and its first and second derivatives, all as functions of time) are susceptible to noise and other sources of error, so the correction is likely to contain large errors itself. More importantly, however, because the effect of specimen inertia is to move the system away from a uniaxial state of stress, it is likely that this will fundamentally modify the specimen behaviour, which cannot be corrected.

The effects of inertia on three representative classes of materials will now be considered.

8.4.1 Metals and Engineering Polymers

A number of authors have presented experimental data showing the effect of strain rate on the mechanical properties of copper. In common with some other metals, and many polymers, the yield or flow stress of copper is observed to increase linearly with the logarithm of strain rate, up to some critical value above which it increases more rapidly. Gorham [17] surveyed the literature on the behaviour of copper at high strain rates and observed that in some cases the strain rates were close to those at which the term $\ddot{\varepsilon}$ in the inertia equation would be 12.5 MPa (his chosen criterion).

These calculations show the importance of considering inertia in specimen design. As the strain rate is increased for a given material, the specimen size must be reduced, and this has led to the design of miniaturised Hopkinson bars for very high strain rates (e.g. [18, 19]). Papers in which careful consideration has been given to specimen dimensions have shown the same form of rate dependence in both copper [20] and polymers [21], and in the latter case it is now well established that this is a result of lower order transitions in the materials.

Ultimately, the experimentalist must decide how much error may be tolerated in their experiments, and choose specimen dimensions accordingly.

8.4.2 Rubber and Biomaterials

Soft polymeric materials, with densities approximately one tenth of steel, but only capable of supporting stresses one thousand times smaller, or less, often exhibit inertial stresses of similar magnitudes to the material strength of the specimen. In these materials, the effect of inertia can be seen as a spike or peak in the output bar signal during the early stages of the experiment. This has been demonstrated particularly effectively by Song et al. [11, 12] who show that by using an annular specimen the effect of inertia can be significantly reduced. Such a specimen design results directly from the stress-analysis based calculations of specimen inertia distributions presented above, and their papers contain an analytical solution for inertia in annular specimens (see Eq. 8.19). A similar approach has also been proposed for the reduction of frictional effects [22]. Further discussion of the effect of inertia on measurements from very soft specimens is given by Casem et al. [23], and in tension by Zhang et al. [13].

8.4.3 Concrete

It has long been known that confinement increases the apparent compressive strength of brittle materials. Experimental observations of increasing apparent strength of concrete with increasing applied strain rate are well known: the phenomenon is often referred to as the dynamic increase factor (DIF). A comprehensive survey of data obtained between 1917 and 1985 was given by Bischoff and Perry [24]. They also discussed the effect of inertia in concrete specimens. Tang et al. [25] gave a simple expression for the lateral inertial confinement, from which they concluded that it was not significant in their experiments up to strain rate of 200 s^{-1}. However, at higher rates, it is generally (but not universally) accepted that inertial confinement plays an important role in the DIF; see for example the detailed discussion by Li and Meng [26], the combined experimental and numerical study of Zhang, Li and co-workers [27, 28], and wide-ranging papers on concrete by other authors [29–31]. The combined effects of inertia and friction in rocks and concrete were studied by Hao et al. [32–34]. The use of pulse shaping to reduce acceleration in the early stages of loading was investigated by Heard et al. [35]. In these papers, both the energy- and stress-analysis based results discussed above are often quoted as evidence for the importance of appropriately chosen specimen dimensions. It should be noted that crack growth in concrete, and therefore the macroscopic behaviour, is also affected by the phenomenon of micro-inertia within the material, as distinct from specimen inertia as discussed here [36]. The importance of inertia has also been demonstrated through numerical modelling, e.g. [37]. Unsurprisingly, other ceramic materials are also affected by the same phenomenon [38].

8.5 Numerical Modelling of Inertial Effects

In addition to numerical modelling performed in some of the studies above, a number of authors have performed instructive numerical simulations that help elucidate inertial effects in Hopkinson bar experiments. A two-dimensional simulation by Bertholf and Karnes [4] supported earlier analytical investigations of both friction and inertia which concluded that with suitable specimen design and lubrication to minimise these effects, valid materials data could be obtained. Simulations by Zencker and Clos [39] supported the analyses of Samanta [3] and Gorham [6] showing that the optimum specimen design that results from Davies and Hunter's analysis does not remove frictional effects. Jankowiak et al. performed numerical simulations [40] to investigate the combined effects of inertia and friction. Their studies supported the Klepaczko and Malinowski's analysis [41], whilst simulations recently performed by Iwamoto and Yokoyama also demonstrated the effects of inertia on Hopkinson bar measurements [42]. Finally, modelling by Georgin and Reynouard [37] and Hao and Hao [32] demonstrates these effects in concrete and rocks.

8.6 Inertia in Tensile Experiments

Little research has been published on the effects of inertia in tensile specimens. Similar arguments to the annular specimens discussed above form the basis of the analysis of tensile specimens that are formed by wrapping a film around a Hopkinson bar to produce thin-walled tension and torsion specimens [43–45]. Guduru and Freund [46] considered inertial effects in high rate extension of ductile metals, producing solutions that were later applied to the Hopkinson bar [14]. However, this data, and that of other authors, does illustrate that inertia plays a role in the ductility of metals deformed at high rates of strain [47]. Rusinek et al. performed numerical simulations which demonstrated the importance of specimen dimensions and inertia on neck formation and the development of inertia driven oscillations in tensile Hopkinson bar experiments [48]. Although these studies give insights into the importance of, and effects of, inertia in tensile testing, we are only aware of one paper, by Hockly and Siviour, that proposes a guide to specimen design [15]. These authors suggest that the effect of inertia on the output bar measurements in a tensile experiment is bounded by

$$\sigma_I = -\frac{m_2}{\pi b^2}\dot{v} + k\rho\frac{a^2}{b^2}\left[\ddot{\varepsilon}\left(\frac{a^2}{8} - \frac{h^2}{6}\right) - \dot{\varepsilon}^2\left(\frac{a^2}{16} + \frac{h^2}{6}\right) - \frac{\dot{v}h}{2}\right], \tag{8.25}$$

where k = 1,2 provide lower and upper bounds. Here, m_2 is the mass of the endcap on the output end, and b its radius, whilst a is the radius of the gauge section, and h the gauge length.

8.7 Introduction to Friction in 'Simple' Compression (Upsetting)

In elastic deformation, compression of a cylinder of material is straightforward: the reduction in height is accompanied by a proportionate expansion in diameter. The coefficient of proportionality between the height strain and the diameter strain is one of the elastic constants of materials, the Poisson ratio. On the atomic scale, the deformation is due to stretching, compression and rotation of atomic bonds. Larger elastic (recoverable) strains are possible for materials consisting of macromolecules than for say metals and ceramics because polymer molecules loop around reach other forming a network of entanglement points [49] (Fig. 8.3). This combined with the most probable configuration of a macromolecule being a specific shape (spherical for a simple linear polymer) results in a large thermodynamic 'driving force' towards recovery of the original shape so long as the network of entanglement points has not been destroyed by fracture or diffusion of the molecules. Recovery from large strains for polymers is therefore possible and is a macroscopic expression of the large size of polymer molecules.

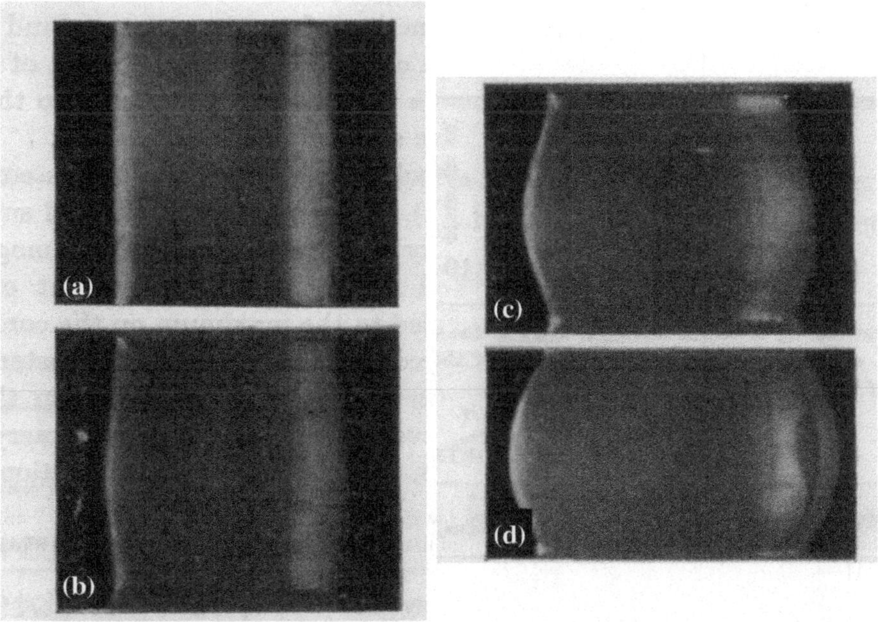

Fig. 8.3 Example of barrelling of epoxy specimen produced by quasistatic loading. From [49]

Fig. 8.4 Cross-section of a cylinder of HRC 40 4340 steel after a 56% reduction in height. From [50]

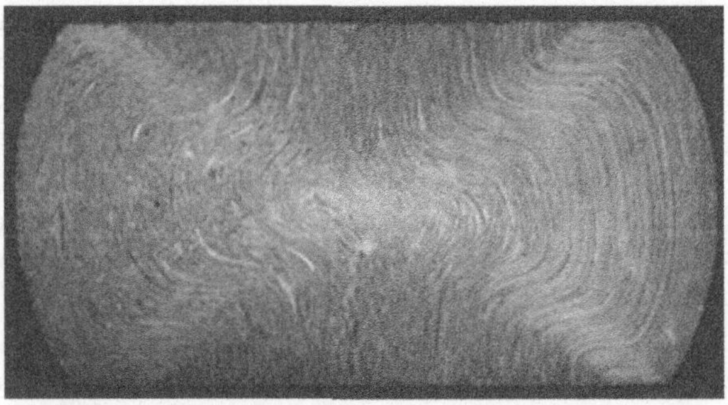

When, however, plasticity is initiated in simpler materials such as metals, atoms and molecules move relative to each other to new positions. True plasticity is irreversible. The material flow patterns even in 'simple' compression can be quite complex (see Fig. 8.4) and are very sensitive to surface stresses associated with friction between the specimen and the anvils.

Friction between specimens and anvils has long been recognised as a problem in compression testing, both at low [51–55] and at high rates of deformation [41, 56–61] (Fig. 8.5).

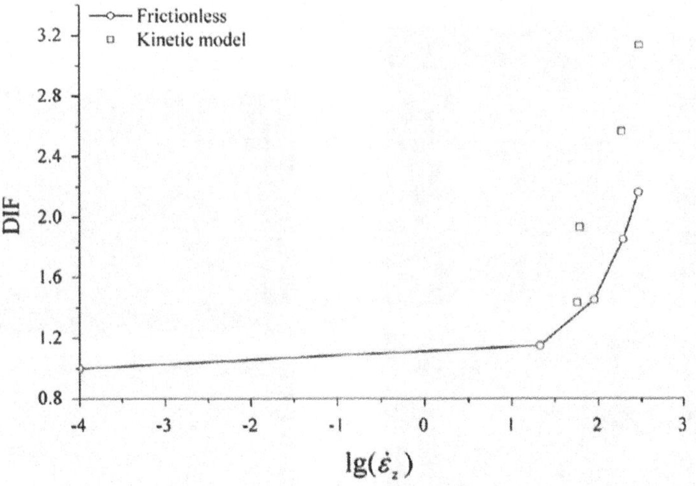

Fig. 8.5 Graph showing effect of friction on the ratio of dynamic to static strength for 75 mm diameter, 37.5 thick mortar specimens deformed in a split Hopkinson pressure bar (SHPB). From [61]

Friction has three major effects on 'simple' compression: (i) it is manifested as a shear stress at the specimen/anvil interface which changes the state of stress in the specimen from uniaxial to triaxial [50]. This leads in turn to the measured stress being higher than the true yield stress of the material [58, 62] (see Figs. 8.5, 8.6, 8.7, 8.8 and 8.9); (ii) the specimen does not remain cylindrical during deformation but bulges out sideways (see Figs. 8.3 and 8.4). This phenomenon is usually termed 'barrelling' e.g. [49, 50, 59, 63–74]; (iii) deformation localizes within the specimen due to frictional locking of the surface resulting in cones of non-deforming material which slide over the unconstrained material at the sides (Figs. 8.4, 8.10 and 8.11) [50, 51, 75–77]. These shear bands are often precursors of fracture [51, 67, 78–86]. Hodgkinson's 1839 paper was stated by Rankine in 1858 [87] to be the first published report of this phenomenon. Walley et al. [84] found that for certain armour steels and tungsten alloys deformed dynamically the aspect ratio affects the outcome: cylinders with a length/diameter (l/d) ratio less than 1 deformed in a ductile manner whereas those with $l/d > 1$ behaved in a brittle manner.

The effect of friction on the bulge factor p for aluminium cylinders loaded at an 'intermediate' strain rate of 0.1 s^{-1} was studied experimentally by Sivaprasad and Davies for three different friction conditions [90] (Fig. 8.12). They defined p as follows:

$$p = \frac{D_t}{D_m}, \tag{8.26}$$

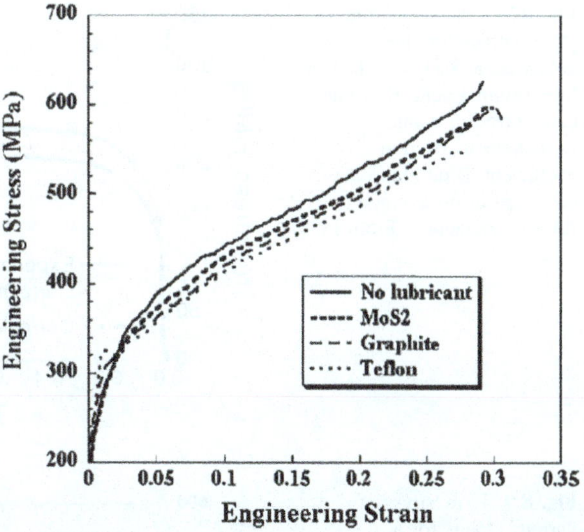

Fig. 8.6 Stress-strain curves for a 6061-T651 aluminium alloy deformed in a split Hopkinson pressure bar using three different solid lubricants and no lubrication. From [58]

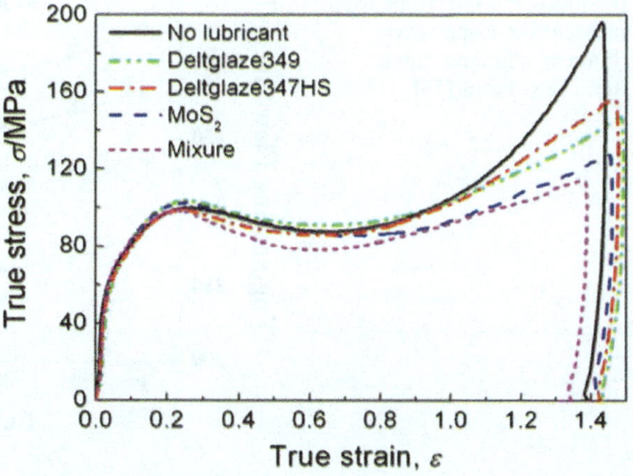

Fig. 8.7 True stress-true strain curves for steel cylinders lubricated with various lubricants at 1000 °C. From [88]

where D_t is the diameter of the specimen at the top or bottom of the cylinder and D_m is the diameter of the specimen at the middle. They modelled the deformation using a simple power law strain hardening description of the metal:

$$\sigma = K\varepsilon^n, \tag{8.27}$$

where σ is the stress, K is the strength coefficient, ε is the strain, and n is the strain hardening exponent. The shape of the bulge can be seen in Fig. 8.13 to depend strongly on the friction coefficient: a high value of the friction factor resulted in a small increase in D_t and a large increase in D_m. Thus the bulge factor p decreased as the friction increased (Fig. 8.14). A comparison of simulations with experimentally

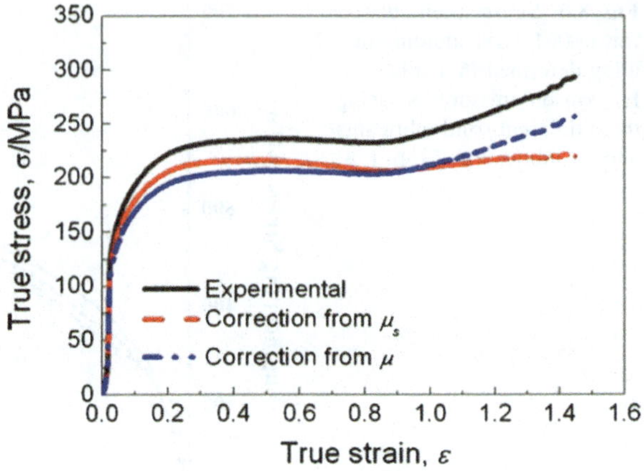

Fig. 8.8 True stress-true strain curves for steel deformed at 800 °C. Black line: original data. Red line: data corrected using instantaneous friction coefficient. Blue line: data corrected using average friction coefficient. From [88]

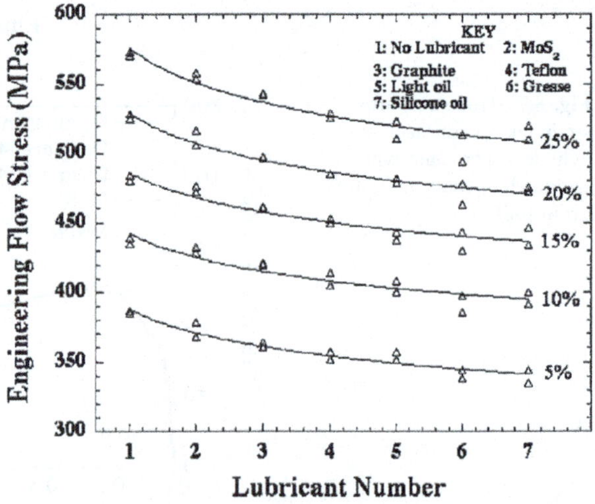

Fig. 8.9 Flow stress at various strains for a 6061-T651 aluminium alloy deformed in a split Hopkinson pressure bar using seven different solid and liquid lubricants. From [58]

determined bulge profiles (e.g. Fig. 8.15) allows the friction factors to be estimated for various lubricants and surface finishes, bearing in mind that this method can only give an average value over the end surfaces and the friction will also change as material from the side of the cylinder 'folds over' onto the ends.

One way of overcoming problems associated with friction in compression testing was reported by G. I. Taylor in 1938. He achieved uniform quasistatic deformation to large strains by compressing specimens 'in very small stages' and regreasing them between each increment of strain [54]. He also commented that an outward flow of grease 'causes an outward tangential force due to viscous drag…[8. which acts] in the opposite direction to the friction'. Although time-consuming and difficult to implement, this method is often used in laboratory compression testing, but is not normally practical in industrial upset forging.

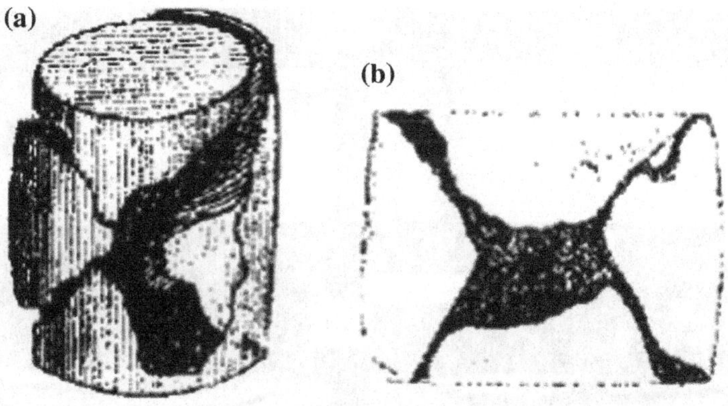

Fig. 8.10 Engraved drawings from what is believed to be the first published report (in 1839) of the cracking produced by quasistatic compression of solid cylinders. The material tested was cast iron. From [51]

Other methods that have been developed to produce compressive deformation under more nearly uniform stress include: (i) compressing the cylinder between two cones of appropriate angle to provide an outwardly directed stress equal to the friction (Fig. 8.16); (ii) machining slight recesses in the two ends which are then filled with lubricant (Fig. 8.17) [91]. Large deformations are possible using this method with the sides of the cylinder remaining straight (no barrelling). Although the state of stress induced can be made uniform by such techniques, the strain that is applied is subject to measurement error due to uncertainty about the positions of the specimen/platen (anvil) interfaces.

Various techniques have been developed to study the flow fields within barrelling cylinders. Examples include building up cylinders using layers of two different aluminium alloys with different etching properties (Fig. 8.18) and (more recently) X-ray tomography (Fig. 8.19).

8.8 Origins of Friction

Whenever two material bodies in contact with each other are moved relative to each other, it is observed that a force resists the motion. As far as we know from the records available to us, this force of friction only began to be studied systematically from the Renaissance onwards, although people had been using lubricants to reduce its magnitude for many thousands of years [95]. The laws of friction we learnt in high-school were worked out by two French scientists Amontons [96] and Coulomb [97]. The most famous of these laws are that the friction force is independent of the apparent area of contact and that the force is proportional to the normal load on the surfaces in contact (Amontons' law). The constant of proportionality is called the coefficient of friction μ. Thus:

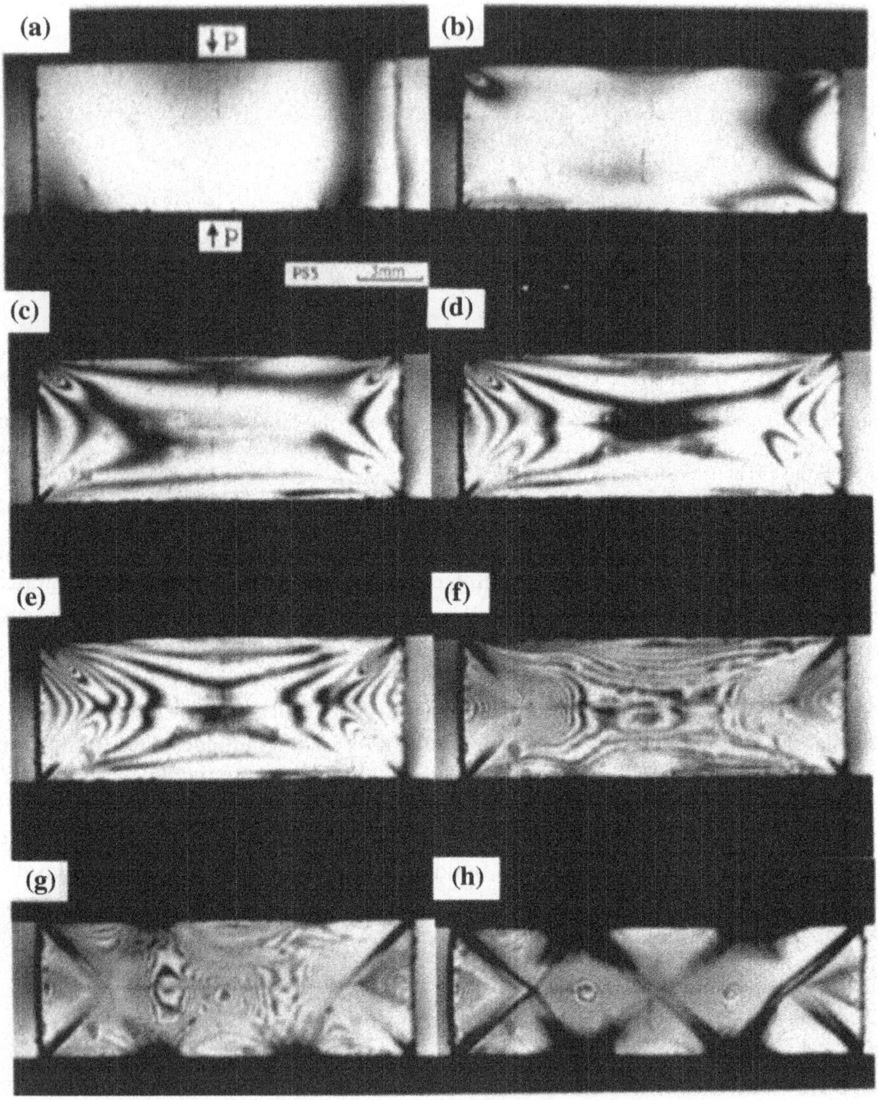

Fig. 8.11 Quasistatic plane strain deformation of a polystyrene specimen showing photoelastic fringes. Original specimen dimensions 17.6 × 7.4 × 3 mm. From [89]

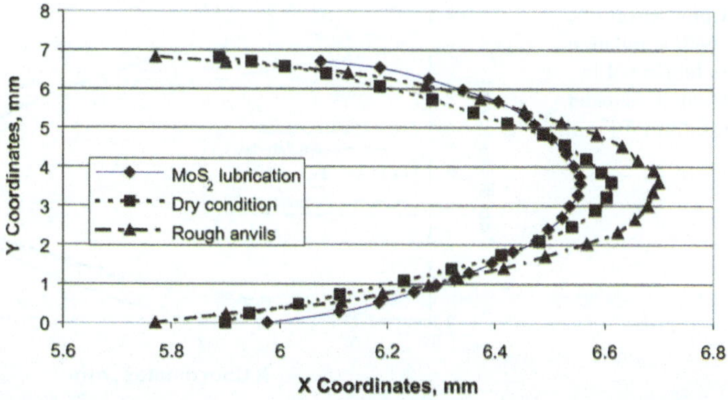

Fig. 8.12 Experimental bulge profiles for compression of 'commercially pure' aluminium cylinders 9 mm in diameter and 13.5 mm in height for three different lubrication conditions. From [90]

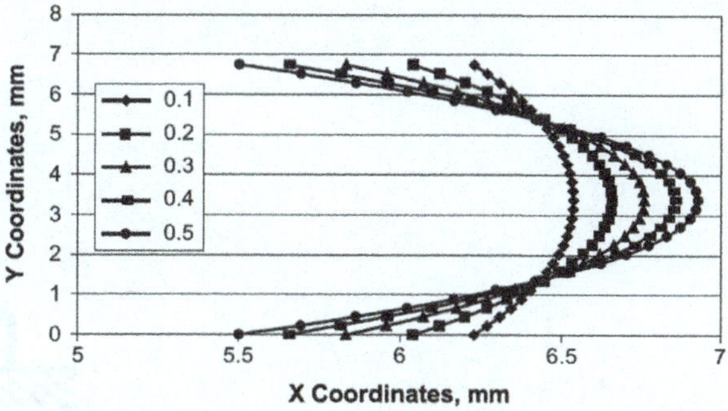

Fig. 8.13 Simulated bulge profiles for the same aluminium cylinders for five values of interface friction. From [90]

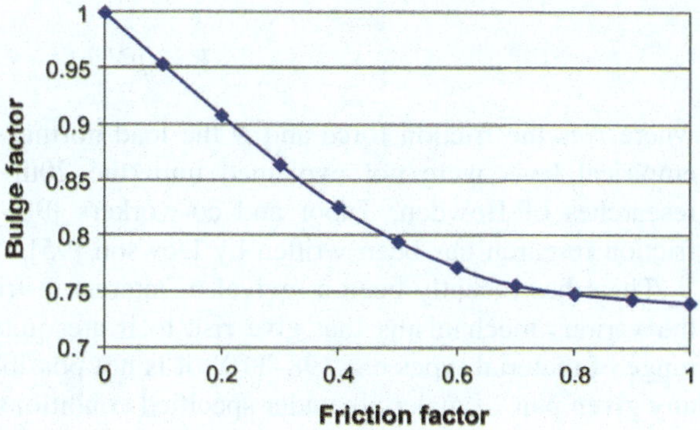

Fig. 8.14 Plot of bulge factor against friction factor. From [90]

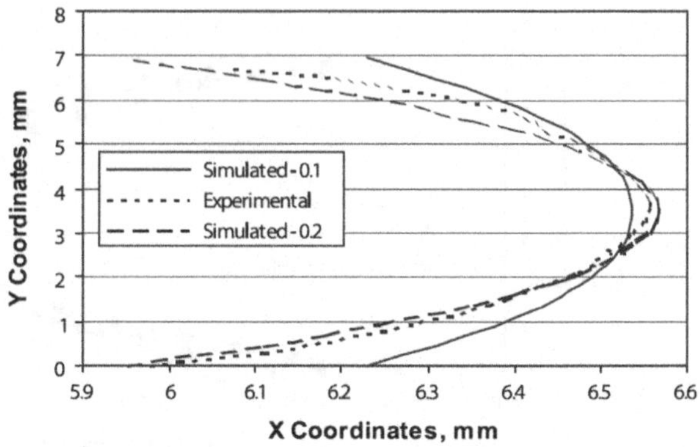

Fig. 8.15 Comparison of experiment with simulation for cylinders lubricated by MoS_2 for a height/diameter ratio of 1.5. From [90]

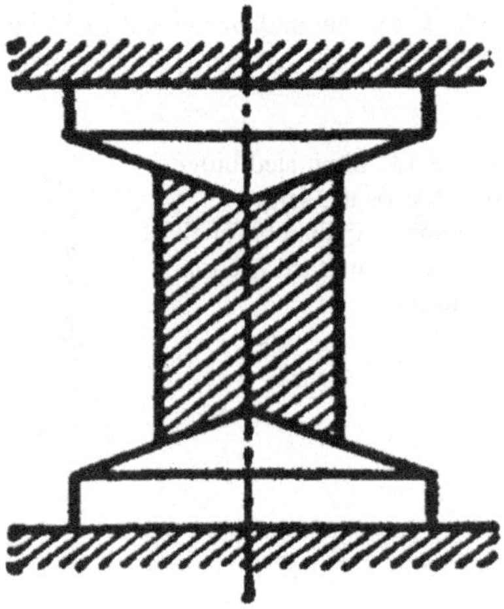

Fig. 8.16 Use of conical anvils to produce a uniform distribution of stress. From [63]

$$F = \mu N, \tag{8.28}$$

where F is the friction force and N the load normal to the sliding surfaces. These empirical laws were not explained until the 20th century, largely through the researches of Bowden, Tabor and co-workers [98, 99]. An excellent history of friction research has been written by Dowson [95].

There has recently been a revival of interest in friction research since although the various mechanisms that give rise to it are quite well-understood for a wide range of material types e.g. [98–119], it is not possible to predict its magnitude for any given pair of materials under specified conditions as Richard Feynman pointed out back in 1963 [120], though there have been some attempts since then at doing so e.g. [121–126]. To put it another way, constitutive equations have not yet been developed which describe friction and this is a major hindrance to the numerical

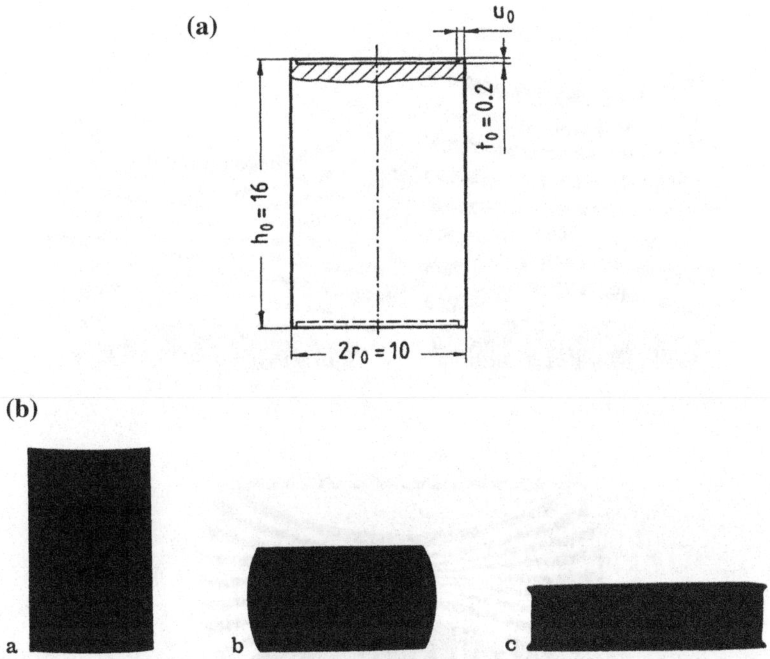

Fig. 8.17 Rastegaev method to produce uniform deformation to large strains. **a** Dimensions of Rastegaev specimen. **b** Comparison of deformation of two specimens of initial dimensions a. **b** without lubricant (strain ca. 0.7). **c** Rastegaev specimen (strain ca. 1.3). From [92]

modelling of sliding contact problems [127, 128]. However, the development of scanning tunnelling and atomic force microscopy [129, 130] and computer simulation techniques [131, 132] have made it possible to perform fundamental studies of friction at the atomic level as well as with atomic resolution e.g. [110, 133–148]. Such studies are proving particularly important in various 'high-tech' applications such as hard disk drives e.g. [149, 150], magnetic tape media e.g. [151–153], and microelectromechanical systems (MEMS) e.g. [154–156], where the sliding bodies interact over very small areas, but it has yet to be demonstrated that these studies are relevant to larger engineering structures.

The main problems are that the surfaces of two materials in contact (i) have different mechanical properties to the bulk due to surface oxide layers, absorbed dirt or lubricants [157], (ii) are rough [158], and (iii) change with time during the deformation due to wear processes [159, 160]. The surfaces can even melt if the sliding speed is high enough [161–165]. In addition, the interaction between asperities on the contacting surfaces may take place at shear strain rates similar to shock conditions [166–168].

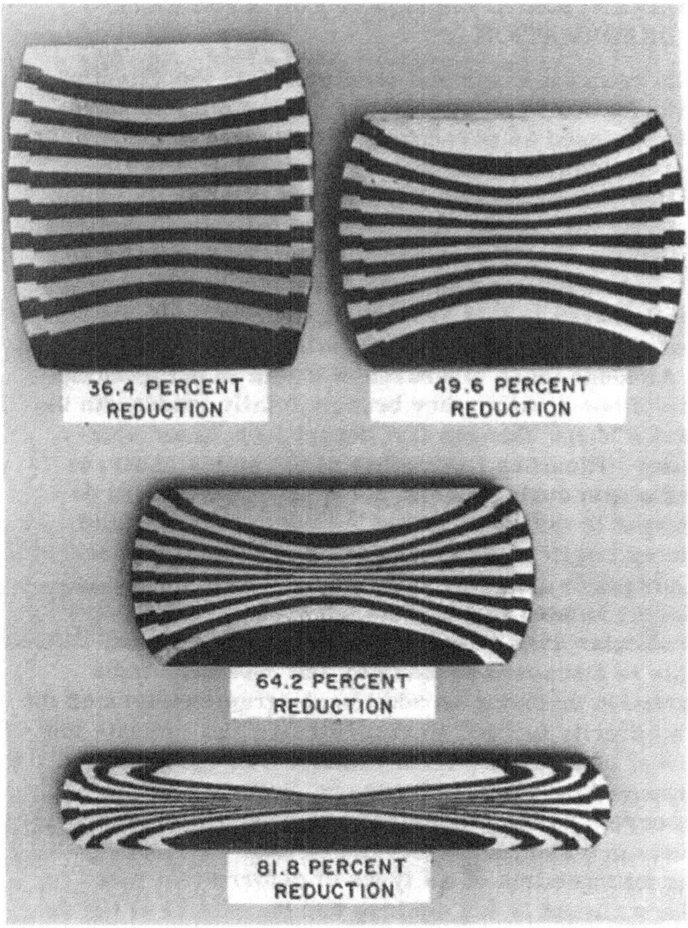

Fig. 8.18 Photograph of anodized cross-sections of upset cylinders formed by stacking alternating discs of two aluminium alloys with similar stress-strain curves but differing anodizing properties. From [93]

8.9 Friction in Upsetting

As outlined in Sect. 8.7, although most studies of friction have been concerned with sliding, there is also a considerable literature on its effects in 'upset forging' due to the importance of this technique in the forming of materials into desired shapes e.g. [90, 169–184].

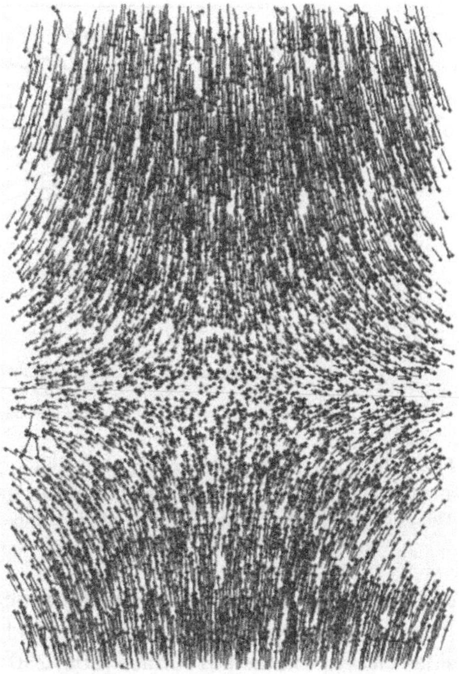

Fig. 8.19 Map of displacement vectors measured using X-ray tomography for a cylindrical specimen originally 2 mm long 1 mm diameter formed by hot compaction of tungsten and aluminium powders. Total field of view 1.5 × 1.5 mm. From [94]

8.9.1 The Ring Test

The standard test for measuring friction in the upset forging geometry is to axially deform an annulus of the test material e.g. [18, 41, 171, 174, 175, 178, 184–224]. If the ratio of the inner to outer diameter remains constant then the lubrication conditions are perfect (zero friction). However, perfect lubrication has only been achieved for low strength materials such as polymers [202, 203, 225], and even then the lubrication is found to break down after a certain strain is exceeded [202]. No combination of surface preparation and lubrication system has been found which reduces friction to zero for metals in this geometry [58, 226, 227], though at high rates of deformation it can be reduced to low values (3–4% of the shear yield strength) [57]. Several studies have shown that at ambient temperatures molybdenum disulphide is the best lubricant out of a number that have been studied [70, 228].

Conventionally the ring test is performed by deforming the annulus to a certain strain, removing it from the apparatus and measuring the radius of the hole and the thickness of the specimen. This measurement is then compared with a set of theoretical curves to read off a value for the friction e.g. [188, 194, 201, 229]; see also Fig. 8.20. There are several assumptions made in this procedure: (i) the specimen dimensions do not change between the time the deformation stopped and the measurement was made. This is probably reasonable for metals but it is certainly not true for polymers [202, 203]; (ii) the friction remains constant during the

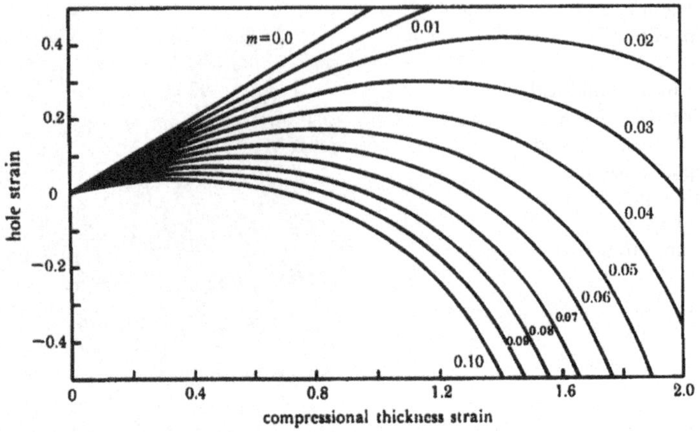

Fig. 8.20 Plots of the hole strain as a function of compression strain for an annulus of original dimensions: OD 5 mm, ID 2.5 mm, thickness 1 mm for various values of the friction parameter m. The curves were calculated using Avitzur's upper bound solution [188, 229]. From [202]

deformation. This last assumption is probably reasonable for small strains but not for large plastic deformation as: (a) 'foldover' occurs in which material from the sides ends up on the top surface [206, 230, 231] (see also Figs. 8.21 and 8.22). This new surface material will have a different roughness and hence a different friction coefficient; (b) lubricant is squeezed out during the deformation (two methods to mitigate this are shown in Figs. 8.17 and 8.22); (c) the theory used is true (Avitzur made a number of simplifications to make the problem tractable such as: the material obeys the von Mises yield criterion, the friction stress is some constant fraction of the shear strength of the material, and the shear stress does not vary through the thickness of the ring).

Walley et al. checked Avitzur's analysis dynamically using high-speed photography for both isotropic polymers [202] and metals [206]. Polymer and copper annuli were observed to follow curves following Avitzur's analysis under certain conditions but not in others (Figs. 8.23 and 8.24). However, over the years, the tendency has been to move away from analytic approaches towards numerical simulation of this problem e.g. [196, 206, 232–236]. Walley et al. found that the friction coefficient needed to obtain agreement between modelling and experiment for the rapid deformation of a steel annulus agreed with that previously determined experimentally on the basis of Avitzur's upper bound solution [206]; see also Figs. 8.25 and 8.26. Note that for numerical methods to make successful predictions of the effect of friction, an adequate constitutive model of the material under consideration must be available.

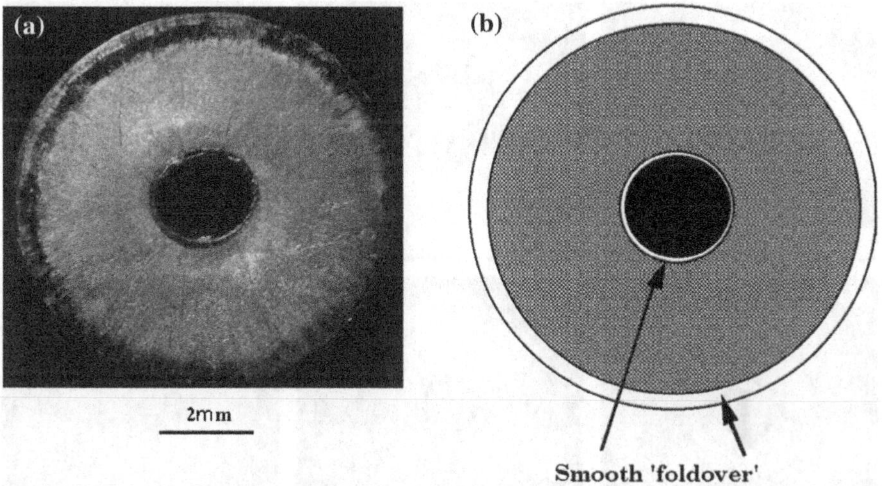

Fig. 8.21 (a) Photograph of deformed annulus of Rolled Homogeneous Armour (RHA) lubricated using MoS_2. Original dimensions: OD = 5.00 mm, ID = 2.5 mm, t = 0.99 mm. (b) Schematic drawing of Fig. 8.19a showing the 'foldover' zones. From [206]

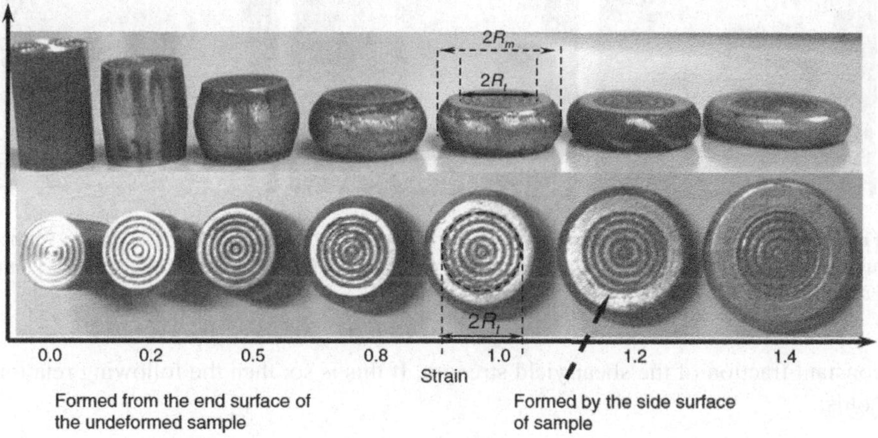

Fig. 8.22 Photographs of steel specimens after compression at 1000 °C. Note that cylindrical grooves were machined in the cylinder's end to hold lubricant in place during deformation. From [88]

8.9.2 The Aspect Ratio Test

An alternative method of determining friction in upsetting is to deform discs of various aspect ratios. Again this technique implicitly makes the assumption that the material obeys a yield criterion (e.g. von Mises) and the friction stress is some

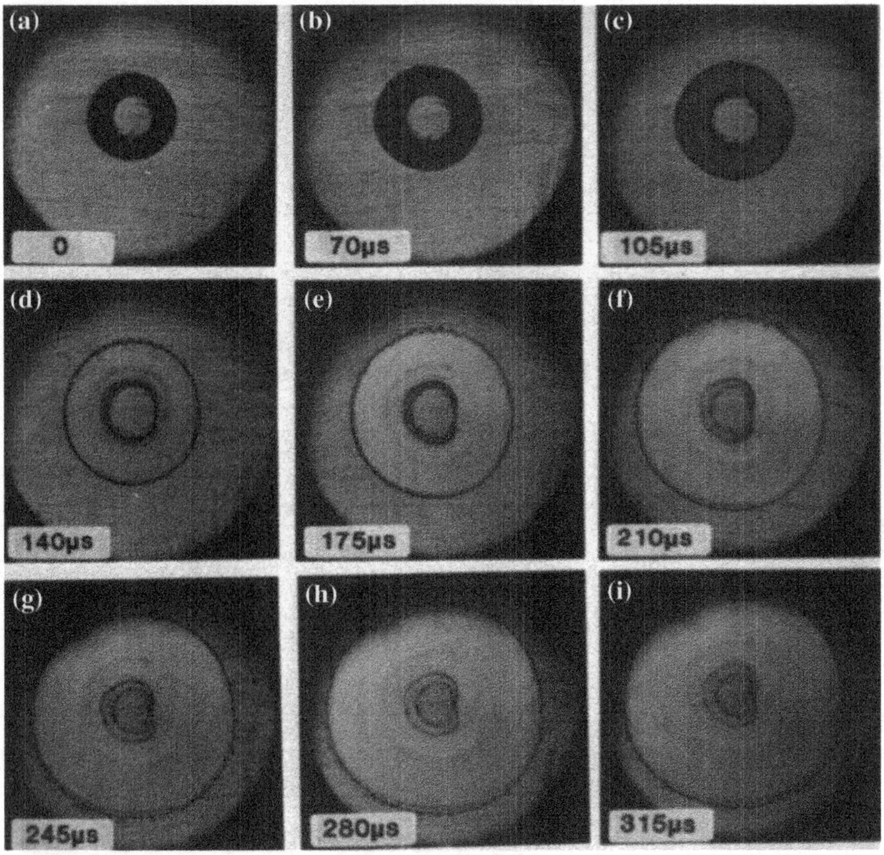

Fig. 8.23 High-speed photographic sequence of the rapid deformation of a nylon washer lubricated with petroleum jelly. Original dimensions: 6.3 mm OD, 3.1 mm ID, thickness 0.8 mm. From [202]

constant fraction of the shear yield strength. If this is so, then the following relation holds:

$$\bar{P} = (1 + \mu d/nh)\sigma_y, \tag{8.29}$$

where $\bar{P}$ is the measured pressure on the specimen, d the diameter of the disc, h its thickness, n a number (see discussion below), and σ_y its yield stress [75]. Clearly if the lubrication is perfect, μ is zero and the measured pressure equals the yield stress for discs of all aspect ratios. If the friction is finite, the thinner specimens exert a higher pressure on the anvils (Fig. 8.27).

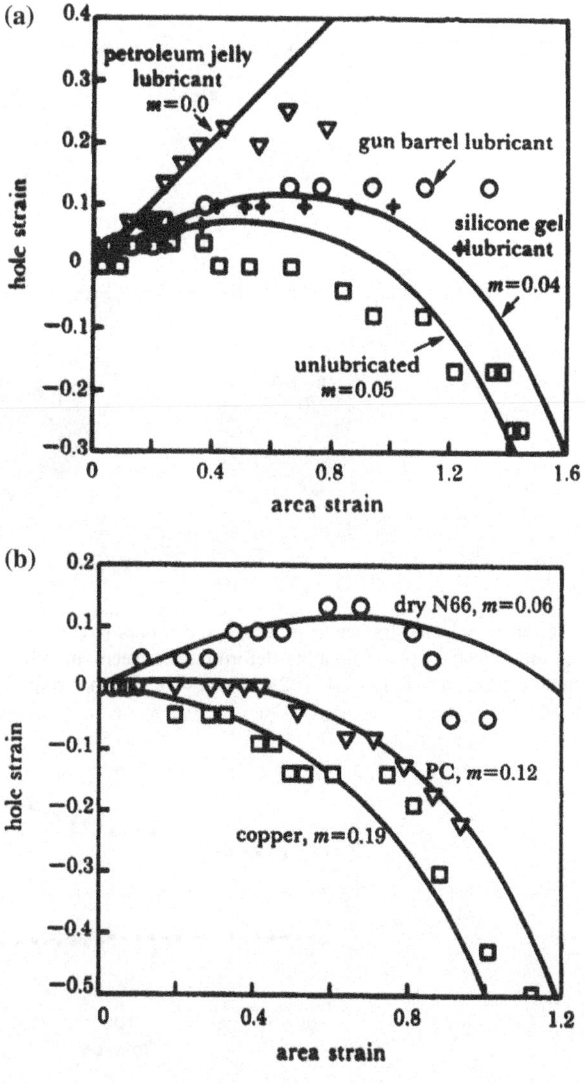

Fig. 8.24 Plots of the hole strain against area strain for **a** nylon washers for various lubrication conditions, **b** various materials. The solid lines are computed from Avitzur's theory to give the closest fit to the experimental data. From [202]

Perfect lubrication has been observed for polymers (Figs. 8.23, 8.24a and 8.28) using hydrocarbon lubricants but a perfect lubrication system has not been found for metals [58]. The value of n depends on the relationship between friction and the shear strength of the material [57, 192]. Theory suggests that for a von Mises solid,

$$\mu = m/\sqrt{3}, \tag{8.30}$$

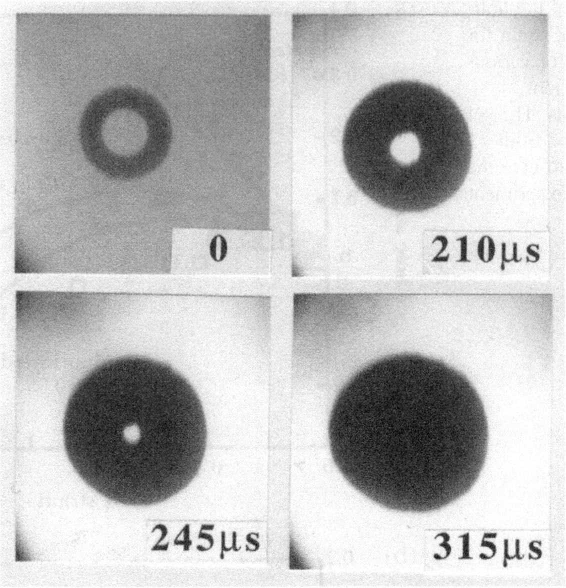

Fig. 8.25 Selected frames from a high-speed photographic sequence of the rapid deformation of an unlubricated copper annulus deforming between smooth anvils. Original dimensions of the annulus: OD = 5.00 mm, ID = 2.4 mm, t = 0.87 mm. From [206]

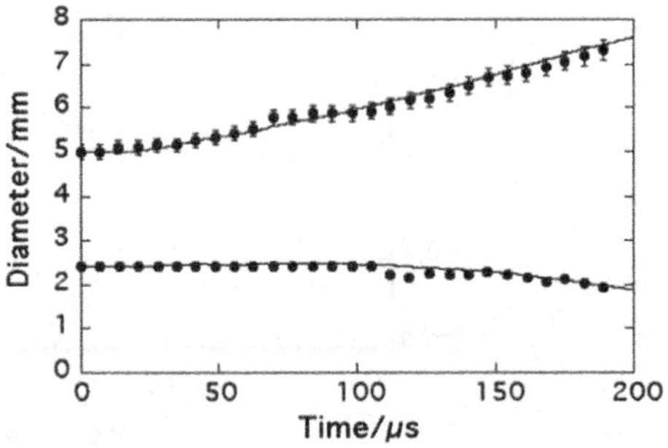

Fig. 8.26 Comparison of simulation (solid line) with experiment (filled dots) for the dynamic deformation of an annulus of RHA deformed between smooth glass anvils and lubricated by molybdenum disulphide. Upper line is the outer diameter; lower line is the inner diameter. Original thickness = 0.99 mm. Error bars 3%. Friction coefficient in the simulation was 0.04. Figure 8.8 courtesy of P.D. Church. Note that the simulation published in [206] was calculated using a friction coefficient of 0.1

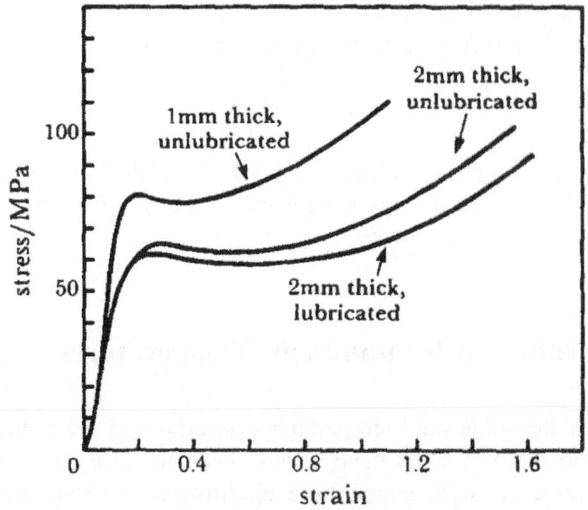

Fig. 8.27 Quasistatic stress-strain curves for 5 mm diameter discs of polyvinylidene difluoride of various thicknesses and lubrication conditions. From [202]

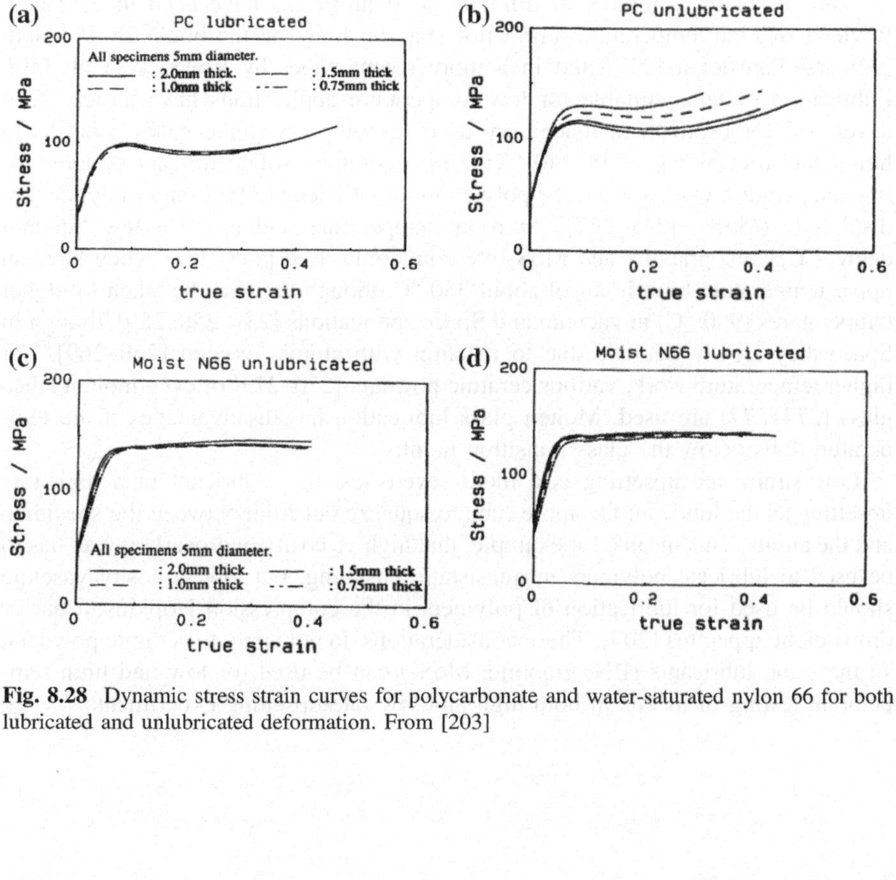

Fig. 8.28 Dynamic stress strain curves for polycarbonate and water-saturated nylon 66 for both lubricated and unlubricated deformation. From [203]

where m is the ratio of the friction stress to the bulk shear strength. However, Male and de Pierre [192] found experimentally that the relation

$$\mu = m/2\sqrt{3} \tag{8.31}$$

fitted the data better for metals. Walley et al. found neither expression worked perfectly for polymers [203]. This is probably because polymers obey a different yield criterion to that of von Mises.

8.10 Lubrication at Non-ambient Temperatures

Metals are often shaped at high temperatures because they are softer and easier to work. Solid lubrication systems have usually been found to be better than liquid lubricants or greases in high temperature upsetting as: (i) they are less prone to being squeezed out from between the specimen and the anvils, and (ii) most liquids and greases would burn away under prolonged contact with hot metal (greases can, however, be used in rolling mills where the contact time between the billet and the rollers is short).

Over the years a number of different solid lubricants have been investigated. Reviews of high temperature lubrication may be found in the books by Pöhlandt [92] and Lansdown [237] and in a more recent paper by Roebuck et al. [59]. Lubrication systems suitable for low temperature applications have largely been developed for bearings in machinery used to pump cryogenic gases, such as in liquid fuel rockets e.g. [238–240]. The most common solid lubricant systems for low temperature work seem to be polytetrafluoroethylene (PTFE) and molybdenum disulphide (MoS_2) [241–247]. At room temperature and up to a few hundred degrees Celsius, graphite and MoS_2 are commonly used [248–257]. They have an upper temperature limit in air of about 350 °C, though they can be taken to higher temperatures (900 °C) in vacuum and Space applications [237, 258, 259], though in Space they slowly degrade due to reaction with atomic oxygen [260–269]. For higher temperature work, various ceramic powders [270–273] or even molten silica glass [274–277] are used. Molten glass lubrication has disadvantages if the temperature falls below the glass transition point.

Low strain rate upsetting is a more severe test of a lubricant than high rate upsetting as the lubricant has more time to squeeze out from between the specimen and the anvils. This means, for example, that high viscosity hydrocarbon wax has to be used to lubricate polymers in quasistatic upsetting, but low viscosity vaseline should be used for lubrication of polymers in the compression Hopkinson bar or dropweight apparatus [203]. These considerations do not apply to ceramic powders, so the same lubricants (BN, graphite, MoS_2) can be used for low and high temperature testing of metals in both high and low rate upsetting experiments.

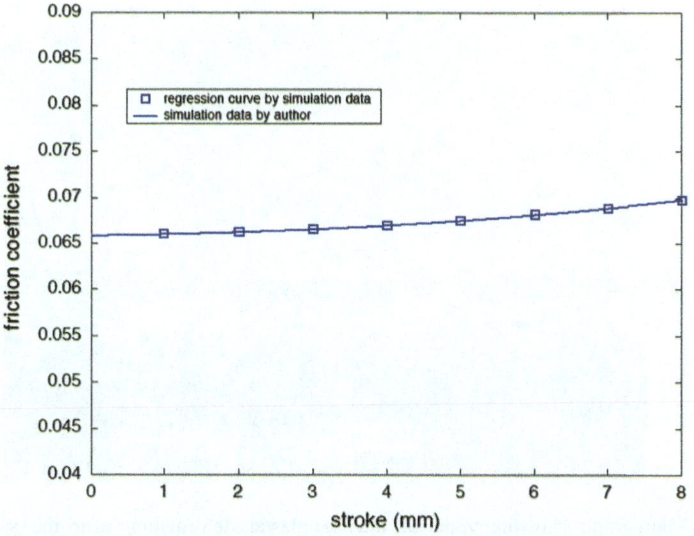

Fig. 8.29 Variation of friction coefficient during upsetting of cylinders of molybdenum. From [284]

Another component of the lubrication system is the surface finish of the specimen. For lubrication to large strains, it is important that the surface is not too smooth or the lubricant will be squeezed out and the friction rise [278–280]. Various methods have been adopted in different studies from simple roughening of the surface [281–283], to machining grooves [88, 226] (Fig. 8.22) or other profiles (Figs. 8.16 and 8.17) [63, 91].

A comprehensive study of this problem for low strain rate upsetting was conducted at Los Alamos by Lovato and Stout [226]. They concluded that boron nitride (BN) powder was the best for high temperature, colloidal graphite for room temperature, and molybdenum disulphide (MoS_2) in a resin was the best for cryogenic upsetting. They also found a grooved surface finish gave good results. As mentioned before, however, material from the sides folds over onto the top and bottom of the cylinder (see Figs. 8.21 and 8.22). So the friction conditions are bound to vary as the test proceeds because unroughened material comes into contact with the anvils (Fig. 8.29). Figure 8.30 presents a micrograph showing the typical depth of material affected by the surface shear stresses that are give rise to friction in cylinder upsetting.

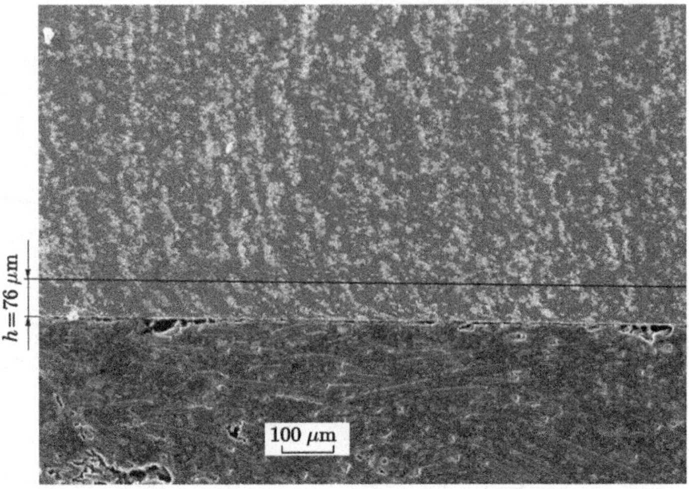

Fig. 8.30 Micrograph showing zone of intense plastic deformation near the surface of a quasistatically deformed 25 mm diameter, 25 mm long cylinder of steel. The radial distance from the centre of the cylinder was 13 mm. The upper surface of the cylinder was displaced 13 mm. The lower surface was held stationary. From [285]

Acknowledgements We would like to thank the UK Ministry of Defence who for many years have funded research at Oxford and Cambridge on dynamic testing of materials using Hopkinson pressure bars. We would also like to thank Philip Church for Fig. 8.26 and Bradley Dodd and Hugh MacGillivray for constructive criticism of this review.

References

1. Kolsky H (1949) An investigation of the mechanical properties of materials at very high rates of loading. Proc Phys Soc Lond B 62:676–700
2. Davies EDH, Hunter SC (1963) The dynamic compression testing of solids by the method of the split Hopkinson pressure bar (SHPB). J Mech Phys Solids 11:155–179
3. Samanta SK (1971) Dynamic deformation of aluminium and copper at elevated temperatures. J Mech Phys Solids 19:117–135
4. Bertholf ID, Karnes CH (1975) 2D analysis of the split Hopkinson pressure bar system. J Mech Phys Solids 23:1–19
5. Malinowski JZ, Klepaczko JR (1986) A unified analytic and numerical approach to specimen behaviour in the split Hopkinson pressure bar. Int J Mech Sci 28:381–391
6. Gorham DA (1989) Specimen inertia in high strain-rate compression. J Phys D Appl Phys 22:1888–1893
7. Sen O, Tekalur SA, Maity P (2011) On the use of non-cylindrical specimens in a split Hopkinson pressure bar. J Strain Anal Eng Des 46:866–878
8. Haddow JB (1965) On the compression of a thin disk. Int J Mech Sci 7:657–660
9. Dharan CKH, Hauser FE (1970) Determination of stress-strain characteristics at very high strain rates. Exper Mech 10:370–376

10. Forrestal MJ, Wright TW, Chen W (2007) The effect of radial inertia on brittle samples using the split Hopkinson pressure bar test. Int J Impact Eng 34:405–411
11. Song B, Chen W, Ge Y, Weerasooriya T (2007) Dynamic and quasistatic compressive response of porcine muscle. J Biomech 40:2999–3005
12. Song B, Ge Y, Chen WW, Weerasooriya T (2007) Radial inertia effects in Kolsky bar testing of extra-soft specimens. Exp Mech 47:659–670
13. Zhang M, Li QM, Huang FL, Wu HJ, Lu YB (2010) Inertia-induced radial confinement in an elastic tubular specimen subjected to axial strain acceleration. Int J Impact Eng 37:459–464
14. Warren TL, Forrestal MJ (2010) Comments on the effect of radial inertia in the Kolsky bar test for an incompressible material. Exp Mech 50:1253–1255
15. Hockly M, Siviour CR (2015) Specimen inertia in high strain rate tensile testing. EPJ Web Conf 94:01050
16. Malinowski Z, Klepaczko J (1972) Estimation of the coefficient of friction on the interfaces of a plastically deformed specimen (in Polish). Mech Teor Stos 10:561–576
17. Gorham DA (1991) The effect of specimen dimensions on high strain rate compression measurements of copper. J Phys D Appl Phys 24:1489–1492
18. Gorham DA, Pope PH, Field JE (1992) An improved method for compressive stress–strain measurements at very high strain rates. Proc R Soc Lond A 438:153–170
19. Jia D, Ramesh KT (2004) A rigorous assessment of the benefits of miniaturization in the Kolsky bar system. Exp Mech 44:445–454
20. Jordan JL, Siviour CR, Sunny G, Bramlette C, Spowart JE (2013) Strain rate-dependant mechanical properties of OFHC copper. J Mater Sci 48:7134–7141
21. Siviour CR, Walley SM, Proud WG, Field JE (2005) The high strain rate compressive behaviour of polycarbonate and polyvinylidene fluoride. Polymer 46:12546–12555
22. Alves M, Karagiozova D, Micheli GB, Calle MAG (2012) Limiting the influence of friction on the split Hopkinson pressure bar tests by using a ring specimen. Int J Impact Eng 49: 130–141
23. Casem DT, Dwivedi AK, Mrozek RA, Lenhart JL (2014) Compression response of a thermoplastic elastomer gel tissue surrogate over a range of strain-rates. Int J Solids Struct 51:2037–2046
24. Bischoff PH, Perry SH (1991) Compressive behaviour of concrete at high strain rates. Mater Struct 24:425–450
25. Tang T, Malvern LE, Jenkins DA (1992) Rate effects in uniaxial dynamic compression of concrete. ASCE: J Eng Mech 118:108–124
26. Li QM, Meng H (2003) About the dynamic strength enhancement of concrete-like materials in a split Hopkinson pressure bar test. Int J Solids Struct 40:343–360
27. Zhang M, Wu HJ, Li QM, Huang FL (2009) Further investigation on the dynamic compressive strength enhancement of concrete-like materials based on split Hopkinson pressure bar tests. 1: experiments. Int J Impact Eng 36:1327–1334
28. Li QM, Lu YB, Meng H (2009) Further investigation on the dynamic compressive strength enhancement of concrete-like materials based on split Hopkinson pressure bar tests. 2: numerical simulations. Int J Impact Eng 36:1335–1345
29. Mu ZC, Dancygier AN, Zhang W, Yankelevsky DZ (2012) Revisiting the dynamic behavior of concrete-like materials. Int J Impact Eng 49:91–102
30. Song Z, Lu Y (2012) Mesoscopic analysis of concrete under excessively high strain rate compression and implications on interpretation of test data. Int J Impact Eng 46:41–55
31. Chen T, Li QB, Guan JF (2013) Effect of radial inertia confinement on dynamic compressive strength of concrete in SHPB tests. Appl Mech Mater 438:215–219
32. Hao Y, Hao H (2013) Numerical investigation of the dynamic compressive behaviour of rock materials at high strain rate. Rock Mech Rock Eng 46:373–388
33. Hao Y, Hao H, Jiang GP, Zhou Y (2013) Experimental confirmation of some factors influencing dynamic concrete compressive strengths in high-speed impact tests. Cement Concr Res 52:63–70

34. Hao Y, Hao H, Li ZX (2013) Influence of end friction confinement on impact tests of concrete material at high strain rate. Int J Impact Eng 60:82–106
35. Heard WF, Martin BE, Nia X, Slawson T, Basu PK (2014) Annular pulse shaping technique for large-diameter Kolsky bar experiments on concrete. Exp Mech 54:1343–1354
36. Klepaczko JR (1990) Behavior of rock-like materials at high strain rates in compression. Int J Plast 6:415–432
37. Georgin JF, Reynouard JM (2003) Modeling of structures subjected to impact: concrete behaviour under high strain rate. Cement Concr Compos 25:131–143
38. Holland CC, McMeeking RM (2015) The influence of mechanical and microstructural properties on the rate-dependent fracture strength of ceramics in uniaxial compression. Int J Impact Eng 81:34–49
39. Zencker U, Clos R (1999) Limiting conditions for compression testing of flat specimens in the split Hopkinson pressure bar. Exper Mech 39:343–348
40. Jankowiak T, Rusinek A, Logygowski T (2011) Validation of the Klepaczko-Malinowski model for friction correction and recommendations on split Hopkinson pressure bar. Finite Elem Anal Des 47:1191–1208
41. Klepaczko J, Malinowski Z (1978) Dynamic frictional effects as measured from the split Hopkinson pressure bar. In: Kawata K, Shioiri J (eds) High velocity deformation of solids. Springer, Berlin, pp 403–416
42. Iwamoto T, Yokoyama T (2012) Effects of radial inertia and end friction in specimen geometry in split Hopkinson pressure bar tests: a computational study. Mech Mater 51: 97–109
43. Lim J, Hong J, Chen WNW, Weerasooriya T (2011) Mechanical response of pig skin under dynamic tensile loading. Int J Impact Eng 38:130–135
44. Nie X, Cheng J-I, Chen WW, Weerasooriya T (2011) Dynamic tensile response of porcine muscle. Trans ASME: J Appl Mech 78:021009
45. Nie X, Prabhu R, Chen WW, Caruthers JM, Weerasooriya T (2011) A Kolsky bar technique for characterization of dynamic shear response of soft materials. Exp Mech 51:1527–1534
46. Guduru PR, Freund LB (2002) The dynamics of multiple neck formation and fragmentation in high rate extension of ductile materials. Int J Solids Struct 39:5615–5632
47. Ma H, Huang L, Tian Y, Li J (2014) Effects of strain rate on dynamic mechanical behavior and microstructure evolution of 5A02-O aluminum alloy. Mater Sci Eng A 606:233–239
48. Rusinek A, Zaera R, Klepaczko JR, Cheriguene R (2005) Analysis of inertia and scale effects on dynamic neck formation during tension of sheet steel. Acta Mater 53:5379–5386
49. Bernatskii AD, Rabinovich AL (1964) Strain in certain crosslinked polymers. Polym Sci USSR 6:1166–1175
50. Mescall J, Papirno R, McLaughlin J (1983) Stress and deformation states associated with upset tests in metals. In: Chait R, Papirno R (eds) Compression testing of homogeneous materials and composites (ASTM STP 808). American Society for Testing and Materials, Philadelphia, pp 7–23
51. Hodgkinson E (1839) On the relative strength and other mechanical properties of cast iron obtained by hot and cold blast. J Franklin Inst 28:184–195 (plate XXIV)
52. Gulliver GH (1908) On the cohesion of steel, and on the relation between the yield points in tension and compression. Proc R Soc Edinb 28:374–381
53. Gulliver GH (1909) On the friction at the extremities of a short bar subjected to a crushing load, and its influence upon the apparant compressive strength of the material. Proc R Soc Edinb 29:432–444
54. Taylor GI (1938) Plastic strain in metals. J Inst Metals 62:307–324
55. Hartley P, Sturgess CEN, Rowe GW (1980) Influence of friction on the prediction of forces, pressure distributions and properties in upset forging. Int J Mech Sci 22:743–753
56. Sato Y, Takeyama H (1979) An extrapolation method for obtaining stress-strain curves at high rates of strain in uniaxial compression. Technol Rep Tohoku Univ 44:287–302
57. Gorham DA, Pope PH, Cox O (1984) Sources of error in very high strain rate compression tests. Inst Phys Conf Ser 70:151–158

58. Hall IW, Guden M (2003) Split Hopkinson pressure bar compression testing of an aluminum alloy: effect of lubricant type. J Mater Sci Lett 22:1533–1535
59. Roebuck B, Lord JD, Brooks M, Loveday MS, Sellars CM, Evans RW (2006) Measurement of flow stress in hot axisymmetric compression tests. Mater High Temp 23:59–83
60. Wang ZJ, Cheng LD (2009) Experimental research and numerical simulation of dynamic cylinder upsetting. Mater Sci Eng A 499:138–141
61. Lu Y, Zhang S (2013) Study on interface friction model for engineering materials testing in split Hopkinson pressure bar tests. Mod Mech Eng 3(3):27–33
62. Schey JA, Venner TR, Takomana SL (1982) The effect of friction on pressure in upsetting at low diameter-to-height ratios. J Mech Work Technol 6:23–33
63. Nadai A (1950) Theory of flow and fracture of solids, vol 1. McGraw-Hill, New York
64. Semiatin SL, Jonas JJ (1984) Formability and workability of metals: plastic instability and flow localization. American Society for Metals, Metals Park
65. Bannerjee JK (1985) Barreling of solid cylinders under axial compression. Trans ASME: J Eng Mater Technol 107:138–144
66. Bannerjee JK, Cárdenas G (1985) Numerical analysis on the barreling of solid cylinders under axisymmetric compression. Trans ASME: J Eng Mater Technol 107:145–147
67. Papirno R (1985) Axial compression testing. In: (eds) Metals handbook (9th edn) vol 8, American Society of Metals, Metals Park, OH, pp 55–58
68. Narayanasamy R, Pandey KS (1997) Phenomenon of barrelling in aluminium solid cylinders during cold upset-forming. J Mater Process Technol 70:17–21
69. Sathiyanarayanan S, Narayanasamy R, Pandey KS (1999) Phenomenon of barrelling in zinc solid cylinders during cold upset forming. Indian J Eng Mater Sci 6:22–26
70. Malayappan S, Narayanasamy R (2003) Observations of barrelling in aluminium solid cylinders during cold upsetting using different lubricants. Mater Sci Technol 19:1705–1708
71. Baskaran K, Narayanasamy R, Arunchalam S (2007) Effect of friction factor on barrelling in elliptical shaped billets during cold upset forging. J Mater Sci 42:7630–7637
72. Narayanasamy R, Baskaran K, Arunachalam S, Krishna DM (2008) An experimental investigation of barreling of aluminium billets during extrusion forging using different lubricants. Mater Des 29:2076–2088
73. Misirili C (2014) On materials flow using different lubricants in upsetting process. Ind Lub Tribol 66:623–631
74. Krishna CH, Davidson MJ, Nagaraju C, Kumar PR (2015) Effect of lubrication in cold upsetting using experimental and finite element modeling. J Test Eval 43:53–61
75. Siebel E (1923) Grundlagen zur Berechnung des Kraft- und Arbeitsbedarf beim Schmieden und Walzen. Stahl u Eisen 43:1295–1298
76. Johnson W, Baraya GL, Slater RAC (1964) On heat lines or lines of thermal discontinuity. Int J Mech Sci 6:409–414
77. Bai YL, Dodd B (1992) Adiabatic shear localization: occurrence, theories and applications. Pergamon, Oxford
78. Marchal M (1891) Study of certain phenomena of compression. Trans Inst Naval Archit 32:81–97, plates XVI–XXI
79. Jenner A, Bai Y, Dodd B (1981) A thermo-plastic shear instability criterion applied to surface cracking in upsetting and related processes. J Strain Anal 16:159–164
80. Jenner B, Dodd B (1981) Cold upsetting and free surface ductility: a review. J Mech Work Technol 5:31–43
81. Bai YL, Dodd B (1985) On plastic instability and fracture in upset forging. Res Mech 13:227–241
82. Dodd B, Stone R, Bai YL (1985) Fracture in upset-forging of cylinders of a magnesium alloy. Res Mech 13:265–273
83. Dodd B, Bai YL (1987) Ductile fracture and ductility with applications to metalworking. Academic, London
84. Walley SM, Radford DD, Chapman DJ (2006) The effect of aspect ratio on the compressive high rate deformation of three metallic alloys. J Phys IV France 134:851–856

144. Braun OM (2010) Bridging the gap between the atomic scale and macroscopic modeling of friction. Tribol Letts 39:283–293
145. Spijker P, Anciaux G, Molinari J-F (2011) Dry sliding contact between rough surfaces at the atomistic scale. Tribol Lett 44:279–285
146. Jeng Y-R, Peng S-R (2012) Atomic simulation of contact behavior during sliding inception of an asperity. Wear 276:70–74
147. Spijker P, Anciaux G, Molinari J-F (2013) Relations between roughness, temperature and dry sliding friction at the atomic scale. Tribol Int 59:222–229
148. Krylov SY, Frenken JWM (2014) The physics of atomic-scale friction: basic considerations and open questions. Phys Stat Sol B 251:711–736
149. Tani H, Mitsutome Y, Tagawa N (2013) Adhesion and friction behavior of magnetic disks with ultrathin perfluoropolyether lubricant films having different end-groups measured using pin-on-disk test. IEEE Trans Magn 49:2638–2644
150. Mate CM, Deng H, Lo GJ, Boszormenyi I, Schreck E, Marchon B (2015) Measuring and modeling flash temperature at magnetic recording head-disk interfaces for well-defined asperity contacts. Tribol Lett 57:https://doi.org/10.1007/s11249-015-0505-6
151. Li X, Bhushan B, Inoue M (2001) Time-dependent mechanical properties and tribological behavior of magnetic tapes. Wear 251:1150–1158
152. Wang JH, Talke FE (2005) Investigation of tape edge wear of magnetic recording tapes. Tribol Int 38:606–615
153. Palacio M, Bhushan B (2009) Nanotribological properties of novel lubricants for magnetic tapes. Ultramicroscopy 109:980–990
154. Singh RA, Satyanarayana N, Sinha SK (2011) Surface chemical modification for exceptional wear life of MEMS materials. AIP Adv 1:042141
155. Subhash G, Corwin AD, de Boer MP (2011) Evolution of wear characteristics and frictional behavior in MEMS devices. Tribol Lett 41:177–189
156. Achanta S, Celis JP (2015) Nanotribology of MEMS/NEMS. In: Gnecco E, Meyer E (eds) Fundamentals of friction and wear on the nanoscale, 2nd edn. Springer, Berlin, pp 631–656
157. Dhinojwala A, Cai L, Granick S (1996) Critique of the friction coefficient concept for wet (lubricated) sliding. Langmuir 12:4537–4542
158. Thomas TR (ed) (1982) Rough surfaces. Longman, Harlow, Essex
159. Archard JF (1958) The temperature of rubbing surfaces. Wear 2:438–455
160. Zipin RB (1984) The effect of wear on surface texture. Appl Surf Sci 18:123–145
161. Bowden FP, Freitag EH (1958) The friction of solids at very high speeds. 1: metal on metal. 2: metal on diamond. Proc R Soc Lond A 248:350–367
162. Bowden FP, Persson PA (1961) Deformation heating and melting of solids in high-speed friction. Proc R Soc Lond A 260:433–458
163. Montgomery RS (1976) Surface melting of rotating bands. Wear 38:235–243
164. Montgomery RS (1976) Projectile lubrication by melting rotating bands. Wear 39:181–183
165. Montgomery RS (1985) Evidence for the melt-lubrication of projectile bands. ASLE Trans 28:117–122
166. Ramesh KT, Clifton RJ (1987) A pressure-shear plate impact experiment for studying the rheology of lubricants at high pressures and high shearing rates. Trans ASME: J Tribol 109:215–222
167. Frutschy KJ, Clifton RJ (1997) Plate-impact technique for measuring dynamic friction at high temperatures. Trans ASME: J Tribol 119:591–593
168. Prakash V (1998) Time-resolved friction with applications to high-speed machining: experimental observations. Tribol Trans 41:189–198
169. Dudder GB (1985) Drop tower compression test. In: Metals handbook (9th edn) vol 8, American Society of Metals, Metals Park, Ohio, pp 196–198
170. Malkani HG, Wilson WRD (1988) Entrapment of laminated solid lubricant films in plane strain upsetting. In: Weinmann KJ (ed) Proceedings of 16th North American Manufacturing Research Conference, Society of Manufacturing Engineers, Dearborn, MI, pp 77–83

171. Avitzur B (1992) Friction during metal forming. In: Friction, lubrication and wear technology. ASM International, Metals Park OH, pp 59–69
172. Hsu T-C, Lee C-H (1997) Realistic friction modeling for simple upsetting. Tribol Trans 40:367–373
173. Hsu T-C, Lee C-H (1997) Refined friction modeling for simple upsetting. Trans ASME: J Manuf Sci Eng 119:563–570
174. Lin SY (1997) Investigation of the effect of dissimilar interface frictional properties on the process of hollow cylinder upsetting. J Mater Process Technol 66:204–215
175. Lin SY (1997) An investigation of unequal interface frictional conditions during the upsetting process. J Mater Process Technol 65:292–301
176. Kuhn HA (2000) Uniaxial compression testing. In: Kuhn H, Medlin D (eds) ASM Handbook, vol 8. Mechanical testing and evaluation. ASM International, Materials Park, pp 143–151
177. Wang JP (2002) An investigation into friction in dynamic plane upsetting. J Mater Process Technol 123:323–328
178. Matsumoto R, Osakada K (2004) Measurement of friction in cold upsetting with mist lubrication. Mater Trans 45:2891–2896
179. Li YP, Onodera E, Chiba A (2010) Evaluation of friction coefficient by simulation in bulk metal forming process. Metall Mater Trans A 41:224–232
180. Solhjoo S (2010) A note on 'barrel compression test': a method for evaluation of friction. Comput Mater Sci 49:435–438
181. Azushima A, Yanagida A, Tani S (2011) Permeation of lubricant trapped within pocket into real contact area on the end surface of cylinder. Trans ASME: J Tribol 133:011501
182. Azushima A, Yoneyama S, Utsunomiya H (2012) Coefficient of friction at interface of lubricated upsetting process. Wear 286:3–7
183. Yao Z-H, Mei D-Q, Shen H, Chen Z-C (2013) A friction evaluation method based on barrel compression test. Tribol Lett 51:525–535
184. Hu C, Ou H, Zhao Z (2015) An alternative evaluation method for friction condition in cold forging by ring with boss compression test. J Mater Process Technol 224:18–25
185. Kunogi M (1954) On plastic deformation of hollow cylinders under axial compression (in Japanese). Rep Sci Res Inst Tokyo 30:63–92
186. Kudo H (1960) Some analytical and experimental studies of axi-symmetric cold forging and extrusion. 1. Int J Mech Sci 2:102–127
187. Kudo H (1961) Some analytical and experimental studies of axi-symmetric cold forging and extrusion. 2. Int J Mech Sci 3:91–117
188. Avitzur B (1964) Forging of hollow discs. Isr J Technol 2:295–304
189. Male AT, Cockcroft MG (1964) A method for the determination of the coefficient of friction of metals under conditions of bulk plastic deformation. J Inst Metals 93:38–46
190. Male AT (1966) Variations in friction coefficients of metals during compressive deformation. J Inst Metals 94:121–125
191. Hawkyard JB, Johnson W (1967) An analysis of the changes in geometry of a short hollow cylinder during axial compression. Int J Mech Sci 9:163–182
192. Male AT, Depierre V (1970) The validity of mathematical solutions for determining friction from the ring compression test. Trans ASME: J Lub Tech 92:389–397
193. Lee CH, Altan T (1972) Influence of flow stress and friction upon metal flow in upset forging of rings and cylinders. Trans ASME: J Eng Ind 94:775–782
194. de Pierre V, Gurney F (1974) A method for determining constant and varying friction factors during ring compression tests. Trans ASME: J Lub Technol 94:482–488
195. Avitzur B, van Tyne CJ, Umaña C (1978) Analytical determination of the sensitivity of the ring test as an experimental study of friction. In: Alexander JM (ed) Proceedings of 18th international machine tool design and research conference, Macmillan, London, pp 305–313
196. Hartley P, Sturgess CEN, Rowe GW (1979) Friction in finite-element analyses of metalforming processes. Int J Mech Sci 21:301–311

197. Gorham DA (1980) Measurement of stress-strain properties of strong metals at very high strain rates. Inst Phys Conf Ser 47:16–24
198. Gomide HA, Burger CP (1981) Three-dimensional strain distribution in upset rings by photoplastic simulation. Exp Mech 21:361–370
199. Martorell IA (1983) On the ring compression test. In: Chait R, Papirno R (eds) Compression testing of homogeneous materials and composites (ASTM STP 808). American Society for Testing and Materials, Philadelphia, pp 78–94
200. Dieter GE (1985) Bulk workability testing. In: Metals handbook (9th edn) vol 8, American Society of Metals, Metals Park, OH, pp 571–597
201. Hansen PH, Bay N, Christiansen P (1988) Analysis of the ring compression test using a general friction model and the upper bound elemental technique. In: Proceedings of 16th North American manufacturing research conference, Society of Manufacturing Engineers, Dearborn, pp 41–47
202. Walley SM, Field JE, Pope PH, Safford NA (1989) A study of the rapid deformation behaviour of a range of polymers. Phil Trans R Soc Lond A 328:1–33
203. Walley SM, Field JE, Pope PH, Safford NA (1991) The rapid deformation behaviour of various polymers. J Phys III France 1:1889–1925
204. Wang F, Lenard JG (1992) An experimental study of interfacial friction-hot ring compression. Trans ASME: J Eng Mater Technol 114:13–18
205. Lichtenberger A, Lach E, Bohmann A (1994) Analyse du frottement dans les essais de compression sur barres d'Hopkinson. J Phys IV France 4(C8):29–34
206. Walley SM, Church PD, Furth M, Field JE (1997) A high-speed photographic study of the rapid deformation of metal annuli: a comparison of theory with experiment. J Phys IV France 7(C3):317–322
207. Dutton RE, Seetharaman V, Goetz RL, Semiatin SL (1999) Effect of flow softening on ring test calibration curves. Mater Sci Eng A 270:249–253
208. Sofuoglu H, Rasty J (1999) On the measurement of friction coefficient utilizing the ring compression test. Tribol Int 32:327–335
209. Li LX, Peng DS, Liu JA, Liu ZQ, Jiang Y (2000) An experimental study of the lubrication behavior of A5 glass lubricant by means of the ring compression test. J Mater Process Technol 102:138–142
210. Han H (2002) Influence of material anisotropy and friction on ring deformation. Trans ASME: J Tribol 124:637–644
211. Robinson T, Ou H, Armstrong CG (2004) Study on ring compression test using physical modelling and FE simulation. J Mater Process Technol 153:54–59
212. Cloete TJ, Hartley RS, Nurick GN (2006) A microstructural investigation of 'machining rings' and deformation uniformity for dynamic ring compression tests. J Phys IV France 134:591–596
213. Hartley RS, Cloete TJ, Nurick GN (2007) An experimental assessment of friction effects in the split Hopkinson pressure bar using the ring compression test. Int J Impact Eng 34:1705–1728
214. Sahin M, Çetinarsian CS, Akata HE (2007) Effect of surface roughness on friction coefficients during upsetting processes for different materials. Mater Des 28:633–640
215. Wang J-P, Lin F-L, Huang B-C, Yun C-C (2008) A new experimental approach to evaluate friction in ring test. J Mater Process Technol 197:68–76
216. Yeh W-C, Wu M-C, Hong J-J (2010) A general upper-bound formulation of stream function of upset forging of ring using a variational approach. Mater Sci Eng B 172:55–67
217. Cristino VAM, Rosa PAR, Martins PAF (2011) Surface roughness and material strength of tribo-pairs in ring compression tests. Tribol Int 44:134–143
218. Noh JH, Min KH, Hwang BB (2011) Deformation characteristics at contact interface in ring compression. Tribol Int 44:947–955
219. Zhu Y-C, Zeng W-D, Ma X, Tai Q-G, Li Z-H, Li X-G (2011) Determination of the friction factor of Ti6Al4 V titanium alloy in hot forging by means of ring-compression test using FEM. Tribol Int 44:2074–2080

220. Lawanwong K (2012) An investigation of thin film coating for evaluating friction coefficient using ring compression test. Appl Mech Mater 152:1281–1286
221. Martin F, Sevilla L, Camacho A, Sebastian MA (2013) Upper bound solutions of ring compression test. Procedia Eng 63:413–420
222. Trabelssi M, Joseph PF (2014) Ring compression test for high-temperature glass using the generalized Navier law. J Amer Ceram Soc 97:3257–3262
223. Cristino VAM, Rosa PAR, Martins PAF (2015) The role of interfaces in the evaluation of friction by ring compression testing. Exp Tech 39(4):47–56
224. Li WQ, Ma QX (2015) Evaluation of rheological behavior and interfacial friction under the adhesive condition by upsetting method. Int J Adv Manuf Technol 79:255–263
225. Briscoe BJ, Nosker RW (1984) The influence of interfacial friction on the deformation of HDPE in an SHPB. Wear 95:241–262
226. Lovato ML, Stout MG (1992) Compression testing techniques to determine the stress/strain behavior of metals subject to finite deformation. Metall Trans A 23:935–951
227. Mirzadeh H, Najafizadeh A, Moazeny M (2009) Flow curve analysis of 17-4 PH stainless steel under hot compression test. Metall Mater Trans A 40:2950–2958
228. Manisekar K, Narayanasamy R, Malayappan S (2006) Effect of friction on barrelling in square billets of aluminium during cold upset forging. Mater Des 27:147–155
229. Avitzur B (1968) Metal forming: processes and analysis. McGraw-Hill, New York, pp 81–102
230. Pearsall GW, Backofen WA (1963) Frictional boundary conditions in plastic compression. Trans ASME: J Eng Ind 85:68–76
231. Hou J, Ståhlberg U, Keife H (1992) Bulging and folding-over in plane-strain upset forging. J Mater Process Technol 35:199–212
232. Nagpal V, Lahoti GD, Altan T (1978) A numerical method for simultaneous prediction of metal flow and temperatures in upset forging of rings. Trans ASME: J Eng Ind 100:413–420
233. Carter WT Jr, Lee D-Y (1985) A finite element analysis of cylinder and ring compression and its experimental verification. Comput Struct 21:1–19
234. Lee JH, Burger CP (1990) Finite element modeling of photoplastic processes: upset forging of rings. Int J Mech Sci 32:19–34
235. Rudkins NT, Hartley P, Pillinger I, Petty D (1996) Friction modelling and experimental observations in hot ring compression tests. J Mater Process Technol 60:349–353
236. Noh JH, Seo JM, Hwang BB (2011) Finite element analysis of the sensitivity of friction calibration curves to dimensional changes in a ring compression test. Metals Mater Int 17:187–197
237. Lansdown AR (1994) High temperature lubrication. Mechanical Engineering Publications Ltd., London
238. Maurer RE (1986) Friction, wear and corrosion control in rolling bearings through coatings and surface modification: a review. J Vac Sci Technol A 4:3002–3006
239. Marchal N, Caisso P, Alliot P, Souchier A (1992) Dry lubrication applied to the HM7 engine hydrogen bearing. J Propul Power 8:968–970
240. Nosaka M, Oike M, Kikuchi M, Kamijo K, Tajiri M (1993) Tribo-characteristics of self-lubricating ball bearings for the LE-7 liquid hydrogen rocket-turbopump. Tribol Trans 36:432–442
241. Nosaka M, Kikuchi M, Kawai N, Kikuyama H (2000) Effects of iron fluoride layer on durability of cryogenic high-speed ball bearings for rocket turbopumps. Tribol Trans 43:163–174
242. McCook NL, Burris DM, Dickrell PL, Sawyer WG (2005) Cryogenic friction behavior of PTFE based solid lubricant composites. Tribol Lett 20:109–113
243. Subramonian B, Kato K, Adachi K, Basu B (2005) Experimental evaluation of friction and wear properties of solid lubricant coatings on SUS440C steel in liquid nitrogen. Tribol Lett 20:263–272
244. Trautmann A, Siviour CR, Walley SM, Field JE (2005) Lubrication of polycarbonate at cryogenic temperatures in the split Hopkinson pressure bar. Int J Impact Eng 31:523–544

245. Burris DL (2008) Investigation of the tribological behavior of PTFE at cryogenic temperatures. Tribol Trans 51:92–100
246. Kauffmann A, Freudenberger J, Geissler D, Yin S, Schillinger W, Sarma VS, Bahmanpour H, Scattergood T, Khoshkhoo MS, Wendrock H, Koch CC, Eckert J, Schultz L (2011) Severe deformation twinning in pure copper by cryogenic wire drawing. Acta Mater 59:7816–7823
247. Descartes S, Godeau C, Berthier Y (2015) Friction and lifetime of a contact lubricated by a solid third body formed from an $MoS_{1.6}$ coating at low temperature. Wear 330:478–489
248. Kaur RG, Heshmat H (2002) 100 mm diameter self-contained solid/powder lubricated auxiliary bearing operated at 30,000 rpm. Tribol Trans 45:76–84
249. Hampson MR, Roberts EW, Cropper MD, Watters RB, Forster DJ (2008) Towards the effective solid lubrication of ball bearings operating at high temperature. Proc Inst Mech Eng Part J: J Eng Tribol 222:1041–1049
250. Ouyang JH, Murakami T, Sasaki S, Li YF, Wang YM, Umeda K, Zhou Y (2008) High temperature tribology and solid lubrication of advanced ceramics. Key Eng Mater 368:1088–1091
251. Jia XM, Chai JR (2011) The experimental study on a new kind of high temperature solid lubricant coating. Adv Mater Res 148:575–578
252. Chen F-Y, Feng Y, Shao H, Zhang X-B, Chen J, Chen N-N (2012) Friction and wear behaviors of Ag/MoS_2/G composite in different atmospheres and at different temperatures. Tribol Lett 47:139–148
253. Zhang W, Demydov D, Jahan MP, Mistry K, Erdemir A, Malshe AP (2012) Fundamental understanding of the tribological and thermal behavior of Ag-MoS_2 nanoparticle-based multi-component lubricating system. Wear 288:9–16
254. Khare HS, Burris DL (2013) The effects of environmental water and oxygen on the temperature-dependent friction of sputtered molybdenum disulfide. Tribol Lett 52:485–492
255. Shtansky DV, Bondarev AV, Kiryukhantsev-Korneev PV, Rojes TC, Godinho V, Fernandez A (2013) Structure and tribological properties of MoCN-Ag coatings in the temperature range of 25–700 & #xB0;C. Appl Surf Sci 273:408–414
256. Yang J, Liu Z, Grey F, Xu Z, Li X, Liu Y, Urbakh M, Cheng Y, Zheng Q (2013) Observation of high-speed microscale superlubricity in graphite. Phys Rev Lett 110:255504
257. Tormala A, Hernandez S, Ripoll MR, Badisch E, Prakash B (2014) Tribological performance of some solid lubricants for hot forming through laboratory simulative tests. Tribol Int 74:164–173
258. Seki K, Suzuki M, Nishimura M, Hasegawa M, Moriyama M (1995) Performance of ball-bearings operated at temperatures up to 500 °C in vacuum. Lub Eng 51:753–763
259. Ohno N, Mia S, Morita S, Obara S (2010) Friction and wear characteristics of advanced space lubricants. Tribol Trans 53:249–255
260. Lewis JH, McDaniel RH (1969) A test of molybdenum disulfide lubricants in radiation and vacuum environments. Trans ASME: J Lub Technol 91:390–396
261. Arita M, Yasuda Y, Kishi K, Ohmae N (1992) Investigation of tribological characteristics of solid lubricants exposed to atomic oxygen. Tribol Trans 35:374–380
262. Sharma SK, Snyder CE Jr, Gschwender LJ (1993) Tribological behavior of some candidate advanced space lubricants. Tribol Trans 36:321–325
263. Wei R-H, Wilbur PJ, Buchholz BW, Kustas FM (1995) *In situ* tribological evaluation of greases and solid lubricants in a simulated atomic oxygen environment. Tribol Trans 38:950–958
264. Tagawa M, Yokota K, Ohmae N, Matsumoto K, Suzuki M (2004) Hyperthermal atomic oxygen interaction with MoS_2 lubricants relevance to space environmental effects in low earth orbit-atomic oxygen-induced oxidation. Tribol Lett 17:859–866
265. Tagawa M, Yokota K, Matsumoto K, Suzuki M, Teraoka Y, Kitamura A, Belin M, Fontaine J, Martin JM (2007) Space environmental effects on MoS_2 and diamond-like carbon lubricating films: atomic oxygen-induced erosion and its effect on tribological properties. Surf Coat Technol 202:1003–1010

266. Tagawa M, Yokota K, Ochi K, Akiyama M, Matsumoto K, Suzuki M (2012) Comparison of macro and microtribological property of molybdenum disulfide film exposed to LEO Space environment. Tribol Lett 45:349–356
267. Gao XM, Hu M, Sun JY, Fu YL, Yang J, Liu WM, Weng LJ (2015) Changes in the composition, structure and friction property of sputtered MoS_2 films by LEO environment exposure. Appl Surf Sci 330:30–38
268. Lv M, Yang LJ, Wang QH, Wang TM, Liang YM (2015) Tribological performance and lubrication mechanism of solid-liquid lubricating materials in high-vacuum and irradiation environments. Tribol Lett 59. https://doi.org/10.1007/s11249-015-0532-3
269. Wang P, Qiao L, Xu J, Li WX, Liu WM (2015) Erosion mechanism of MoS_2-based films exposed to atomic oxygen environments. ACS Appl Mater Interfaces 7:12943–12950
270. Heshmat H (1991) High-temperature solid-lubricated bearing development: dry powder-lubricated traction testing. J Propul Power 7:814–820
271. Heshmat H, Dill JF (1992) Traction characteristics of high-temperature powder-lubricated ceramics (Si_3Ni_4/aSiC). Tribol Trans 35:360–366
272. Heshmat H (1993) Rolling and sliding characteristics of powder-lubricated ceramics at high temperature and speed. Lub Eng 49:791–797
273. Heshmat H, Godet M, Berthier Y (1995) On the role and mechanism of dry triboparticulate lubrication. Lub Eng 51:557–564
274. Alder JF, Phillips VA (1954) The effect of strain rate and temperature on the resistance of aluminium, copper and steel to compression. J Inst Metals 83:80–86
275. Hillery MT, Symmons GR (1996) The hydrodynamic drawing of wire at elevated temperatures. J Mater Process Technol 56:302–312
276. Li LX, Peng DS, Liu ZQ (2000) Study on the lubricities of glass and graphite in the deformation processing of the alloy TC4 at high temperatures (in Chinese). Rare Metal Mater Eng 29:239–242
277. Matsumoto K, Izawa M, Nakanishi T, Tsubouchi K (2009) Tribological properties of water glass lubricant for hot metalworking. Tribol Trans 52:553–559
278. Kudo H (1965) A note on the role of microscopically trapped lubricant at the tool-work interface. Int J Mech Sci 7:383–388
279. Osakada K (1977) A mechanism of lubricant trapping in slow speed compression. Int J Mech Sci 19:413–421
280. Wilson WRD, Lak S (1977) The transport and breakdown of solid lubricants in a simple forging operation. Trans ASME: J Lub Technol 99:230–235
281. Krokha VA (1974) An efficient experimental test piece shape for compressive tests. Strength Mater 6:634–637
282. Wilson WRD, Delmolino WP (1974) The influence of surface roughness on the lubrication breakdown in upsetting between overhanging dies. Wear 29:1–10
283. Krokha VA (1978) Stresses induced in upsetting of specimens with cylindrical recesses in the end faces filled with grease. Strength Mater 10:1467–1471
284. Lin Z-C, Chen C-K (2005) Inverse calculation of the friction coefficient during the warm upsetting of molybdenum. Int J Mech Sci 47:1059–1078
285. Alexandrov SE, Vilotich D, Lyamina EA, Shidzhanin L (2011) Thickness of the layer of intense plastic deformation in the vicinity of the friction surface during upsetting of a cylinder with flat dies. J Appl Mech Tech Phys 52:491–495

Chapter 9
Very High Strain Rate Range

Thomas Heuzé, Xiaoli Guo and Ramzi Othman

9.1 Introduction

The classical Split Hopkinson Pressure Bar (SHPB) system is considered to be able to perform tests at strain rates ranging from 10^2 to 10^4 s^{-1} [1]. However, some modifications can be carried out to extend the reachable strain rate within the specimen. The mean strain rate defined within the specimen:

$$\dot{\varepsilon}_s = \frac{V_{out} - V_{in}}{l_s}, \tag{9.1}$$

where V_{out} and V_{in} stand for the velocity of output and input cross-sections of the specimen respectively, shows that the achievable strain rate varies inversely proportionally to the length of the specimen l_s, while the achievable stress is confined by the elastic limit of the bars [2], especially by the incident bar sustaining the entire impacting energy. From this viewpoint, extending the strain rate in the test can be either achieved by scaling down the size of the specimen and consequently that of the entire device, or by dispensing with the limit on the stress of the incident bar by removing it. Two modified Hopkinson devices are widely adopted to test the material at the strain rates beyond 10^4 s^{-1}, referred to as the miniaturized Hopkinson bar and Direct-Impact (DI) devices.

T. Heuzé (✉) · X. Guo
Research Institute in Civil and Mechanical Engineering
(GeM, UMR 6183 CNRS), École Centrale de Nantes, 1 rue de la Noë,
44321 Nantes, France
e-mail: thomas.heuze@ec-nantes.fr

R. Othman
Mechanical Engineering Department, Faculty Engineering,
King Abdulaziz University, P.O. Box 80248, Jeddah 21589, Saudi Arabia

R. Othman (ed.), *The Kolsky-Hopkinson Bar Machine*,
https://doi.org/10.1007/978-3-319-71919-1_9

9.2 Two Solutions

9.2.1 *Miniaturized Hopkinson Bar Devices*

Miniaturized Hopkinson devices have the same configuration than that of the classical SHPB device, but are scaled down. Thus the wave propagation in the bars and the deduction of the flow stress-strain relationship remain the same. Theoretically, the extension of the strain rate is unlimited. However, largely scaling down the equipment rises difficulties on the manufacturing of the components and on achieving a good alignment of the bars. First, the manufacturing becomes difficult with the traditional means such as turning; it can thus be replaced by Electrical Discharge Machining to get a sufficient accuracy of the geometry. Second, the specimen has to be considered as a homogeneous continuous medium, hence a characteristic length of the microstructure should remain far smaller than the length of the specimen. Lower lengths of specimen have already been experienced, Kamler et al. [3] have used a length of 0.7 mm to reach strain rates of the order of 10^5 s^{-1} on copper. Moreover, miniaturized Hopkinson device also rises difficulties on the strain measuring by the traditional strain gauges. The length of the gauge has to be taken into account once the dimension of the device has been decreased. However, smaller gauges need lower bridge excitations, and this decreases the sensitivity [4]. Othman [5] showed that the strain gauge length acts as a low-pass filter, and this filter effect is more important when the bar diameter is lower than 8 mm. In other words, the gauge may filter out the useful signals as the bar diameter becomes relatively small with respect to the length of the gauge. Thus smaller gauges or alternative means may be required to measure the strain in the bar. Othman's study shows that for a specific material of the bar, the cut-off frequency introduced by the length of the gauge l_g rises more rapidly as l_g decreases, as shown in Fig. 9.1.

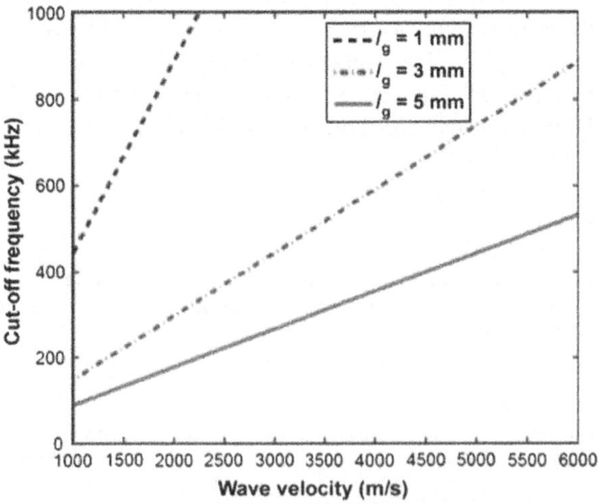

Fig. 9.1 Cut-off frequencies of gauge lengths [5]

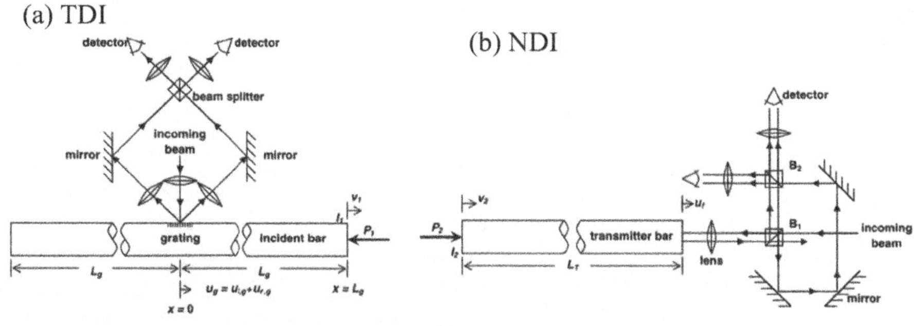

Fig. 9.2 Optical system for displacement measurement [7]

More precisely, for a bar material with a sound speed of about 5000 m/s, the cut-off frequency of a gauge of 1 mm length reaches the large value of 2216 kHz. Jia and Ramesh [6] have used a very small strain gauge of 0.79 mm length to measure the strain in a thin bar of 3.155 mm diameter. According to Othman's research, the signal travelling in this thin bar has a frequency not more than 1000 kHz, while the cut-off frequency of this small gauge is much higher than that of the signal. Hence the major part of the signal is not filtered. With this miniaturized device, they attained strain rates ranging from 10^3 to 5×10^4 s^{-1} on an aluminium 6061-T651 specimen of 1 mm length and 1.818 mm diameter. Casem et al. [7] have also used a very small bar of 1.6 mm in diameter to test specimens made of 6061-T6 aluminium, at a very high strain rate beyond 10^5 s^{-1}. In their research, the strain gauge measurement system is here replaced by an optical measurement techniques where the movement of the incident and transmitted bars were measured by the Transverse Displacement Interferometer (TDI) and Normal Displacement Interferometer (NDI), as shown in Fig. 9.2.

Besides the reduction of the specimen length, the Eq. (9.1) says that the strain rate can also be extended by increasing the impact velocity of the projectile. However, the increase of the impact velocity is restricted by the elastic limit of the incident bar. A direct-impact Hopkinson device can reach higher strain rate by removing the incident bar. But, the modification of the Hopkinson arrangement changes the processing of wave propagation and consequently the deduction of the stress-strain curve of the specimen as well. This point is discussed below.

9.2.2 Direct-Impact Hopkinson Bar

Since the incident bar has been removed in the direct-impact system, the projectile impacts the specimen directly at high velocity in order to reach a higher strain rate within the tested material. The loading pulse is generated at the impacting interface, that is the interface between the projectile and the specimen, then two waves

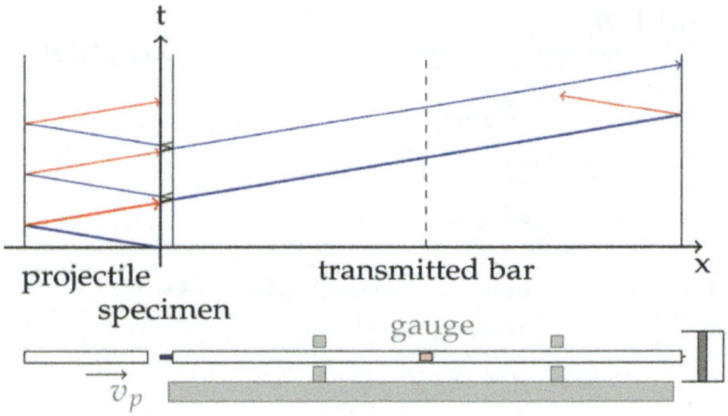

Fig. 9.3 Direct impact Hopkinson bar

propagate simultaneously in opposite directions in the specimen and the projectile, as shown in Fig. 9.3.

On the one hand, the specimen yields under the impact loading and then the pulse is both reflected within the specimen and transmitted to the transmitted bar, owing to the mismatch of material impedance. The pulse is reflected within the specimen. On the other hand, the wave in the projectile is reflected at the free section, and then propagates back towards the impacting interface as an unloading wave. Furthermore, additional difficulties also rise in the deduction of the material behaviour. In the SHPB device, the incident bar plays the role of a load transducer and allows to check the force equilibrium of the specimen [6]. In the direct-impact system, the achievement of the force equilibrium is also usually assumed, and is used to compute the stress. However, it cannot be checked anymore. Besides, alternative techniques are requested to complete the strain and strain rate in the specimen.

Despite the difficulties induced by the absence of an incident bar, the direct-impact Hopkinson device is still of great interest for attaining higher strain rate than classical SHPB system. With such a device, Shioiri et al. [8] tested polycrystalline aluminium, iron, copper and niobium alloys at the strain rate of $2 \times 10^4\,\mathrm{s}^{-1}$. Wulf [9] achieved a strain rate of about $2.5 \times 10^4\,\mathrm{s}^{-1}$ on the 7039 aluminium alloy. Gorham et al. [10] extended the strain rate to $4 \times 10^4\,\mathrm{s}^{-1}$ on Ti–6Al–4V and tungsten alloys. Impacting a very thin specimen of pure aluminium, Dharan and Hauser [11] have achieved strain rates up to $1.2 \times 10^5\,\mathrm{s}^{-1}$. Furthermore, Kamler et al. [3] established a miniaturized direct-impact Hopkinson device where the bar diameter is just 1.5 mm. In his test, a very high strain rate of $2.5 \times 10^5\,\mathrm{s}^{-1}$ was claimed on a copper specimen of 0.3 mm length and 0.7 mm diameter.

This chapter focuses on the direct impact Koslky device as used in the very high strain rate testing. A dedicated Hopkinson system allows to reach the expected levels of strain and strain rate while enforcing the basic assumptions required to deduce explicitly the stress-strain curve of the specimen. The design of such a system often relies on a set of empirical confinement equations used in order to fulfill the required assumptions. However, though it allows to restrict the range of possibilities, additional constraints built on a physical basis permit to clarify and complete these empirical bounds. Moreover, the design process should be adapted to the design of a direct impact Hopkinson system of interest here. The general design requirements set for the design are first introduced in Sect. 9.3. Next, additional criteria defined on a physical basis are introduced in Sect. 9.4 to design a direct impact Hopkinson device. Then, it is shown in Sect. 9.5 that the design process can be written as an optimization problem submitted to equality and inequality type constraints. At last in Sect. 9.6, the processing of direct-impact Hopkinson experiments is discussed.

9.3 General Design Requirements

A direct impact device consists of the projectile, the transmitted bar, the specimen and the accessory devices such as the canon, the recording device, etc. Designing a dedicated experimental device comes down to the design of the bar, the specimen, the projectile, and to determine its impact velocity. Most of the constraints used to design a conventional SHPB device are adoptable to design a direct-impact system. This leads to two topics classified as the *system design* and the *experimental design* [12]. The *system design* is independent of the specific experiment carried out, and involves constraints on the dimensions of the bar and the specimen by the way of three ratios: the ratio of the length of the bar to its diameter (l_b/ϕ_b), the ratio of the length of the specimen to its diameter (l_s/ϕ_s), and the ratio of the specimen diameter to that of the bar (ϕ_s/ϕ_b).

These ratios will be referred in the sequel to as the first, second and third ratios of the *system design* respectively. The indices ($\cdot_p$, $\cdot_s$ and $\cdot_b$) will refer in the sequel to the projectile, the specimen and the transmitted bar respectively, as shown in Fig. 9.4. The *experimental design* determines the specimen dimensions (l_s, ϕ_s), the length (l_p) and the impact velocity (v_p) of the projectile, to deform the specimen in such a way that a given strain rate be reached at a given level of strain.

Fig. 9.4 Geometric schema of the direct-impact Hopkinson system

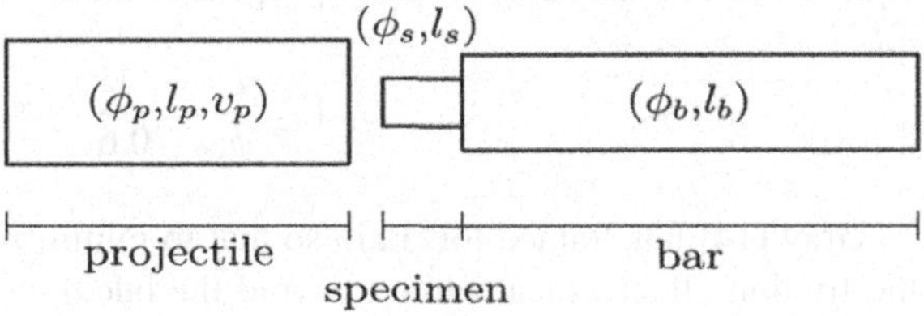

9.3.1 System Design

The three ratios of the system design aim primarily to ensure the enforcement of the unidimensional wave propagation assumption, the sustainability of the system and the reduction of possible disturbances that may alter the quality of the results. This condition is of primary importance for the identification of the specimen behaviour in this test.

9.3.1.1 First Ratio (l_b/ϕ_b)

The assumption of unidimensional wave propagation in the transmitted bar requires a uniform distribution of the stress throughout the entire cross-sections. Provided a given diameter ϕ_b, this suggests a lower bound on the length l_b to ensure a quasi-unidimensional wave propagation along the bar. Davies [13] proved in his work that a bar of the length l_b being greater than $20\phi_b$ can satisfy this requirement. On the other hand, the diameter of the transmitted bar ϕ_b should be large enough to withstand the loading pulse without buckling or being plastically compressed. However, the errors on the stress-strain curve identification induced by dispersion and lateral effects become more important as the diameter increases. A wider range has been suggested by Ramesh [12] and is adopted in practice:

$$\frac{l_b}{\phi_b} \geq 100. \tag{9.2}$$

9.3.1.2 Second Ratio (l_s/ϕ_s)

The design of the specimen comes down to determine its diameter ϕ_s and its length l_s. The constraint on the geometry of the specimen is usually given in the form on the length-to-diameter ratio (l_s/ϕ_s). On the one hand, a small value of this ratio leads to greater lateral inertia and friction effects. On the other hand, a too large value could lead to buckling. The restrictions on this ratio are not unique or equivalent. For instance, Ramesh [12] recommends it to be framed as follows:

$$1 \leq \frac{l_s}{\phi_s} \leq \frac{1}{0.6}. \tag{9.3}$$

Gray [14] has framed this ratio so that to minimize the lateral inertia effects and the friction effects, meanwhile to avoid the buckling of the specimen.

$$0.5 \leq \frac{l_s}{\phi_s} \leq 1. \tag{9.4}$$

The lower bound in Eq. (9.4) is determined in such a way that the lateral inertia effects be minimized. According to Gray's work, the inertia effects are minimized for a ratio:

$$\frac{l_s}{\phi_s} = \sqrt{\frac{3\nu_s}{4}} \tag{9.5}$$

where ν_s, the Poisson's ratio, is taken equal to 1/3, yields the ratio $l_s/\phi_s = 1/2$, which is the lower bound of in Eq. (9.4). Furthermore, a ratio ranging from 1.5 to 2 permits a minimal friction at the contact interface between the bar and the specimen [2], as demonstrated in [15]. A ratio value within this range may generate buckling at high strain rate. Thus Gray suggested the value of one for the upper bound of the ratio.

Davies and Hunter [16] have used a ratio of approximately 0.5 to minimize the interface friction in their experiments. Kamler et al. [3] have impacted a copper specimen designed with a ratio (l_s/ϕ_s) slightly smaller than 0.5, around 0.43 [6]. In the limit case, a thin plate specimen is also adopted in the literature to reach higher strain rate. Dharan and Hauser [11] performed tests on aluminium workpieces with the ratio of 0.25 and achieved the strain rate of $1.2 \times 10^5\ \mathrm{s}^{-1}$. Edington [17] used a thin plate of ratio 0.16 to study the dynamic behaviour of the copper. The lubrication of the interfaces and the numerical correction of lateral inertia effects are generally mandatory in the tests. For a non-cylindrical specimen, the geometry effect is studied and discussed in [18]. All these authors do not exactly agree on the allowable bounds of this ratio, but combining all these works, Guo et al. [19, 20, 25] used the in Eq. (9.4), i.e., $0.5 \le l_s/\phi_s \le 1$.

9.3.1.3 Third Ratio (ϕ_s/ϕ_b)

A small value of the third ratio (ϕ_s/ϕ_b) enables to ensure a good contact at the specimen/bar interface even if ϕ_s dilates largely during the plastic compression of the specimen, and allows to test much more ductile materials. The decrease of this ratio is meanwhile bounded since it can lead to the punch effect. The error, between the measured strain in the specimen and the expected one, which is induced by the non-uniform distribution of the stress, increases with the decrease of this ratio.

Buchar et al. [21] indicate that the specimen diameter must be large enough compared with that of the bars and give a lower bound of this ratio of $(\phi_s/\phi_b > 0.9)$ [22]. Gray [14] has suggested the ratio $(\phi_s/\phi_b > 0.8)$ in order to minimize the mismatch of material impedance and the diameters of the bar and the specimen and consequently reduced the inertia and friction effects [2]. In order to adapt for large strain in the specimen, an empirical range can be adopted [12] as follows:

$$\frac{1}{4} \leq \frac{\phi_s}{\phi_b} \leq \frac{1}{2}. \tag{9.6}$$

9.3.2 Experimental Design

The *system design* has suggested a general frame for the dimension of each component of the Hopkinson device. The *experimental design* determines the specific dimensions of the specimen and the projectile, and the impact velocity, fulfilling the requirement to reach the expected strain rate and the allowable strain, and thus complete the dimensions of the whole device.

The achievable strain rate is related to the dimensions of the specimen, these of the projectile and its impact velocity. When the projectile impacts the specimen, a loading pulse is generated so that two waves propagate simultaneously in opposite directions within the specimen and the projectile. In the projectile, the first wave propagates to the free end and is then reflected back to the impacting interface. Likewise, the second wave propagates through the specimen and is both reflected and transmitted to the transmitted bar. When the first wave, reflected at the end side of the projectile, arrives at the projectile/specimen interface at time $\Delta t = 2l_p/c_p$, referred to as the *characteristic time*, the impact is considered to be terminated. In this definition, c_p denotes the sound speed in the projectile and l_p the projectile length. In order to achieve a high strain rate $\dot{\varepsilon}_s$ in the specimen, the projectile should be accelerated to a sufficient velocity v_p to deform the specimen, though the capacity of the canon may limit it. The maximum engineering strain induced in the specimen can be estimated by:

$$\varepsilon_{s,max} = \int_0^{2l_p/c_p} \dot{\varepsilon}_s \, dt \tag{9.7}$$

where the specimen strain rate $\dot{\varepsilon}_s$ can be estimated by Eq. (9.1).

The experimental design has thus to simultaneously consider the expected strain rate and the allowable strain in the specimen.

9.4 Specific Requirements of the Direct Impact Kolsky/Hopkinson Device

As mentioned above, the combination of the *system design* and the *experimental design* allows to reduce the range of possibilities for the design. Confinement criteria based on less empirical statements can be introduced to narrow down this range. These additional criteria involve quantities that either pertain to each component of the direct impact system (projectile, specimen and bar) or combine quantities related to different ones [19, 20]. They will be presented in the sequel as the criteria involving quantities relative to both the projectile and the bar, to both the specimen and the bar, and to both the projectile and the bar. Constraints on the diameter of the bar and on the projectile are also introduced.

9.4.1 Criteria Involving Quantities Relative to Both the Projectile and the Specimen

First of all, the level of strain achieved within the specimen should be bounded. A sufficient desired level of strain $\varepsilon^d_{s,min}$ is required to correctly characterize the dynamic behaviour of the material, whereas a maximum desired strain $\varepsilon^d_{s,max}$ is required to avoid to crush the specimen. Two criteria are then involved. First, these bounds on the level of strain permit to assess bounds on the length of the projectile l_p. Indeed, assuming a given average strain rate $\dot{\varepsilon}_{s,avg}$ during the characteristic time $\Delta t = 2l_p/c_p$, the maximum strain achieved in the specimen reads:

$$\varepsilon_{s,max} = \dot{\varepsilon}_{s,avg} \frac{2l_p}{c_p} \tag{9.8}$$

which yields the following bounds on the length of the specimen (Fig. 9.5):

$$\frac{c_p}{2} \left| \frac{\varepsilon^d_{s,min}}{\dot{\varepsilon}_{s,avg}} \right| \leq l_p \leq \frac{c_p}{2} \left| \frac{\varepsilon^d_{s,max}}{\dot{\varepsilon}_{s,avg}} \right| \tag{9.9}$$

Second, an approximate upper bound of the impact velocity of the projectile can be assessed in order to avoid to exceed the allowable level of strain. Let's consider a system that consists of the projectile plus the specimen at two instants denoted t_0 and t_1, corresponding to their impact and depart times, as shown in Fig. 9.6.

Writing the conservation of energy of this system between these two instants assuming a unidimensional system, a rigid projectile, a rigid perfectly plastic

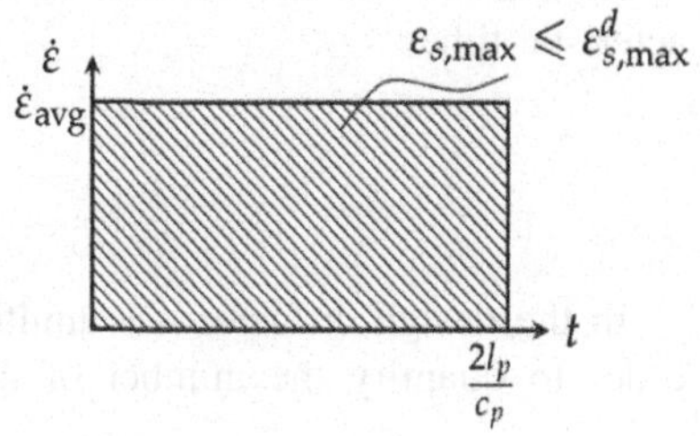

Fig. 9.5 Strain assessed to derive bounds on the length of the specimen

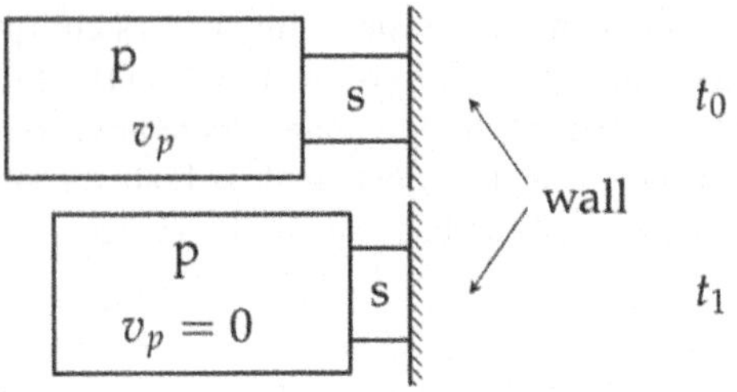

Fig. 9.6 The unidimensional system consisting of a projectile and a specimen

behaviour of the specimen and a vanishing velocity of the projectile at the end time, yields:

$$K_1 - K_0 = W_{\int,0\to1} + W_{ext,0\to1}, \tag{9.10}$$

where K_1 and K_0 denote the kinetic energy of the projectile and specimen together at times t_1 and t_0, respectively, and $W_{\int,0\to1}$ and $W_{ext,0\to1}$ are the internal and external works, respectively. Equation (9.10) leads to:

$$\frac{1}{2}\rho_p l_p A_p v_p^2 = \sigma_{s,y}|\varepsilon_{s,max}|l_s A_s, \tag{9.11}$$

where $\sigma_{s,y}$ is the yield stress of the specimen, l_s and A_s are the length and the cross-sectional area of the specimen, ρ_p, A_p and l_p the mass density, the cross-sectional area and the length of the projectile, respectively.

This leads to the following upper bound on the velocity of the projectile [20, 25]:

$$v_p \leq \sqrt{\frac{2\sigma_{s,y}|\varepsilon_{s,max}|l_s A_s}{\rho_p l_p A_p}}. \tag{9.12}$$

Of course, a refined bound could be assessed using a more complex constitutive model to compute the strain energy of the specimen, and considering the projectile as deformable.

Afterwards, since the test is assumed to be terminated at the end of the characteristic time, the force equilibrium within the specimen should be reached before this time, so that its writing can be used in the post-processing to extract directly the stress-strain curve a posteriori. It is generally considered that a great number of round trips of the wave within the specimen should be achieved during the characteristic time:

$$\frac{2l_s}{c_s} \ll \frac{2l_p}{c_p}. \tag{9.13}$$

In the design, a factor n is multiplied to the left hand side of in Eq. (9.13) in order to quantify the number of these round trips needed to achieve the force equilibrium. For the value of n, Davies and Hunter [16] have proposed that the achievement of the stress equilibrium requires three reflections of the loading pulse within the specimen for ductile metal material. A greater value $n = 10$ is usually adopted to ensure the achievement of the force equilibrium.

9.4.2 *Criteria Involving Quantities Relative to Both the Specimen and the Bar*

During the test, the transmitted bar should remain elastic, while it has to be sufficiently strained to record a usable signal in post-processing. On the one hand, strength criteria of the bar pertain to its resistance to buckling and to plasticity. The former can be assessed in a first approximation through the critical load obtained in quasi-static:

$$F_b \leq \frac{\pi^2 E_b I_b}{L^2} \tag{9.14}$$

where F_b denotes the axial force supported by a cross-section of the bar, E_b is its Young's modulus, L is the length between two bearings (It is almost equal to $l_b/2$ if three bearings are used) and I_b is the cross-section moment of inertia about the bar axis.

Since the bar should behave elastically, the stress in the bar should remain lower than the elastic yield stress of the bar material. Assuming the force equilibrium is achieved within the specimen, the axial stress in the bar should fulfill the following inequality:

$$|\sigma_b| = |\sigma_s| \frac{A_s}{A_b} \leq \frac{|\sigma_{y_b}|}{\alpha} \tag{9.15}$$

where α denotes a safety factor (>1). Provided a maximum expected level of stress $|\sigma_{s_{max}}|$ within the specimen, an upper bound to the third ratio of the system design is given:

$$\left|\frac{\phi_s}{\phi_b}\right| \leq \sqrt{\frac{|\sigma_{y_b}|}{\alpha |\sigma_{s_{max}}|}}. \tag{9.16}$$

On the other hand, a minimum signal-to-noise ratio is required in order to obtain usable signals for post-processing. This implies the normal force in the bar should most of the time be higher than a minimum value which can be roughly assessed by the force in the bar higher than the force corresponding to a stress in the specimen equal to a third of its yield stress.

$$|\sigma_{b_{min}}| \geq \frac{A_s}{A_b} \frac{|\sigma_{y_b}|}{3}. \tag{9.17}$$

Provided a minimum level of strain recorded by gauges on the bar $\varepsilon_{b_{min}}$,

$$|\varepsilon_{b_{min}}| \geq \frac{A_s}{A_b} \frac{|\sigma_{y_b}|}{3E_b}. \tag{9.18}$$

Thus another upper bound to the third ratio of system design can be given:

$$\left|\frac{\phi_s}{\phi_b}\right| \leq \sqrt{\frac{3E_b|\varepsilon_{b_{min}}|}{|\sigma_{y_b}|}}. \tag{9.19}$$

9.4.3 Criteria Involving Quantities Relative to Both the Bar and the Projectile

The length of the bar has to be determined so that, on the one hand, a uni-dimensional propagation of the wave is ensured, which requires a minimum slenderness; the length should be at least ten times the diameter. On the other hand, no wave reflection should occur at the free end of the bar during the characteristic time. Combining both items, one gets:

$$l_b \geq \frac{2l_p c_b}{c_p} + 10\phi_b. \tag{9.20}$$

This inequality comes in addition to that of the first ratio [Inequality (9.2)] of the system design that couples the length and the diameter of the bar.

9.4.4 Diameter of the Transmitted Bar

The loading pulse propagates in the transmitted bar as a plane wave, which consists of a superposition of an infinite number of monochromatic waves with different amplitudes and frequencies:

$$u(r_b, z, t) = \frac{1}{2\pi} \int_{-\infty}^{+\infty} u(r_b, \omega) e^{i(\pm\xi(\omega)z + \omega t)} d\omega \tag{9.21}$$

where $\xi(\omega)$ is the wave number, ω the angular frequency, r_b the radius of the bar and z the coordinate along axis direction. If we want to compute directly the stress in the bar from the strain which is recorded at a different position, we have to make sure that the sole first mode of the bar will be excited by the loading pulse. Thus, the profile of this loading signal has first to be assessed. Second, the spectrum of the bar is needed, and more precisely the frequency of the second mode. The well-known Pochhammer-Chree [23, 24] analytical solution enables to relate the radius of the bar to the angular frequency of a given mode. Determining an upper bound for the bar diameter thus comes down to compare the cut-off frequency of the exciting signal with respect to the frequency of the second mode of the bar.

In order to assess the profile of the loading signal that propagates within the bar, a constitutive model can be postulated to describe the behaviour of the specimen, and therefore to assess its response to the initial pulse. The cut-off frequency of this signal is then computed in the frequency space through a Fourier transform. The Johnson-Cook model has here been used in a first approximation with parameters calibrated for the Ti–6Al–4V alloy at the strain rate of 20 s^{-1} extracted from [25] as summarized in Table 9.1. In order to convert the stress-strain curve into time space, a constant strain rate of 10^5 s^{-1} is chosen. According to Eq. (9.9), the total strain in the specimen can be approximately assessed by $\varepsilon = \dot{\varepsilon}_{avg} t$. Thus a time period of 5 ms is needed to attain a strain of 0.5.

The unloading stage is also plotted at an arbitrary (but negative) strain rate in order to form a complete pulse. The time interval of the signal is set at 0.1 ms according to the sampling frequency of the gauges. The profile of the exciting signal is plotted in Fig. 9.7a. The exciting signal is then converted from the time space to the (angular) frequency space by the Fourier Transformation. The stress after transformation is denoted $\tilde{\sigma}$, whose profile is shown in the Fig. 9.7b. In order to determine the cut-off frequency, we define the total energy of the transformed pulse as:

$$E_\infty = \int_0^{+\infty} |\tilde{\sigma}(\omega)|^2 d\omega \tag{9.22}$$

while the truncated one is:

Table 9.1 Physical and mechanical properties

	ρ (kg/m^3)	E (GPa)	σ_y (MPa)	V
Ti–6A1–4V	4430	113	970	0.34
Marval	8000	186	1750	0.3
	A (MPa)	B (MPa)	n	C
Ti–6A1–4V	950	603	0.20	0.02

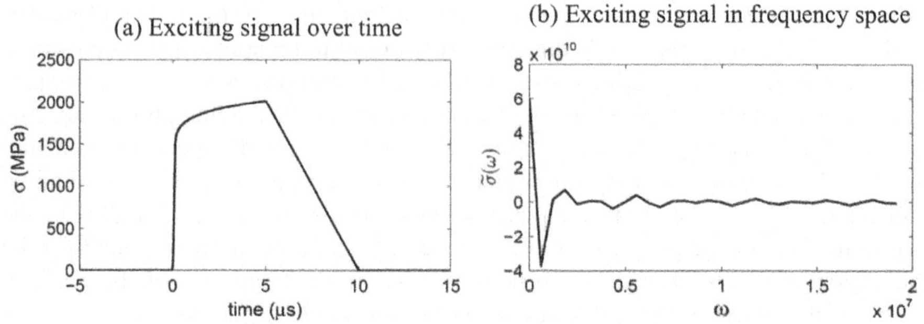

Fig. 9.7 Exciting signal. **a** Exciting signal over time, **b** Exciting signal in frequency space

$$E_c = \int_0^{\omega_c} |\tilde{\sigma}(\omega)|^2 d\omega. \tag{9.23}$$

The cut-off angular frequency ω_c is defined as the critical frequency that allows the relative error between the total energy and the truncated one to remain within a prescribed tolerance:

$$\left| \frac{E_\infty - E_c}{E_\infty} \right| \leq TOL \tag{9.24}$$

Consequently, the cut-off frequency $\omega_c = 1.237 \times 10^6$ rad s^{-1} is found for a tolerance of 0.1.

The Pochhammer-Chree solution is obtained by solving the set of elastodynamic equations for an infinite cylindrical bar. Non-trivial solutions are given when the Pochhammer-Chree equation [5, 23, 24] vanishes, i.e.,

$$\begin{aligned} \frac{2\alpha}{r_b}(\beta^2 + \xi^2)J_1(\alpha r_b)J_1(\beta r_b) - (\beta^2 - \xi^2)^2 J_0(\alpha r_b)J_1(\beta r_b) \\ - 4\alpha\beta\xi^2 J_1(\alpha r_b)J_0(\beta r_b) = 0 \end{aligned} \tag{9.25}$$

where $\alpha^2 = \rho\omega^2/(\lambda + 2\mu) - \xi^2$ and $\beta^2 = \rho\omega^2/\mu - \xi^2$, $J_n(.)$ is the Bessel function of the first kind at order n, ξ is the wave number, ω is the angular frequency, r_b is the bar radius, and v, λ, μ, E, ρ denote the Poisson's ratio, the Lamé's constants, the Young's modulus and the mass density respectively. This equation gives an implicit relation between the wave number ξ and the angular frequency ω, but also involves the radius and material properties of the bar. The first mode propagates for every frequencies. Higher modes (starting from the second one) only propagates above a critical frequency $\omega_c^{(m)}$. If we denote by ξ_m the mth solution of Eq. (9.25), the displacement field reads:

$$u(r_b, z, \omega) = \sum_{m=1}^{+\infty} u_m^{\pm}(r_b, \omega) e^{\pm i\xi_m(\omega)z} \tag{9.26}$$

where $\xi_m(\omega)$ is the mth solution of Eq. (9.25) and well known as the dispersion relation of the mth mode. Below these critical frequencies, the dispersion relation has no real part and is purely imaginary complex number:

$$\Re(\xi_m(\omega)) = 0 \quad if|\omega| \leq \omega_c^{(m)}. \tag{9.27}$$

The limit angular frequencies $\omega_c^{(m)}$ of the modes are such that the wave number vanishes [5, 13]:

$$\xi\left(\omega_c^{(m)}\right) = 0. \tag{9.28}$$

Thus, we obtain an equation satisfied by the limit frequencies. The solution of the obtained equation allows to relate the angular frequency of the second mode to the radius of the bar. Figure 9.8 depicts an example calculated assuming a bar made of a maraging steel. The frequency of the second mode is plotted as a function of the bar radius. In the Fig. 9.8, the cut-off frequency ω_c is also calculated for the two strain rates 10^4 s^{-1} and 10^5 s^{-1}, and is then plotted as two horizontal lines; ω_c increases with the targeted strain rate. With regards to the bar frequency, the second-mode frequency $\omega_c^{(2)}$ decreases as the bar radius increases. The $\omega_c^{(2)}$ curve intersects the dashed line of ω_c calculated with a strain rate of 10^5 s^{-1} at a bar radius of about $r_b = 9$ mm, while it approaches the solid line of ω_c calculated with a strain rate of 10^4 s^{-1} at a bar radius of about $r_b = 40$ mm. Thus, 40 mm can be taken as the upper bound of the bar radius if the targeted strain rate is 10^4 s^{-1} or lower. However, 9 mm has to be taken as the upper bound of the bar radius if the targeted strain rate can be as high as 10^5 s^{-1}. This ensures that the propagating frequencies of the second mode and higher are not excited. In that case, the second and higher modes can be neglected.

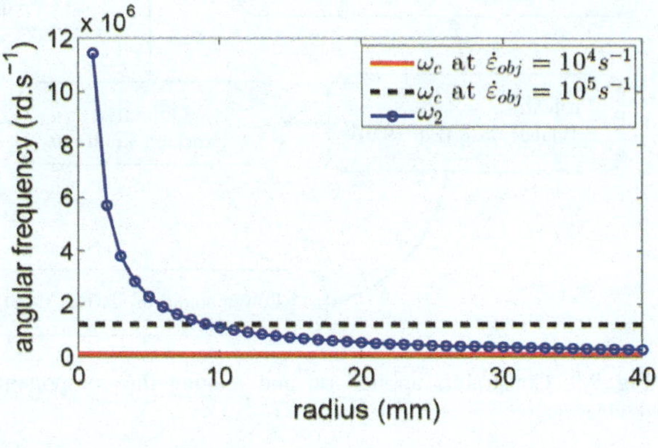

Fig. 9.8 Cut-off frequency of the loading pulse and angular frequency of the second mode of the bar

9.5 Design Procedure

Guo [25] summarized the constraints applied on the components of the direct impact device in the flow chart represented in Fig. 9.9. Moreover, provided some input data, the design problem of the direct-impact configuration was formulated by Guo et al. [20, 25–27] as an optimization problem submitted to equality and inequality constraints.

First, Guo et al. [20] assumed to be given a family of materials to be tested at a targeted strain rate $\dot{\varepsilon}_{s_{obj}}$, so that the level of strain in the specimen be framed between its given minimum $\varepsilon_{s_{min}}$ and maximum $\varepsilon_{s_{max}}$ values, and so that a minimum level of recorded bar strain $\varepsilon_{b_{min}}$ be reached. Second, Guo et al. [20] assumed that it is possible to choose the material of the projectile and the bar, according to the family of material to be tested, so that their Young's modulus and mass density (and thus their sound speed), and the yield stress of the bar σ_{b_y} are known. Finally, Guo et al. [20] assumed a coarse constitutive model of the specimen material in order to be able to assess the yield stress σ_{s_y}, the maximum level of stress reached $\sigma_{s_{max}}$ and the sound speed c_s.

The unknown vector $\mathbf{x}$ associated to the optimization problem consists of the length of the projectile l_p, its impact velocity v_p, the dimensions of the specimen (l_s, ϕ_s), and those of the bar (l_b, ϕ_b):

$$\mathbf{x} = \{l_p, v_p, l_s, \phi_s, l_b, \phi_b\}. \tag{9.29}$$

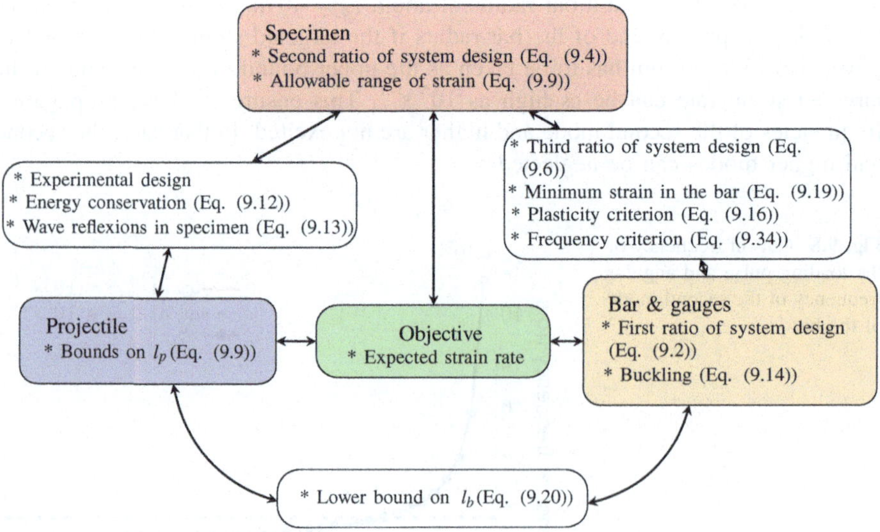

Fig. 9.9 Constraints applied on and relating the components of the direct-impact Hopkinson device

A solution of $\boldsymbol{x}$ may be sought by comparing the computed value of the strain rate $\dot{\varepsilon}_s(\mathbf{x})$ during the design process to the objective one $\dot{\varepsilon}_{s_{obj}}$. The cost function is thus defined as follows:

$$f(\mathbf{x}) = \left|\dot{\varepsilon}_s(\mathbf{x}) - \dot{\varepsilon}_{s_{obj}}\right| \tag{9.30}$$

where the strain rate $\dot{\varepsilon}_s(\mathbf{x})$ within the specimen reads:

$$\dot{\varepsilon}_s = \frac{-v_p + \frac{A_p + A_b}{A_p} c_b \varepsilon_b}{l_s}. \tag{9.31}$$

This equation allows for different diameters of the projectile and the bar and assumes that they are made of the same material. We recall that A_p and A_b are the cross-sectional areas of the projectile and the bar, respectively. The strain in the bar ε_b is related to the stress in the specimen and to the second ratio of the system design through the equilibrium of the specimen/bar interface:

$$\varepsilon_b = \frac{\sigma_s}{E_b}\left(\frac{\phi_s}{\phi_b}\right)^2. \tag{9.32}$$

The optimization problem submitted to equality and inequality constraints is thus formulated as follows [20, 25]:

$$\min_{\mathbf{x}} f(\mathbf{x})\,;\mathbf{G}(\mathbf{x}) \leq 0\,;\mathbf{h}(\mathbf{x}) = 0 \tag{9.33}$$

where $(\mathbf{h}(\mathbf{x}) = 0)$ is a set of equality constraints that consists of Eqs. (9.31), (9.32) and the equality between the projectile and bar diameters. The set of inequality constraints $(\mathbf{G}(\mathbf{x}) \leq 0)$ consists of inequalities (9.2), (9.4), (9.6), (9.9), (9.12)–(9.14), (9.16), (9.19) and (9.20). Another inequality is added to the above set of inequalities which is related to the upper bound prescribed on the bar diameter given as discussed in Sect. 9.4.4. and summarized through the following implicit relation:

$$g\left(\phi_b, \omega_c, \omega_b^{(2)}\right) \leq 0 \tag{9.34}$$

where $\phi_b, \omega_c, \omega_b^{(2)}$ denote the bar diameter, the cut-off frequency of the loading pulse and the angular frequency of the second mode of the bar respectively. Notice also that the cost function defined by (9.30) may a priori exhibit several minima.

Guo et al. [20] used the above approach to design a direct-impact Hopkinson bar rig for testing materials at strain rates between 5000 and 30,000 s^{-1}. They were interested in characterizing Ti–6Al–4V material. To withstand the impact force required to test the titanium alloy, the high-strength maraging steel MARVAL X2NiCoMo18-8-5 is adopted to manufacture the bar and the projectile. This steel has a yield stress of 1800 MPa after sustaining an aging treatment. The diameter of the projectile is determined by that of the canon and is set at $\phi_p = 15.8$ mm, being given a diametral clearance of 0.2 mm. Consequently, the unknown vector $\mathbf{x} = \{l_p, v_p, l_s, \phi_s, l_b, \phi_b\}$ that consists of five parameters remains to be determined. They selected four strain rates of 5000, 10,000, 20,000 and 30,000 s^{-1} are given as the objective strain rates. Consequently, Guo et al. [20] proposed a bar diameter, a bar length and a specimen diameter of 10, 1.2 and 5 mm, respectively. They also selected ranges for the specimen length and the impact velocity, namely, [1.5 mm, 5 mm] for the former and [15 m/s, 40 m/s] for the latter. Moreover, they selected the projectile lengths: 30, 60 and 125 mm. The shortest projectile is used for the highest strain rate.

9.6 Processing of Direct-Impact Hopkinson Bar Experiment

The common and main assumption usually performed in the post-processing of Kolsky experiments is to consider the force equilibrium achieved within the specimen. The stress is then given by:

$$\sigma_s(t) = \frac{E_b A_b}{A_s} \varepsilon_b(t) \tag{9.35}$$

where $\varepsilon_b(t)$, A_b and A_s denote the strain recorded on the bar, and the cross-sections of the projectile and the specimen respectively. The absence of the incident bar rises some difficulties on the deduction of the strain rate and thus the strain in the specimen. As a consequence, the classical equations of an SHPB do not work any more in the case of the direct-impact tests.

Alternative methods are required to bypass this difficulty. Intending to deduct the strain in the specimen, several approaches have been followed so far. A first method is to measure the strain directly during the test. A high speed camera permits to record the variation of the geometry of the specimen so that the strain can be calculated by the image processing. Gorham [10] developed an optical system for the measurement. In the optical system, he employed a refraction element. The light from the flash passes through the specimen and get refracted by the refraction element, so that the profile of the transverse edges of the specimen is shot by a high-speed camera. Thus the longitudinal strain in the specimen is deduced from the transverse deformation assuming the incompressibility of the tested material.

The calculation of the strain rate within the specimen requires the difference between the velocity of the specimen/bar (V_{out}) and that of the projectile/specimen interface (V_{in}) [see Eq. (9.1)], the latter being expressed as a function of the strain in the transmitted bar according to the unidimensional wave theory as:

$$V_{out}(t) = -c_b \varepsilon_b(t). \tag{9.36}$$

Thus the calculation of the strain and strain rate in a direct impact device comes down to approach the velocity of the projectile/specimen interface during the contact. Additional measurement/deduction techniques are proposed to obtain (V_{in}).

Malinowski [28] used two separated light systems to measure the displacement of the projectile/specimen interface and the impact velocity of the projectile in the tests. Two laser diodes accompanied with two photodiodes are equipped close to the emitting end of the accelerating tube for the projectile with a certain distance between them. When the projectile is travelling through this distance, a time counter records the period so that the velocity of the projectile is computed by the distance over the time period. The velocity of the interface can be deduced by assuming a constant deceleration. Another setup consisting of laser diodes and a photodiode is instrumented to measure the displacement of the projectile/specimen interface at different instants, using the principle of shadow. A gap is reserved between the projectile and the deceleration tube. By scaling the light passing through the gap, the evolution in time of the displacement of the projectile/specimen interface can be deduced. Besides these experimental setups used to assess (V_{in}), the impact velocity of the projectile (V_p) can be quite easily measured, and can thus also be used to approach (V_{in}) and hence to compute the strain rate $(\dot{\varepsilon}_s)$.

Dharan and Hauser [11] developed a diagram method to assess the strain and the strain rate in the specimen as indicated in Fig. 9.10. In this figure, v_1, v_2 and v_x denote the velocities on the both sides of the specimen and that of the projectile before impact, while ε, $\dot{\varepsilon}$, σ_x and a are the strain, strain rate, stress and the length of the specimen. The subscript E refers to the transmitted bar. The polygonal line starting from the original point O denotes the velocity of the projectile/specimen interface identified to V_{in} in Eq. (9.1). It indicates that after a short constant rising time, the velocity of the projectile/specimen interface equals the impact velocity of the projectile (v_x in Fig. 9.10), supposing that the projectile sections are rigid. The rising time is obtained from many tests by impacting the transmitted bar with the projectile directly. The segment OA represents the time period needed for the loading pulse to transmit through the specimen. Thus the curve starting from the point A denotes the stress σ_x in the specimen deduced from the strain ε_E measured in the transmitted bar. Then the velocity on the right side of the specimen is expressed as $v_2 = \sigma_x / \rho_E c_E$ and plotted. Consequently, the strain in the specimen can be deduced by calculating the shaded area and then dividing by the length a. This procedure supposes a constant impact velocity since the work done to strain the specimen is negligible when compared to the kinetic energy of the comparative massive projectile.

Gorham et al. [29] has used an expression proposed by Pope and Field [30], to compute the strain in the specimen from the measurements of v_p and ε_b. Let l_{s0} be the initial length of the specimen, the current length of the specimen reads at a time t:

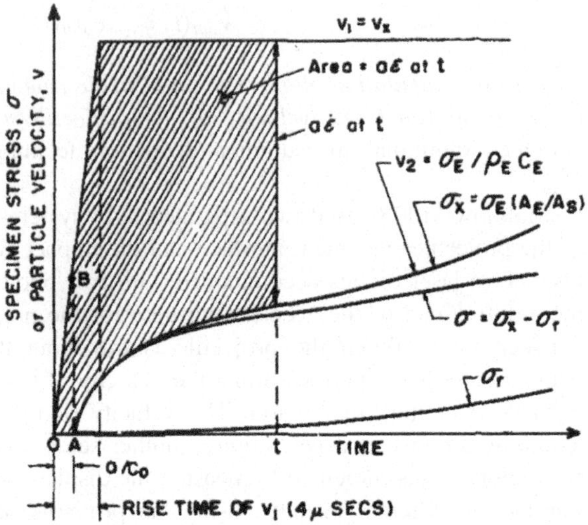

Fig. 9.10 Determination of stress, strain and strain rate from measurements [11]

$$l_s(t) = l_{s0} - v_0 + x_1(t) + x_2(t), \tag{9.37}$$

where v_0 is the velocity of projectile, $x_1(t)$ and $x_2(t)$ are the displacement of the projectile and the transmitted bar due to elastic deformation. According to $v = c\varepsilon$, the displacement is given by the integration of the velocity with respect to time:

$$x(t) = \frac{1}{ZA}\int_0^t f(\tau)d\tau, \tag{9.38}$$

where f, $Z = \rho c$ and A are the force, the acoustic impedance and the cross-section of the bar, respectively.

Substituting Eqs. (9.38) into (9.37), the length of the specimen reads:

$$l_s(t) = l_{s0} - v_0 t + \frac{Z_b + Z_p}{Z_p Z_b A_b}\int_0^t f(\tau)d\tau. \tag{9.39}$$

The above equation assumes that the projectile and the transmitted bar have the same cross-section but different acoustic impedance. Gorham et al. [29] emphasized that, the expression (9.39) is only valid within the characteristic time, that is until the reflected wave in the projectile reaches the specimen. Moreover, Pope and Field [30] have also discussed the possible errors arisen by the propagation delay in the recorded signal and by the force rising and oscillation due to the dispersion. They clarify that, the equilibrium cannot be achieved until at least two round trips of the wave in the specimen (Fig. 9.11).

Guo et al. [25] used a striker and a bar made of the same material but having different diameters. Thus Eq. (9.39) is changed to:

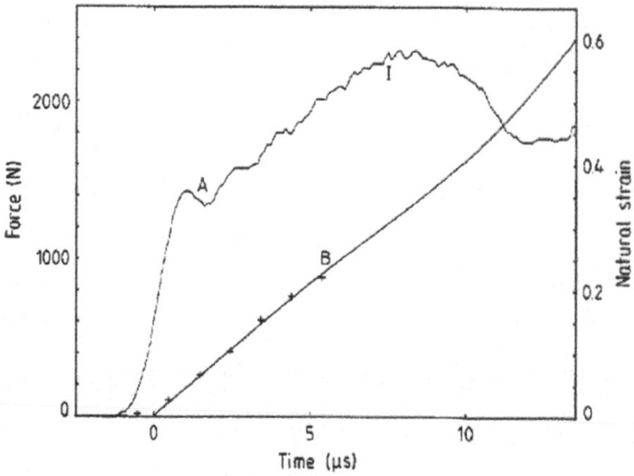

Fig. 9.11 Force (*A*) and calculated strain (*B*) curves for Ti–6Al–4V alloy [30]

Fig. 9.12 Specimen deformation capture by the high-speed camera during the test T4 [25]

$$l_s(t) = l_{s0} - v_0 t + \frac{1}{Z}\frac{A_b + A_p}{A_p}\int_0^t E_b \varepsilon_b(\tau) d\tau, \tag{9.40}$$

where $Z = Z_b = Z_p = \rho_b c_b$. Guo et al. [20, 25] used two laser diodes with photodiodes to measure the impact velocity of the projectile v_p. Thus, the strain rate in the specimen reads:

$$\dot{\varepsilon}_s = \frac{-v_p + \frac{A_p + A_b}{A_p} c_b \varepsilon_b}{l_s}. \tag{9.41}$$

Guo et al. [25] proposed a second alternative for the measurement of the strain and strain rate in the specimen. The direct-impact bar is then equipped with a high-speed camera to monitor the response of the specimen. Namely, the high-speed camera allows to film the deformation of the specimen during the tests. The camera records the motion of the projectile/specimen and specimen/bar interfaces. The Fig. 9.12 presents the sequential images recorded by the high-speed camera at 1.8×10^5 frames per second. A tracking technique was applied to obtain

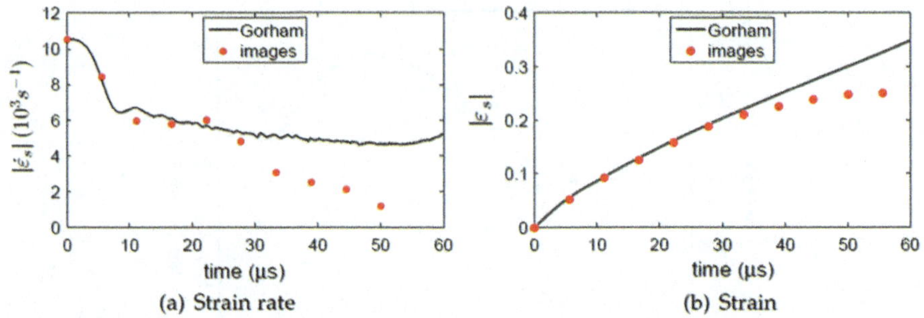

Fig. 9.13 Strain rate and strain calculated in the direct-impact tests by different manners for test T4 [25]

the displacement of the projectile/specimen interface $u_{in}(t)$ and that of the specimen/bar interface $u_{out}(t)$. The strain rate is estimated as

$$\dot{\varepsilon}_s = \frac{\dot{u}_{out}(t) - \dot{u}_{in}(t)}{l_s} \tag{9.42}$$

The Fig. 9.13 depicts a comparison between Eq. (9.41) and the strain rate deduced from high-speed images. Denoting the coordinates of the sections of the projectile and the bar as $x_n^{(p)}$ and $x_n^{(b)}$ at the instant $t = (n-1)\Delta t$, then the strain rate can be assessed through the following expression:

$$\dot{\varepsilon}_s = \left(\frac{x_{n+1}^{(b)} - x_n^{(b)}}{\Delta t} - \frac{x_{n+1}^{(p)} - x_n^{(p)}}{\Delta t} \right) \frac{1}{x_1^{(b)} - x_1^{(p)}} \tag{9.43}$$

where Δt is the time interval and is a constant given by the frequency of the camera. The dots, extracted from the images, fits well the solid curve within the period 25 μs, corresponding to the characteristic time. This supports the hypothesis that Eq. (9.41) is only valid within the period of the characteristic time.

Integrating the strain rate over time, one gets an approximately linearly rising strain as shown in the Fig. 9.13b, that superposes well with the data extracted from the images during the characteristic time.

9.7 Conclusion

The design of direct impact Kolsky/Hopkinson bar devices for very high strain rate ranges has been presented in this chapter, as well as the processing of experimental data. It has been shown that the design of this device is constrained by a set of empirical statements supplemented by others embedding more physics, pertaining and linking quantities associated to all the components of the direct impact system.

The design process can also be formulated as an optimization problem submitted to equality and inequality constraints, the solution of which can involve many projectiles of different length to span a large range of strain rate of the tested material as Guo [20, 25] did for the Ti–6Al–4V.

The removal of the input bar introduces an additional difficulty with respect to the classical SHPB device that consists in assessing the velocity of the projectile/specimen interface, in order to obtain the strain rate as well as the strain induced within the specimen. Then, two solutions have been adopted so far. Either an additional measurement technique is used (optical system for Gorham [10]) to deduce the strain within the specimen, or an additional equation based on a further assumption can be introduced (Gorham [29], Guo [20, 25]) to obtain a formulae defining the strain rate induced within the specimen during the test.

References

1. Zhao H, Gary G (1996) The testing and behaviour modelling of sheet metals at strain rates from 10^{-4} to $10^{4}s^{-1}$. Mater Sci Eng A 207:46–50
2. Gama BA, Lopatnikov SL, Gillespie JW (2004) Hopkinson bar experimental technique: a critical review. Appl Mech Rev 57:223–250
3. Kamler F, Niessen P, Pick RJ (1995) Measurement of the behaviour of high-purity copper at very high rates of strain. Can J Phys 73:295–303
4. Casem DT (2009) A small diameter Kolsky bar for high-rate compression. In: Proceedings of the SEM annual conference
5. Othman R (2012) Cut-off frequencies induced by the length of strain gauges measuring impact events. Strain 48:16–20
6. Jia D, Ramesh KT (2004) A rigorous assessment of the benefits of miniaturization in the Kolsky bar system. Exp Mech 44:445–454
7. Casem DT, Grunschel SE, Schuster BE (2011) Normal and transverse displacement interferometers applied to small diameter Kolsky bars. Exp Mech 52:173–184
8. Shioiri J, Sakino K, Santoh S (1996) Strain rate sensitivity of flow stress at very high rates of strain. In: Kozo K, Jumpei S (eds) Constitutive relation in high/very high strain rates, IUTAM Symposia. Springer, Japan, pp 49–58
9. Wulf GL (1978) The high strain rate compression of 7039 aluminium. Int J Mech Sci 20:609–615
10. Gorham DA (1979) Measurement of stress-strain properties of strong metals at very high rates of strain. In: Proceedings of the third conference on the mechanical properties at high rates of strain, conference series no. 47, pp 16–24
11. Dharan CKH, Hauser FE (1970) Determination of stress-strain characteristics at very high strain rates. Exp Mech 10:370–376
12. Ramesh KT (2008) High strain rate and impact experiments. In: Springer handbook of experimental solid mechanics, Part D: Chapter 33, pp 1–31
13. Davies RM (1948) A critical study of Hopkinson pressure bar. Philos Trans Roy Soc A 240:375–457
14. Gray GT (2000) Classic split-Hopkinson pressure bar testing. ASM International, Materials Park, pp 462–476
15. ASTM-E9-09. ASTM Standard E9-09 (2009) Standard test methods of compression testing of metallic materials at room temperature. ASTM International, West Conshohocken

16. Davies EDH, Hunter SC (1963) The dynamic compression testing of solids by the method of the split Hopkinson pressure bar. J Mech Phys Solids 11:155–179
17. Edington JW (1969) The influence of strain rate on the mechanical properties and dislocation substructure in deformed copper single crystals. Philos Mag 19:1189–1206
18. Tekalur SA, Sen O (2011) Effect of specimen size in the Kolsky bar. Procedia Eng 10:2663–2671
19. Guo X, Heuzé T, Othman R, Racineux G (2012) Dynamic test with modified Hopkinson-bar at high-strain rate. In: Centrale - Beihang workshop 2012. Ecole Centrale Lille, Lille, France
20. Guo X, Heuzé T, Othman R, Racineux G (2014) Inverse identification at very high strain rate of the Johnson-Cook constitutive model on the Ti–6Al–4V alloy with a specially designed direct-impact Kolsky bar device. Strain 50:527–538
21. Buchar J, Bilek Z, Dusek F (1986) Mechanical behavior of metals at extremely high strain rate (Section 3.2). Trans Tech Publications, Switzerland
22. Safa K, Gary G (2010) Displacement correction for punching at a dynamically loaded bar end. Int J Impact Eng 37:371–384
23. Pochhammer L (1876) Uber die fortpflanzungsgeschwindigkeiten kleiner schwingungen in einem unbergrenzten isotropen kreiszylinder. J für die Reine und Angezande Math 81:324–336
24. Chree C (1889) The equations of an isotropic elastic solid in polar and cylindrical coordinates their solution and applications. Camb Philos Soc Trans 14:250–369
25. Guo X (2015) On the direct impact Hopkinson system for dynamic tests at very high stain rates. Ph.D. thesis, École Centrale de Nantes
26. Guo X, Heuzé T, Othman R, Racineux G (2015) Direct-impact Hopkinson tests on the Ti–6Al–4V alloy at very high strain rate and inverse identification of the Johnson-Cook constitutive model. In: Colloque National MECAMAT sur Matériaux sous solicitations intenses grandes vitesses et fortes pressions: de l'impact à la planétologie. Aussois, France
27. Guo X, Heuzé T, Othman R, Racineux G (2015) Direct-impact Hopkinson tests on the Ti–6Al–4V alloy at very high strain rate and inverse identification of elasticviscoplastic constitutive models. In: Congrès Français de Mécanique, Lyon, France
28. Malinowski JZ, Klepaczko JR, Kowalewski ZL (2007) Miniaturized compression test at very high strain rates by direct impact. Exp Mech 47:451–463
29. Gorham DA, Pope PH, Field JE (1992) An improved method for compressive stress-strain measurements at very high strain rates. Philos Trans Roy Soc Lond Ser A Math Phys Sci 438:153–170
30. Pope PH, Field JE (1984) Determination of strain in a dynamic compression test. J Phys E Sci Instrum 17:817–820

Chapter 10
Testing of Adhesively Bonded Joints by Split Hopkinson Bar Technique

Chiaki Sato

10.1 Introduction

Application of adhesively bonded joints has recently been expanding in many fields such as automotive and airspace industries, in which the products may be subjected to impact loading. Therefore, experimental methods of impact for adhesively bonded joints are becoming much more important than previous situations.

For the testing of strain-rate dependent materials, several types of experimental methods are usually applied depending on strain rates [1]. Figure 10.1 shows the relation between types of impact, strain rates occurring in materials subjected to impacts, and typical experimental methods. For instance, quasi-static tests can be carried out using mechanical testing machines. However, in the range of higher strain rate, special testing machines such as hydraulic testing machines, split Hopkinson bar equipment, and powder or gas guns as increasing the required impact velocity as shown in Fig. 10.1. For higher strain rates than $10^2\ s^{-1}$, ordinary load-cells cannot be used due to the free vibration, thus split Hopkinson bar equipment becomes the only one option to measure the load-displacement relations, where powder guns and gas guns cannot measure the applied loads because they do not have any load measurement device.

Split Hopkinson bar equipment can be categorized in two, collision type and clamp type. Collision type split Hopkinson bar methods are popular and suitable for compression tests of materials, although tensile tests are possible with a special set-up. In contrast, clamp type split Hopkinson bar methods, which are also called an energy stored Hopkinson bar, are not popular but can be used for tensile, torsional, and also combined loading.

C. Sato (✉)
Precision and Intelligence Laboratory, Tokyo Institute of Technology,
4259 Nagatsuta, Midori-ku, Yokohama 226-8503, Japan
e-mail: csato@pi.titech.ac.jp

R. Othman (ed.), *The Kolsky-Hopkinson Bar Machine*,
https://doi.org/10.1007/978-3-319-71919-1_10

Yokoyama et al. also conducted tensile tests of butt joints bonded with an epoxy adhesive using hat-shaped specimens and a compressive testing setup shown in Fig. 10.6 [8]. This experiments indicated that the joint strength increased and became two times at a high strain rate condition of 10^6 MPa/s. Finite element analyses were carried out and the stress distribution in the adhesive layer in a specimen was investigated. The result indicated that there is stress distribution in the joint but it is not crucial.

In quasi-static conditions, tensile tests are more popular than compressive tests for adhesively bonded joints. Such tests have been carried out by a modification of ordinary collision type split Hopkinson bar.

Challita et al. have conducted finite element analyses of a double lap joint specimen [9] and carried out the impact test of the specimen using a split Hopkinson bar apparatus [10]. The shape of the specimen is shown in Fig. 10.7. They proposed correction coefficients to compensate the stress concentration and applied their theory to the experimental results. It was concluded that the failure strain decreased with increasing strain rate and bonded assemblies are highly strain rate sensitive regarding both substrates surface treatment and the three overlap length. The sensitivities of joint dimensions on the stress distribution of the double lap joint specimen were also investigated analytically by the authors [11].

Recently, a novel technique, digital image correlation (DIC) method, has been developed for measurement of the strain field on the surface of an object. Neumayer et al. have applied the DIC method to adhesively bonded joints and investigated experimentally the strain distribution in butt joints and single lap joints subjected to impact loading [12]. They determined the traction-separation rule of the adhesive layer directly from the load obtained by a SHPB apparatus and the strain field measured by the DIC method and concluded that inaccuracies in strain measurement could be overcome by measuring the adherends deformation directly by high-speed imaging and the DIC method.

10.3.2 Experimental Results by the Clamped Hopkinson Bar Method

The clamped Hopkinson bar method is suitable for torsional impact tests of materials. Much research on the torsional impact properties of materials has been

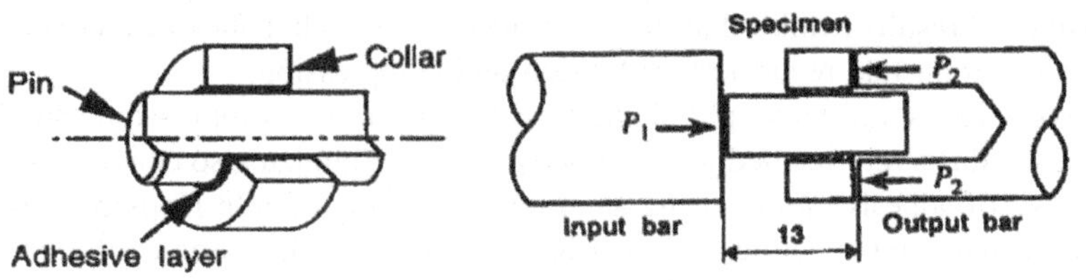

Fig. 10.5 Configuration of pin- and-collar joint specimen [7]

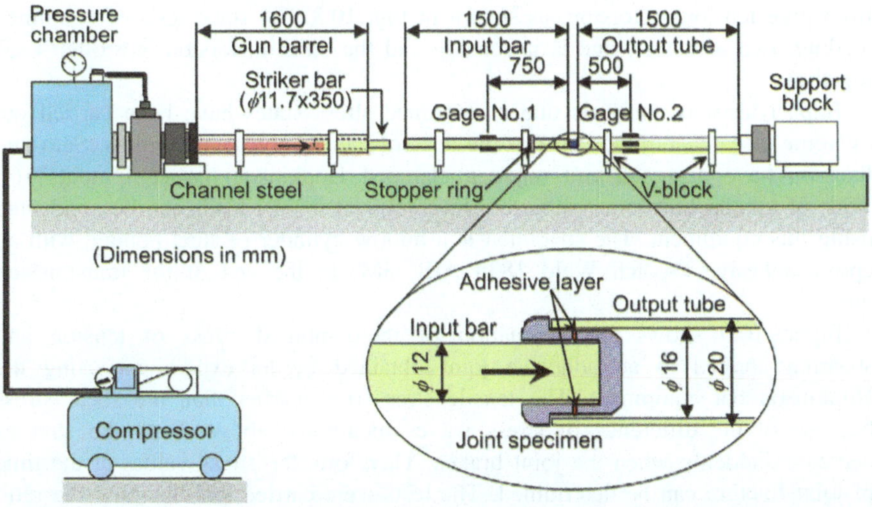

Fig. 10.6 Schematic diagram of a split Hopkinson bar apparatus and a hat-shaped specimen [8]

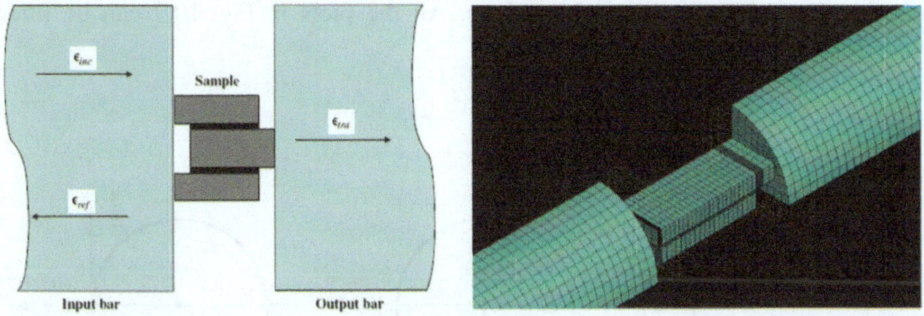

Fig. 10.7 Schematic illustration of double lap joint specimen for impact tests and meshing condition for FEM [9]

conducted using this method. To determine the impact strength of adhesively bonded joints, Raykhere and Kumar et al. have tested tubular butt joints of aluminum alloy—aluminum alloy and aluminum alloy-CFRP bonded adhesively using a clamped split Hopkinson bar equipment [13].

Since combined loading of tension and torsion to tubular specimens is convenient for determination of the strength and constitutive relations of materials, especially metals, combined loaded static tests are widely performed. A wide variety of stress states specimens can occur on the wall of tubular specimens. For instance, tensile loading causes tensile stress, torsional loading shear stress. In addition, combined loading of tension and torsion induces different stress states from pure tension or torsion, as shown in Fig. 10.8. For static tests of combined loading, two actuators, one is in tension and the other in torsion, are often used together.

Impact tests of materials under combined stress states have been carried out by some researchers [14, 15]. For the tests of adhesively bonded joints, Sato and Ikegami have designed and made a clamped Hopkinson bar equipment [16]. Tests of tubular butt joints subjected to combined impact loads can be conducted using this equipment. The specimen is a hollow cylinder of steel bonded with an epoxy adhesive (Scotch Weld 1838 A/B, 3M) to the end of the transmission tube.

Figure 10.9 shows the fluctuation of the combined stress of tension and shearing applied to an adhesive joint obtained by an experiment using the Hopkinson bar equipment. The tensile stress rises earlier than the shear stress because of the difference of stress waves mentioned above. Both the stresses decrease suddenly when the joint brakes. Therefore, the stress values at the time of joint fracture can be determined. The tests were carried out changing the ratio of tension and torsion. The test results are plotted in Fig. 10.10, where the abscissa denotes tensile normal stress to the joint, and the ordinate signifies shear stress. The static strengths of the joints are also shown in the figure. The impact strengths of the joints are two times higher than those of static tests even in tensile, shear, and combined conditions. All the plots in Fig. 10.9 can be fitted with a 2nd order polynomial curve.

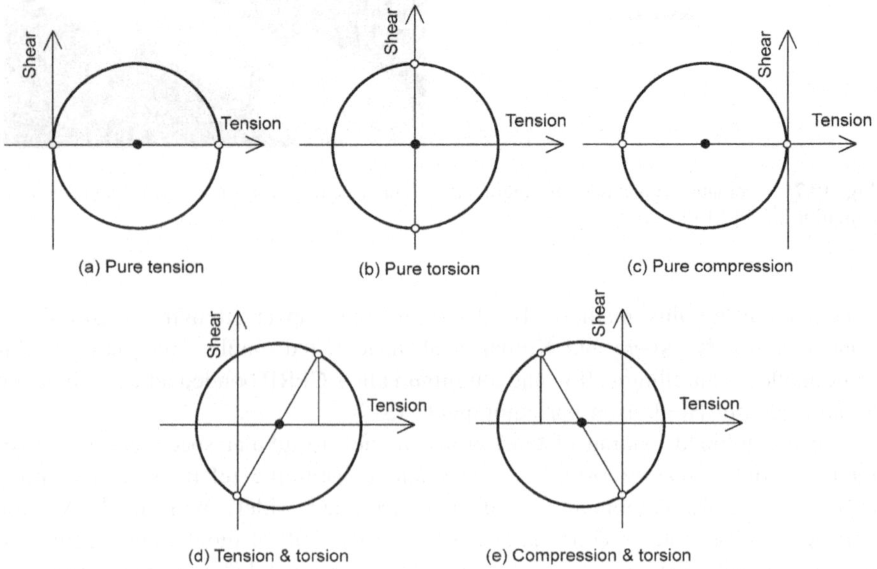

Fig. 10.8 Mohr's stress circles of stress states obtained by tensile, torsional, compression, and combined loading tests [4]

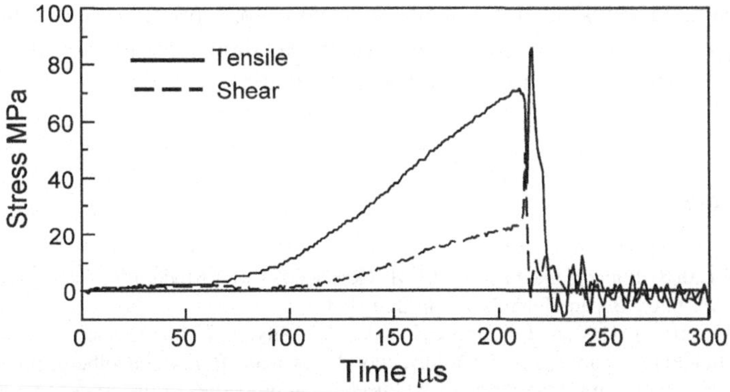

Fig. 10.9 Stress fluctuation applied to tubular butt-joint specimen [16]

Fig. 10.10 Joint strengths of tubular butt-joint specimen under combined loading conditions [16]

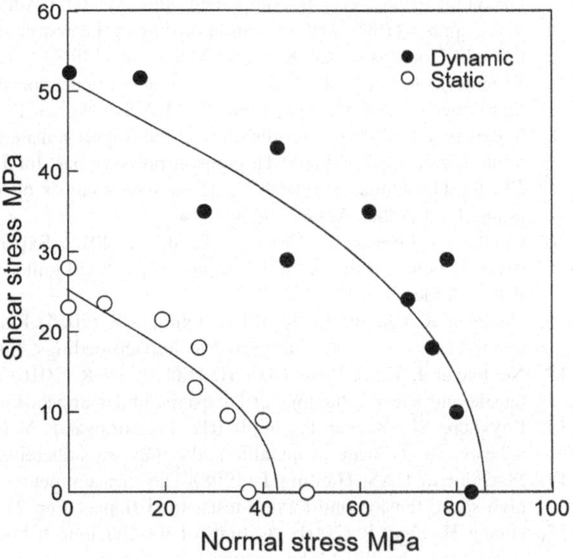

10.4 Conclusion

In this chapter, an outline of the Hopkinson bar method and examples of high-rate tests using the method are explained. Ordinary split Hopkinson bar equipment is suitable for compression tests so that the use is quite limited for adhesive joint evaluation because tensile tests are more useful for the purpose. In this case, additional devices are necessary to carry out tensile tests by the SHPB method.

The clamped Hopkinson bar method is more appropriate to torsional or combined impact tests than the split Hopkinson bar method. This method is also useful especially for tubular butt joints bonded adhesively.

References

1. Sato C (2008) Impact. In: da Silva LFM, Ochsner A (eds) Modeling of adhesively bonded joints. Springer, Berlin Heidelberg, pp 279–304
2. Sato C (2011) Design for impact loads. In: Silva LFM, Ochsner A, Adams, RD (eds) Handbook of adhesion technology, vol 2. Springer, Berlin Heidelberg, pp 743–764
3. Kolsky H (1949) An investigation of the mechanical properties of materials at very high rates of loading. Proc Phys Soc Ser B 62:676–700
4. Sato C (2012) Clamped Hopkinson bar. In: Silva LFM, Dillard DA, Blackman BRK, Adams RD (eds) Testing adhesive joints—Best practice. Wiley, Weinheim, pp 297–308
5. Goglio L, Peroni L, Peroni M, Rossetto M (2008) High strain-rate compression and tension behaviour of an epoxy bi-component adhesive. Int J Adhes Adhes 28:329–339
6. Yokoyama T (1999) Determination of impact shear strength of adhesively bonded joints with the split Hopkinson bar. Key Eng Mater 145–149:317–322
7. Bezemer AA, Guyt CB, Volt A (1998) New impact specimen for adhesives: optimization of high-speed-loaded adhesive joints. Int J Adhes Adhes 18:255–260
8. Yokoyama T (2015) Determination of the impact tensile strength of structural adhesive butt joints with a modified split Hopkinson pressure bar. Int J Adhes Adhes 56:13–23
9. Challita G, Othman R (2010) Finite-element analysis of SHPB tests on double-lap adhesive joints. Int J Adhes Adhes 30:234–244
10. Challita G, Othman R, Casari P, Khalil K (2011) Experimental investigation of the shear dynamic behavior of double-lap adhesively bonded joints on a wide range of strain rates. Int J Adhes Adhes 31:146–153
11. Hazimeh R, Challita G, Khalil K, Othman R (2015) Finite element analysis of adhesively bonded composite joints subjected to impact loadings. Int J Adhes Adhes 56:24–31
12. Neumayer J, Kuhn P, Koerber H, Hinterhölzl R (2016) Experimental determination of the tensile and shear behaviour of adhesives under impact loading. J Adhes 92:503–516
13. Raykhere SL, Kumar P, Singh RK, Parameswaran V (2010) Dynamic shear strength of adhesive joints made of metallic and composite adherents. Mater Des 31:2102–2109
14. Macdougall DAS, Harding J (1998) The measurement of specimen surface temperature in high-speed tension and torsion tests. Int J Impact Eng 21:473–488
15. Huang H, Feng R (2004) A study of the dynamic tribological response of closed fracture surface pairs by Kolsky-bar compression-shear experiment. Int J Solids Struct 41:2821–2835
16. Sato C, Ikegami K (1999) Strength of adhesively-bonded butt joints of tubes subjected to combined high-rate loads. J Adhes 70:57–73

The manufacturer's authorised representative in the EU is Springer Nature Customer Service Centre GmbH, Europaplatz 3, 69115 Heidelberg, Germany. If you have any concerns regarding our products, please contact ProductSafety@springernature.com

Printed and bound by CPI Group (UK) Ltd, Croydon, CR0 4YY
30/09/2026
02225655-0007